BIOTECHNOLOGY: RESEARCH, TECHNOLOGY AND APPLICATIONS

BIOTECHNOLOGY: RESEARCH, TECHNOLOGY AND APPLICATIONS

FELIX W. RICHTER
EDITOR

Nova Science Publishers, Inc.

New York

NOTICE TO THE READER

The Publisher has taken reasonable care in the preparation of this book, but makes no expressed or implied warranty of any kind and assumes no responsibility for any errors or omissions. No liability is assumed for incidental or consequential damages in connection with or arising out of information contained in this book. The Publisher shall not be liable for any special, consequential, or exemplary damages resulting, in whole or in part, from the readers' use of, or reliance upon, this material. Any parts of this book based on government reports are so indicated and copyright is claimed for those parts to the extent applicable to compilations of such works.

Independent verification should be sought for any data, advice or recommendations contained in this book. In addition, no responsibility is assumed by the publisher for any injury and/or damage to persons or property arising from any methods, products, instructions, ideas or otherwise contained in this publication.

This publication is designed to provide accurate and authoritative information with regard to the subject matter covered herein. It is sold with the clear understanding that the Publisher is not engaged in rendering legal or any other professional services. If legal or any other expert assistance is required, the services of a competent person should be sought. FROM A DECLARATION OF PARTICIPANTS JOINTLY ADOPTED BY A COMMITTEE OF THE AMERICAN BAR ASSOCIATION AND A COMMITTEE OF PUBLISHERS.

LIBRARY OF CONGRESS CATALOGING-IN-PUBLICATION DATA
Biotechnology, : research, technology & applications / Felix W. Richter (editor).
p. ; cm.
Includes bibliographical references and index.
ISBN 978-1-60456-901-8 (hardcover)
1. Biotechnology. I. Richter, Felix W.
[DNLM: 1. Biomedical Research--methods. 2. Biotechnology--methods. W 20.5 B6185 2008]
TP248.2.B572   2008
660.6--dc22   2008025310

Published by Nova Science Publishers, Inc. ✦ New York

Contents

Preface

Biotechnology combines disciplines like genetics, molecular biology, biochemistry, embryology and cell biology, which are in turn linked to practical disciplines like chemical engineering, information technology, and robotics. Patho-biotechnology describes the exploitation of pathogens or pathogen derived compounds for beneficial effect. This new book presents the latest research in the field.

Chapter 1 - Polyhydroxyalkanoates (PHAs) are a family of poly-β-hydroxyesters of 3-, 4-, 5- and 6-hydroxyalkanoic acids, produced by a variety of bacterial species under nutrient-limiting conditions with excess carbon. These water-insoluble storage polymers are biodegradable, exhibit thermoplastic properties and can be produced from renewable carbon sources. Hence, these polymers have the potential to replace the petrochemical-based plastics. In addition, PHAs are also known to be biocompatible and hence have the potential to be utilised for a range of biomedical applications.

PHAs are synthesised by a wide variety of bacteria, both Gram negative and Gram positive such as *Pseudomonas, Bacillus, Ralstonia, Aeromonas, Rhodobacter* and certain Archaea, such as members of the Halobactericeae. There are two main types of PHAs, short chain length PHAs that have C_3-C_5 hydroxyacids as monomers and medium chain length PHAs that have C_6-C_{16} hydroxyacids as monomers. The mechanical properties of these PHAs vary from being quite brittle to extremely elastomeric.

Microbial PHA biosynthesis is carried out by the PHA synthase enzyme that catalyses the stereo-selective conversion of (R)-3-hydroxyacyl-CoA substrates to PHAs with the concomitant release of CoA. Based on their primary structures, substrate specificities of the enzymes and the subunit composition, PHA synthases have been classified into four major classes. It is crucial to understand the mechanism of these enzymes in order to develop better polymer production strategies.

There is considerable interest in the commercial exploitation of these biodegradable polyesters. PHAs can be used for a range of different applications such as tissue engineering, drug delivery, and production of medical devices such as urological stents, in the packaging industry, dye industry, textile industry, for manufacturing adhesives, coatings and moulded goods. In addition, all of the monomeric units of PHAs are enantiomerically pure and in R-configuration. R-hydroxyalkanoic acids produced by the hydrolysis of PHAs can be used as chiral starting materials in fine chemical, pharmaceutical and medical industries. The market value of these polymers is estimated to be about £300 billion annually.

In this chapter the authors will cover information regarding the various facets of polyhydroxyalkanoates and trace the evolution of these microbially produced biodegradable polymers as 'new' materials. This information will also include the research findings of their laboratory.

Chapter 2 - With the increasing price of crude oil and the concerns about global warming, petrochemical based plastics now faces two major challenges: their non-degradability in the environment and diminishing supply of resource. Bio-based polymers present several advantages over conventional plastics; first, they originate from sustainable renewable resources and secondly, they are completely biodegradable. Polyhydroxyalkanoates (PHAs) is the most well-known biologically-based polyester. They accumulate naturally as intracellular inclusions in microorganisms grown under imbalanced nutritional conditions. PHAs can contain different and repeating units of carbon 'groups' and this structural diversity confer different physical properties. Thus, they can serve as a reliable substitute to synthetic plastics.

Although more than 90 genera of bacteria have been identified for their ability to accumulate a wide spectrum of PHA, none could serve as a strain suitable for industrial production of PHA. This is partially due to their lack of genetic tractability resulting in an inability for strain improvement. Recent efforts have been focused on reducing the cost for PHAs production in order to make the price of bio-polymers competitive against petrochemical based plastics. Strategies include strain improvement by recombinant DNA technology, high cell-density fermentation and the utilization of inexpensive carbon sources for the production of PHA. The launch of biobased plastics Mirel™ by Metabolix Inc. in 2007 clearly demonstrates that PHAs is starting to be accepted as an alternative to petro-based plastics. With the completion of the genome sequence of the well-known PHA-producing bacterium *Ralstonia eutropha* H16 and the advanced metabolic engineering technology available, opportunities abound for the improvement of PHA quality. This contribution describes the microorganisms (both indigenous and recombinant strains) that are capable of accumulating PHA in their natural habitats, in the laboratory and the feasibility of large scale industrial production.

Chapter 3 - Oligosaccharides play important roles in a number of biological events. To elucidate the biological functions of oligosaccharides, sufficient quantities of structurally defined oligosaccharides are required and one of the main limitation for this, is the availability of these molecules by tradicional purification methods. Hence, chemical and enzymatic synthesis of oligosaccharides are becoming increasingly important in glycobiology and glycotechnology. In addition, oligosaccharides often occur as glycoconjugates attached to proteins or lipids on the cell membranes.

Glycoconjugates can be synthesized by both chemical and enzymatic methods. The former approach involves the careful design of protecting groups, catalysts, reaction conditions, donor leaving groups, and acceptors. The enzymatic approach has several advantages over its chemical counterpart. Enzymatic glycosidation often takes place stereo- and regioselectively under mild reaction conditions without elaborated procedures such as protection and deprotection of hydroxyl groups or activation of anomeric position. Enzymatic approaches can be divided into two major categories based on glycosidases and glycosyltransferases.

Glycosyltransferases, which are responsible for the biosynthesis of oligosaccharides and glycoconjugates in nature, catalyze the efficient and specific transfer of a saccharide from a sugar nucleotide donor to an acceptor. But, both the enzyme and the sugar nucleotide are expensive and the glycosylation pathway is often subject to feedback inhibition from the nucleoside phosphate that is generated. Recently, these problems have been rapidly resolved by the increasing of availability of some recombinant glycosyltransferases and the use of a nucleotide sugar regeneration system and engineered bacteria. Hence, glycosyltransferases are expected to become a useful class of enzymes for the synthesis of oligosaccharides and glycoconjugates.

Glycosidases have been used to synthesize oligosaccharides via transglycosidation and condensation. Although the regioselectivity of glycosidases is rarely absolute and their yields are lower than those glycosyltransferases, glycosides are relatively expensive and are utilized because they are relatively available as glycosyl donor substrates. Therefore, glycosidases have been considered a powerful tool for the practical synthesis of oligosaccharides.

As the authors review here, the development of simple and effective chemoenzymatic methods for the synthesis of oligosaccharides and glycoconjugates, is essential for a better understanding of glycosylation in biology and for the development of new diagnostic and therapeutic strategies.

Chapter 4 - Biotransformations – that is, the use of biocatalysts in organic synthesis –, constitutes a relevant area within White Biotechnology, especially significant in the industrial manufacture of enantiopure drugs or building blocks. In this area, hydrolases are powerful biocatalysts for the production of enantiopure α-hydroxy ketones. These chemicals are valuable structures for the asymmetric synthesis of biologically active compounds, such as antidepressants (i.e., bupropion), inhibitors of amyloid-β protein production – for Alzheimer's disease treatment –, farnesyl-transferase inhibitors Kurasoin A and B, and antitumor antibiotics (Olivomycin A and Chromomycin A$_3$).

Originally, the hydrolase-based kinetic resolution (KR) of α-hydroxy ketones *via* hydrolysis and/or acylation led to positive results using different lipases, i.e., *Candida antarctica* lipase B (CAL-B), *Candida rugosa* lipase (CRL), *Burkholderia cepacia* lipase (BCL), or *Pseudomonas stutzeri* lipase (PSL). Later on, to overcome intrinsic drawbacks derived from the application of KR´s – where reaction yields are obviously limited to 50 % –, chemo-enzymatic dynamic kinetic resolutions (DKR) have been developed as well. In that case the enzymatic resolution is combined with a(n) (*in situ*) racemization of the starting material. Thus, enantiopure α-hydroxy ketones can be obtained in high yields when a lipase-based resolution is properly combined with a ruthenium catalyst racemization *via* hydrogen transfer. For this DKR approach, lipases from *Pseudomonas stutzeri* (PSL) and *Candida antarctica* lipase B (CAL-B) have been reported as promising biocatalysts.

Taking into account the emerging importance of this topic, the present chapter aims to provide a comprehensive view on the field of hydrolases and α-hydroxy ketones, by focussing on the type of substrates, enzymes, and optimum operational conditions used.

Chapter 5 - Human brain pyroglutamyl peptidase (PAPI; EC 3.4.19.3) is an omega exopeptidase which cleaves pyroglutamic acid from the N-terminus of bioactive peptides and proteins. It plays an important role in the processing and degradation of regulatory peptides such as thyrotropin releasing hormone (TRH) and luteinizing hormone releasing hormone (LHRH). To gain further insights into its performance *in vivo* and suggest possible applications, such as peptide processing or sequencing, this study focuses on the *in vitro* stability properties and Michaelis-Menten kinetics of the recombinant wild type enzyme and a single-site mutant, Tyr147→Phe (Y147F).

At 60°C in 50mM potassium phosphate buffer, pH 8.0, recombinant PAPI underwent a first-order decay constant with a *k* value of 0.046 ± 0.002 min^{-1} and a half-life of 15 min. PAPI was unstable to most of the water-miscible solvents tested (dimethyl sulphoxide, methanol, acetone, tetrahydrofuran, acetonitrile, dimethyl formamide and ethanol) even at low v/v concentrations. Methanol and dimethyl sulphoxide were the least injurious to PAPI activity: 56% and 50% residual PAPI activity remained at 10% v/v methanol and DMSO, respectively. Chemical modification with dimethyl suberimidate gave only 20% recovery of initial activity and did not stabilize the enzyme. Polyol and other stabilizing additives were investigated: activity and stability increased with xylitol but not with trehalose, glycerol or ammonium sulphate. PAPI displayed Michaelis-Menten kinetics with the fluorescent substrate pyroglutamyl 7-aminomethylcoumarin at pH 8.0: values for K_m and k_{cat} were 0.132 ± 0.024 mM and $2.68 \pm 0.11 \times 10^{-5}$ s^{-1} respectively.

Mutant Y147F was notably more thermostable, despite differing from wild type only by the absence of a hydroxyl. At 70°C, the Y147F first-order *k* value was 0.0028 ± 0.001 min^{-1} (half-life 25 min) compared with 0.079 ± 0.003 min^{-1} (half life 9 min) for wild type (the higher temperature was required to achieve timely inactivation of Y147F). Values of K_m and k_{cat} for Y147F (0.115 ± 0.019 mM and $2.45 \pm 0.05 \times 10^{-5}$ s^{-1} respectively) closely resembled those of wild type.

It appears that the *in vitro* stability of wild type PAPI might limit its potential applicability in peptide processing or other fields. Additives and chemical modification seem to have limited scope for enhancing its stability but the generation of stabilized mutant variants, or the use of a thermophilic counterpart, should be explored further.

Chapter 6 - Biomass is the only renewable resource that can provide a sufficient fraction of both future transportation fuels and renewable materials at the same time. Synthetic biology is an emerging interdisciplinary area that combines science and engineering in order to design and build novel biological functions and systems. Different from *in vivo* synthetic biology, cell-free *in vitro* synthetic biology is a largely unexplored strategy. Cell-free synthetic enzymatic pathway engineering (SEPE) is to *in vitro* assemble a number of enzymes and coenzymes to implement complicated biotransformations that can mimic natural fermentation or achieve unnatural processes. Recently, a novel synthetic enzymatic pathway composed of 13 enzymes and a cofactor has been demonstrated to produce 12 molecules of hydrogen per molecule of glucose unit of starch and water (PLoS One, 2007, 2:e456). This new sugar-to-hydrogen technology promises to solve several obstacles to the hydrogen economy – cheap hydrogen production, high hydrogen storage density (14.8 H$_2$ mass%), and costly hydrogen infrastructure, and to eliminate safety concerns about mass utilization of hydrogen. Furthermore, the advantages and limitations of producing liquid biofuels -- ethanol and butanol -- from sugars by SEPE are discussed. The research and development of SESE require more efforts, especially in low-cost recombinant thermophilic enzyme building block manufacturing, efficient cofactor recycling, enzyme and cofactor stabilization, and so on.

Chapter 7 - *Artemisia annua* is currently sole herbaceous biomass for industrial manufacture of artemisinin, an antimalarial sesquiterpene lactone with the unique endoperoxide architecture. Due to presence in trace amount, artemisinin has been targeted for *in planta* overproduction by genetic modification of *A. annua*. Beneficial from such pursuits, the overall enzymatic cascades involving artemisinin biogenesis have been elucidated and a dozen of critical artemisinin responsible genes identified. Consequently, transgenic *A. annua* plants with enhanced artemisinin production are available although substantial and profound potentials in artemisinin accumulation expected. Alternatively, due to conservation of the entire terpene pathways among higher plants and eukaryotic or even prokaryotic microbes, re-establishment of extended or diverted pathways toward *de novo* microbial artemisinin production has been eagerly attempted in genetically tractable microbes. In such aspect, a suit of downstream pathway genes specific for artemisinin biogenesis have been transplanted from *A. annua* into *Sacchromyce cerevisiae* and *Escherichia coli*, in which a series of incredible amounts of artemisinin precursors manufactured. The next-step goal is to further accelerate forward the total artemisinin biosynthesis through biotransformation of the artemisinin precursor(s) either *in vivo* or *in vitro*. For this purpose, the putative oxidant sink molecule capable of quenching the reactive oxygen species (ROS), in particular, the singlet oxygen ($^1O_2$), must be produced, in a large scale, in genetically modified microbes or transgenic *A. annua* plants. Whether dihydroartemisinic acid or artemisinic acid is such a $^1O_2$-scavenging direct intermediate has not been convinced, but conversion from dihydroartemisinic acid or artemisinic acid to artemisinin recognized as a bottleneck for artemisinin biosynthesis and versatile strategies aiming at breaking the rate-limited step enthusiastically pursued in *A. annua*, for example, by utilization of the primary abiotic or biotic stress signals or secondary stress signal transducers. These achievements should benefit our future intervention with the homeostatic tempo-spatial regulation mode of genetic background-based and environment-dependent artemisinin accumulations. This article introduces, from the genomics, transcriptomics, proteomics and metabolomics, the updated literatures describing the relationship between artemisinin biosynthetic gene overexpression and subsequent artemisinin overproduction as well. It should shed light on further elucidation of the intrinsic rule and mechanism underlying that artemisinin biochemical synthesis is fine-tuned by the genetic and environmental regulators, and should also urge the researchers all over the world more intensively investigating the intriguing *A. annua* plant that has implications in the medicinal and aromatic industries.

Chapter 8 - Bioethics and public acceptance of biotechnology are important social issues in this century. It appears that people's attitude toward bioethics is closely related to the acceptance of biotechnology. However, few psychological studies have examined this relationship or attitude structure of bioethics in the present circumstances. The following three psychological studies investigated these matters.

In study 1, attitudes toward bioethics and the acceptance of various biotechnologies were investigated with psychological scales. The participants were 231 Japanese undergraduate students. A cognitive map of biotechnology was constructed from these attitude scores. The cognitive map showed that people oppose to biotechnologies that are perceived as unethical, but approve of biotechnologies that are perceived as ethical. This study also indicated that people's attitude toward biotechnology, for example, toward gene recombination technology, differs considerably according to whether the technology is applied to human beings, animals or plants.

In study 2, based on Tanaka's study (2004), a structural equation model was set up, in which the common factor of "bioethics," affects three factors, namely, "human dignity," "ethically right or wrong for researching and developing biotechnologies," and "nature and the natural order." The participants were 154 Japanese undergraduate students who answered a questionnaire. Gene recombination technology toward human beings was taken up as a subject. The results confirmed the validity of the proposed model. There were also some interesting findings concerning the elements that construct each of the three factors. For example, it was suggested from the element that constructs the factor of "nature and the natural order" that people feel it is unnatural to change the human body and human nature during a short time span by using gene technology, because their characteristics have been formed over a very long time period since the beginning of life.

In study 3, based on Tanaka's study (2004) and the study 2 of this chapter, a hypothesis that the factor of bioethics is also important for acceptance of genetically modified foods (GM foods) was tested by using a structural equation model. The participants were 166 Japanese undergraduate students who answered a questionnaire. A causal model in which the factor of bioethics affects the factor of acceptance of GM foods was set up. The results of analysis indicated that the hypothesis was clearly supported. The result of study 3 suggests that the factor of bioethics is important for not only the acceptance of biotechnologies themselves, but also the acceptance of foods and products which are produced by use of biotechnologies.

Chapter 9 - Lactic acid bacteria (LAB) cultures production, as well as milk fermentations, are carried out in bioreactors operated in the batch mode using free -not immobilized-microorganisms. In the recent years however, immobilized cell technologies have been applied successfully in many cases to LAB, bifidobacteria, and probiotic cultures demonstrating the importance and applications potential of this technology in the food area. As a bio-processing strategy, cell immobilization may result in comparative advantages over cell-free cultures, such as increased fermentation rates or metabolite production from lower biomass levels. Other benefits include reduced contamination possibilities, physical and chemical protection of cells and reuse of biocatalysts. Immobilized cell technologies, long-term continuous culture modes and development of controlled-release systems aim to enhance the tolerance of sensitive LAB to environmental stresses and subsequently result in new applications in the food technology, and production of high-value products with positive effect on consumers' health. This review focuses on the current status of the main bio-processing strategies in this field, their technological characteristics, applications and perspectives.

Chapter 10 - Membrane emulsification has received increasing attention over the last 15 years, with applications in many fields. In the membrane emulsification process, a liquid phase is pressed through the membrane pores to form droplets at the permeate side of a membrane; the droplets are then carried away by a continuous phase flowing across the membrane surface. Compared to conventional techniques for emulsification, membrane processes offer advantages such as control of average droplet diameter by average membrane pore size, and low energy input. Under specific conditions, monodispersed emulsions and particles can be produced. The purpose of the present paper is to provide an updated review on the membrane emulsification process including: principles of membrane emulsification, experimental devices, influence of process parameters, and applications such as drug delivery systems, food emulsions, and a large range of particles.

Chapter 11 - Chemical sensors can be used to analyze a wide variety of environmental and biological gases and liquids for properties of interest. For many applications, sensors need to be sensitive and able to detect a target analyte selectively. Sensors are used to detect analytes in a wide range of samples including environmental samples and biological samples. Several different methods including gas chromatography (GC), chemiluminescence, selected ion flow tube (SIFT), and mass spectroscopy (MS) have been used to measure different biomarkers. These methods show variable results in terms of sensitivity and they cannot satisfy all the requirements for a handheld biosensor.

The desired sensors should be small in size, inexpensive and capable of real time detection without consumable carrier gases. One promising new sensing technology utilizes AlGaN/GaN high electron mobility transistors (HEMTs) [1-16]. HEMT structures have been developed for use in microwave power amplifiers as well as gas and liquid sensors due to their high two dimensional electron gas (2DEG) mobility and saturation velocity. The conducting 2DEG channel of GaN/AlGaN HEMTs is very close to the surface and extremely sensitive. HEMT sensors can be used for detecting a variety of analytes including, for example, gases, ions, pH values, proteins, and DNA. In this chapter the authors review recent progress on functionalizing the surface of HEMTs for specific detection of glucose, kidney marker injury molecules, prostate cancer and other common substances of interest in the biomedical field.

Chapter 12 - In order to meet the needs of production in industry and improve the yield and productivity of the industrial strains, the metabolism of microbial cells need to be regulated and controlled. Today, the main regulations are carried out at the DNA level through genetic engineering. Many successful samples have appeared in the past 35 years, but genetic engineering does not satisfy the needs of industry without an external means of regulation. The artificial periodic stimulation theory was proposed based on the maladjustment of the classical chemical reactor theory on bioreactor design and operation. The theory emphasized that the artificial periodic stimulations enhanced the bio-reaction and mass transfer at the cell level. Toward the bio-reaction system the cell level, the periodic input regulation is a generally optimal means of control. Following this theory, the airlift loop bioreactor was built successfully and a novel solid-state fermentation bioreactor: "Gas Double-dynamic Solid-State Fermentation Bioreactor" (GDSFB) was invented. The two types of bioreactors got outstanding practice results. During the further study of GDSFB, it was validated that periodic pressure oscillation reflected as a strong normal force and weak tangential force. This was a better environment for mass and heat transfers in the bio-system. The artificial periodic stimulation can increase the respiratory intensity greatly and affect the key enzyme activity of sugar metabolism, the quantity and abundance of protein expression. The research on solid-state fermentation also discovered the unique respiratory quotient periodic phenomena not found in submerged fermentation. Many types of research are being carried out on perfecting and validating the artificial periodic stimulation theory. This theory will play a more important role in bioprocess regulation.

Chapter 13 - It is well established that soil animals are important contributors to soil health. But keeping in view the paucity of information in harsh climatic conditions. They play a key role in decomposition process and ingesting large volumes of soil containing dead, decaying material, animal debris and associated microorganisms. Soil arthropods favour microbial activity increase enzymatic activity, stimulate root development and maintained soil fertility via biochemical and biomechanical processes. The microbe-grazing arthropods scrape bacteria and fungi of the root surface and complete nutrient cycle processes. Overall, soil faunal populations influence soil biological processes, nutrient cycling, soil structure and environment regulatory functions.

Chapter 14 - Bacterial alkaline phosphatase (BAP) is a useful enzyme for detection in biotechnological researches. There are vast arrays of commercial available substrates, which can be converted to soluble or precipitated products for either colorimetric or chemiluminescent detection. This research article describes the application of bacterial alkaline phosphatase fusion protein as a convenient and versatile molecular probe for direct detection of different molecular interactions. Short peptide, protein binding domain, or single chain variable fragment (scFv) of monoclonal antibody were fused to bacterial alkaline phosphatase and used as one step detection probe for the study protein-protein or antibody-antigen interactions. The BAP-fusion could be generated by cloning a gene of interest in frame of BAP gene in an *Escherichia coli* expression vector. The fusion protein contained N-terminal signal peptide for extracellular secretion and could be induced for over-expression with isopropyl-β-D- thiogalactopyranoside (IPTG), allowing simple harvesting from culture broth or periplasmic extract. The BAP-fusion that was tagged with poly-histidine could be further purified by nickel affinity chromatography. This one step detection probe generated a specific and robust signal, suitable for detection in various formats as demonstrated on microtiter plate, dot blot, or western blot. In addition, it could also be used for an estimation of binding affinity by competitive inhibition with soluble ligand.

In: Biotechnology: Research, Technology and Applications   ISBN 978-1-60456-901-8
Editor: Felix W. Richter   © 2008 Nova Science Publishers, Inc.

Chapter 1

Polyhydroxyalkanoates: A New Generation of Biotechnologically Produced Biodegradable Polymers

Ipsita Roy and Sheryl Philip
University of Westminster, UK

Abstract

Polyhydroxyalkanoates (PHAs) are a family of poly-β-hydroxyesters of 3-, 4-, 5- and 6-hydroxyalkanoic acids, produced by a variety of bacterial species under nutrient-limiting conditions with excess carbon. These water-insoluble storage polymers are biodegradable, exhibit thermoplastic properties and can be produced from renewable carbon sources. Hence, these polymers have the potential to replace the petrochemical-based plastics. In addition, PHAs are also known to be biocompatible and hence have the potential to be utilised for a range of biomedical applications.

PHAs are synthesised by a wide variety of bacteria, both Gram negative and Gram positive such as *Pseudomonas, Bacillus, Ralstonia, Aeromonas, Rhodobacter* and certain Archaea, such as members of the Halobactericeae. There are two main types of PHAs, short chain length PHAs that have C_3-C_5 hydroxyacids as monomers and medium chain length PHAs that have C_6-C_{16} hydroxyacids as monomers. The mechanical properties of these PHAs vary from being quite brittle to extremely elastomeric.

Microbial PHA biosynthesis is carried out by the PHA synthase enzyme that catalyses the stereo-selective conversion of (R)-3-hydroxyacyl-CoA substrates to PHAs with the concomitant release of CoA. Based on their primary structures, substrate specificities of the enzymes and the subunit composition, PHA synthases have been classified into four major classes. It is crucial to understand the mechanism of these enzymes in order to develop better polymer production strategies.

There is considerable interest in the commercial exploitation of these biodegradable polyesters. PHAs can be used for a range of different applications such as tissue engineering, drug delivery, and production of medical devices such as urological stents, in the packaging industry, dye industry, textile industry, for manufacturing adhesives, coatings and moulded goods. In addition, all of the monomeric units of PHAs are enantiomerically pure and in R-configuration. R-hydroxyalkanoic acids produced by the hydrolysis of PHAs can be used as chiral starting materials in fine chemical, pharmaceutical and medical industries. The market value of these polymers is estimated to be about £300 billion annually.

In this chapter we will cover information regarding the various facets of polyhydroxyalkanoates and trace the evolution of these microbially produced biodegradable polymers as 'new' materials. This information will also include the research findings of our laboratory.

Introduction

Polyhydroxyalkanoates (PHAs) are a family of linear polyesters of 3, 4, 5 and 6-hydroxyacids, synthesised by a wide variety of bacteria through the fermentation of sugars, lipids, alkanes, alkenes and alkanoic acids (Figure 1). They are found as discrete cytoplasmic inclusions in bacterial cells. Once extracted from the cells, PHAs exhibit thermoplastic and elastomeric properties. PHAs are recyclable, are natural materials and can be easily degraded to carbon dioxide and water. Hence, they are excellent replacements to petroleum-derived plastics in terms of processability, physical characteristics and biodegradability. In addition, these polymers are biocompatible and hence have several medical applications (Madison and Huisman, 1999; Mochizuki, 2002).

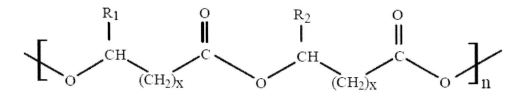

Figure 1. The general structure of polyhydroxyalkanoates. R_1/R_2 = alkyl groups; $C_1 - C_{13}$, x = 1-4, n = 100 – 30000.

All of the monomeric units of PHAs are enantiomerically pure and in the R-configuration. R-hydroxyalkanoic acids produced by the hydrolysis of PHAs can also be widely used as chiral starting materials in fine chemical, pharmaceutical and medical industries.

History

In 1923, Lemoigne at the Institut Pasteur demonstrated that aerobic spore-forming bacillus, formed quantities of 3-hydroxybutyric acid in anaerobic suspensions. He proceeded to investigate further and was successful in estimating quantitatively the amount of 3-hydroxybutyric acid formed. Finally, in 1927, he was able to extract a substance from *Bacillus* using chloroform and prove that the material was a polymer of 3-hydroxybutyric acid (Macrae and Wilkinson, 1958). However, it was not until the early 1960s that the production of poly(3-hydroxybutyrate), P(3HB), was explored on a commercial scale. The pioneering work of Baptist and Werber at W.R. Grace and Co.(U.S.A) earned them several patents to produce and isolate P(3HB). They began using this polymer to fabricate articles like sutures and prosthetic devices. Their efforts had to be terminated as the fermentation yields were low and the polymer was tainted with bacterial residues. Moreover, the solvent extraction process was very costly (Hocking and Marchessault, 1994).

The oil crisis of the 1970s provided a boost in the quest for alternative plastics. ICI (UK) was able to formulate conditions for *Alcaligenes latus* to produce P(3HB) up to 70% of its dry cell weight. The drawbacks were that P(3HB) was brittle, with poor mechanical properties and high production costs. Hence, after the oil crisis subsided, the development of P(3HB) as the future material came to a temporary standstill (Howells, 1982). In the meantime, ICI produced a novel polymer - BIOPOL®- a copolymer of 3-hydroxybutyrate and 3-hydroxyvalerate. BIOPOL® had improved properties, such as lower crystallinity and more elasticity as compared to P(3HB) (King, 1982). ICI split in June 1993 and the Zeneca BioProducts branch of ICI started dealing with BIOPOL®. Zeneca then sold their BIOPOL® technology to Monsanto in April 1996. Metabolix Inc. obtained the license from Monsanto in 1998 (Braunegg, 2003). In 1998, following collaboration between Metabolix Inc. and Children's Hospital, Boston, a new spin off company named Tepha Inc. was initiated. Tepha leads the technology in developing medical devices from biologically derived biodegradable polymers including bioengineered heart valves, surgical sutures, meshes and orthopaedic fixtures.

Occurrence and Biosynthesis of Polyhydroxyalkanoates

A wide variety of bacteria, both Gram negative and Gram positive, such as *Pseudomonas, Bacillus, Ralstonia, Aeromonas, Rhodobacter* and certain Archaea, especially members of the Halobactericeae, like *Haloferax sulfurifontis*, synthesise polyhydroxyalkanoates. They function as energy storage compounds and are present in cells as insoluble granules in the cytoplasm (Anderson and Dawes, 1990). Marine prokaryotes, both bacteria and archae, also produce PHAs that have tremendous commercial potential. PHAs accumulate up to 80% dry cell weight in these organisms when present in a "high-nutrient econiche" (Weiner, 1997). This displays the widespread occurrence of PHA-producing microbes in the environment.

Polyhydroxyalkanoates (PHAs) are divided into two groups based on the number of constituent carbon atoms in their monomer units – short-chain-length (SCL) PHAs and medium-chain-length (MCL) PHAs. The former consists of monomers with 3-5 carbon atoms and the latter contains monomers with 6-14 carbon atoms (Anderson and Dawes, 1990). Recently, there have been reports of several bacteria that are able to synthesise PHAs containing both SCL- and MCL- monomer units (Doi *et al.*, 1995). PHA$_{SCL}$ are stiff and brittle with a high degree of crystallinity, whereas PHA$_{MCL}$ are flexible, have low crystallinity, tensile strength and melting point (Holmes, 1988; Gagnon *et al.*, 1992).

The supply of the substrate monomer and the polymerization of these monomers are the two main steps involved in the biosynthesis of PHAs. The PHA synthesised by a microbe is dependent on the carbon source used. Carbon sources have been classified as 'structurally related' sources that give rise to monomers that are structurally identical to that particular carbon source and 'structurally unrelated' sources that generate monomers that are completely different from the given carbon source. The cause for this difference can be elucidated from the metabolic pathways operating in the microorganism. There are three well-known PHA biosynthetic pathways (Taguchi *et al.*, 2002) (Figure 2).

Pathway I, used by *Cupriavidus necator* (previously known as *Wauterisia eutropha* or *Ralstonia eutrophus*) is the best known among the PHA biosynthetic pathways. In this pathway, 3HB monomers are generated by the condensation of two acetyl-CoA molecules, from the tricarboxylic acid (TCA) cycle to form acetoacetyl-CoA by the enzyme β-ketothiolase. (Senior and Dawes, 1971; Senior and Dawes, 1973). Then acetoacetyl-CoA reductase acts on acetoacetyl-CoA to form 3-hydroxybutyryl-CoA. Finally, the PHA synthase enzyme catalyses the polymerisation via esterification of 3-hydroxybutyryl-CoA into poly(3-hydroxybutyrate), P(3HB).

Pathways involved in fatty acid metabolism generate different hydroxyalkanoate monomers utilised in PHA biosynthesis (Lageveen *et al.*, 1988). The fatty acid β-oxidation pathway (Pathway II) generates substrates that can be polymerised by the PHA synthases of *Pseudomonads* belonging to the ribosomal RNA-homology group I such as *Pseudomonas aeruginosa*. These microbes can synthesise PHA$_{MCL}$ from various alkanes, alkenes and alkanoates. The monomer composition is related to the carbon source used. In *Aeromonas caviae*, the β-oxidation intermediate, trans-2-enoyl-CoA is converted to (R)-3-hydroxyacyl-CoA by a (R)-specific enoyl-CoA hydratase (Fukui and Doi, 1997; Fukui *et al.*, 1998). Tsuge has also found the presence of similar enoyl-CoA hydratases in *Pseudomonas aeruginosa* (Tsuge *et al.*, 2000). Davis *et al.*, (2008) have reported the presence of four enoyl-CoA hydratases that connect the β-oxidation pathway with the PHA biosynthetic pathway in *Pseudomonas aeruginosa* (Davis *et al.*, 2008).

Huijberts *et al.*, 1992 showed that PHA synthases (as that of *Pseudomonas putida*) that catalyse PHA synthesis from fatty acids are also responsible for PHA synthesis from glucose (Huijberts *et al.*, 1992). The intermediates for this channel of synthesis were obtained from the fatty acid *de novo* biosynthetic pathway (Pathway III). Pathway III is of significant interest because it helps generate monomers for PHA synthesis from structurally unrelated, simple and inexpensive carbon sources such as glucose, sucrose and fructose. The (R)-3-hydroxyacyl intermediates from the fatty acid biosynthetic pathway are converted from their acyl carrier protein (ACP) form to the CoA form by the enzyme acyl-ACP-CoA transacylase (encoded by *phaG*). This enzyme is the key link between fatty acid synthesis and PHA biosynthesis (Rehm *et al.*, 1998).

Most bacteria that are capable of synthesising PHAs, are able to degrade the polymer intracellularly. During intracellular degradation, the PHA depolymerase in the cell breaks down P(3HB) to give 3-hydroxybutyric acid. A dehydrogenase acts on the latter and oxidises it to acetylacetate and a β-ketothiolase acts on acetylacetate to break it down to acetyl-CoA. The β-ketothiolase enzyme plays an important role in both the biosynthetic and the biodegradation pathways. Under aerobic conditions, the acetyl-CoA enters the citric acid cycle and gets oxidised to carbon dioxide (Jendrossek, 2001). Very little is known about the intracellular depolymerases since they are always found to be intimately connected to the P(3HB) granules and the overall process is very complex (Tokiwa and Calabia, 2004; Merrick and Doudoroff, 1964).

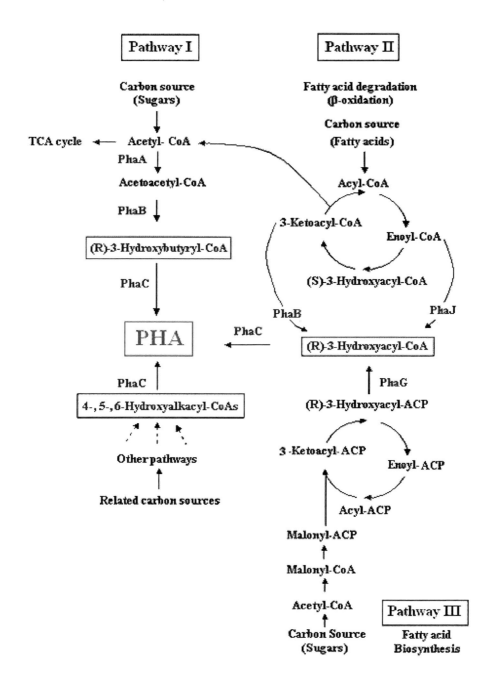

Figure 2. PHA biosynthetic pathways (adapted from Gagnon *et al*., 1992)

Physical Properties of PHAs

Depending on the copolymer composition of the polyhydroxyalkanoate, a range of designer PHAs with desirable properties can be obtained. The polymer can be hard and crystalline or elastic and rubbery [12]. P(3HB) is highly crystalline with a melting temperature of 180^0C. The material is brittle and stiff. However, the introduction of different HA monomers such as 3HV or 3HHx into the chain greatly improves the material properties of P(3HB). P(3HB) is however a biocompatible polymer, is optically pure and possesses piezoelectricity which helps in osteoinduction, the process of inducing osteogenesis (Holmes, 1988; Holmes, 1985; Doi, 1990).

Poly(3-hydroxybutyrate-co-3-hydroxyvalerate), P(3HB-co-3HV), has a lower melting temperature and lower crystallinity than P(3HB). The mole percent of 3HV in the polymer is important in determining the properties of the copolymer. P(3HB-co-3HV) containing more than 20 mole% of 3HV units can be used to make films and fibres with different elasticities, by controlling the processing conditions (Kim and Lenz, 2001).

P(4HB) on the other hand, is a strong and malleable thermoplastic material with a tensile strength closely comparable to that of polyethylene. It has an elongation at break of 100% resulting in extremely elastic properties. When combined with other hydroxyacids, the material properties of P(4HB) can be varied (Martin and Williams, 2003). The material properties of P(3HB), P(3HB-co-3HV) and P(4HB) are summarised below (Table 1):

Polymer microstructure development is an innovative technique of developing novel and desired properties. Block copolymers are one such example. They consist of repeating polymer regions covalently bound to each other, thus confining the properties of the individual polymers as well as giving rise to fresh properties that would have been impossible to achieve by blending the constituent polymers. Studies indicate that introducing a block copolymer-based microstructure in PHAs enables the copolymer to withstand the effects of ageing that lead to the brittle character of the random copolymer samples. Crystallization continues to occur in block copolymers. However, the effects on the elongation at failure are much reduced when compared to that of the random copolymer (McChalicher and Srienc, 2007). There have been recent reports on the bacterial synthesis of PHA block copolymers. In *Cupriavidus necator*, P(3HB) units were synthesised during fructose utilisation and 3-hydroxyvalerate units were synthesised during pulse feeds of pentanoic acid resulting in the formation of P(3HB-co-3HV) random copolymer.

To synthesise a block copolymer, an off-gas mass spectrometry (MS) detection system was used to accurately control pentanoic acid additions to the bioreactor in *Cupriavidus necator*. Differential scanning calorimetry (DSC) and nuclear magnetic resonance spectroscopy (NMR) of the polymer produced showed that about 30% of the total polymer samples reveal melting properties suggestive of block copolymers. Rheology tests showed additional mesophase transitions found only in block copolymers. The block copolymers exhibited secondary transitions in the storage modulus that are not displayed in random copolymers. This study thus depicted the possibility of synthesis of PHA block copolymers within a biological synthesis system. Once the link between the co-feeding time and the resulting properties are established, synthesis of PHA block copolymers with desired and improved properties can be effectively achieved (Pederson *et al.*, 2006).

Table 1. Physical properties of poly(3-hydroxybutyrate), poly(3-hydroxybutyrate-co-3-hydroxyvalerate) and poly(4-hydroxybutyrate)

Properties	Poly(3HB)	Poly(3HB-co-3HV)	Poly(4HB)
Melting temperature (^{0}C)	177	150	60
Glass transition temperature (^{0}C)	4	-7.25	-50
Tensile strength (MPa)	40	25	104
Elongation at break (%)	6	20	1000

PHA Synthases

The biosynthesis of polyhydroxyalkanoates is catalysed by enzymes called PHA synthases. These enzymes use coenzyme A (CoA) thioesters of hydroxyalkanoic acids (HAs) as substrates to catalyse the polymerisation of HAs into PHAs with the simultaneous release of CoA (Steinbuchel and Hein, 2001) (Figure 3).

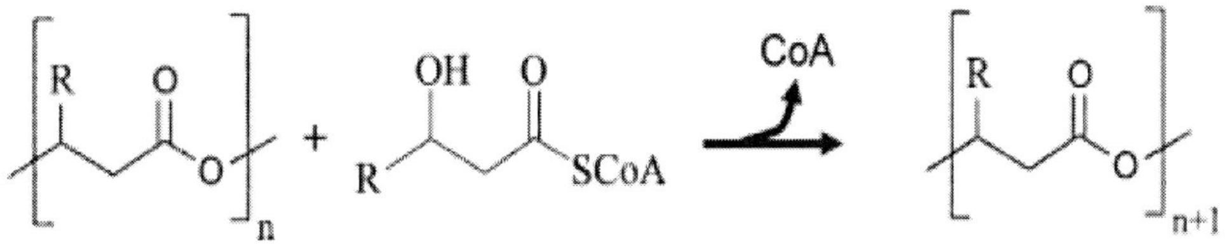

Figure 3. The general reaction catalysed by PHA synthases (Rehm, 2003).

As many as 88 different PHA synthases from 68 microorganisms have been cloned and their primary structures determined (Steinbuchel and Hein, 2001). Based on their primary structures, substrate specificities of the enzymes and the subunit composition, PHA synthases have been classified into four major classes (Table 2).

Class I (e.g. *Cupriavidus necator*) and class II (e.g. *Pseudomonas aeruginosa*) PHA synthases consist of only one subunit each (PhaC) with molecular masses ranging between 61kDa and 73kDa (Wieczorek *et al.*, 1995). Class III PHA synthases (e.g. *Allochromatium vinosum*) consists of two subunits: a 40kDa PhaC subunit displaying similarity to the class I and class II PHA synthases and a 40kDa PhaE subunit with no similarity to PHA synthases at all. Class IV PHA synthases (e.g. *Bacillus cereus*) also possess two subunits like class III but in this case, the PhaE subunit is replaced by the PhaR subunit (Rehm, 2003, Valappil *et al.*, 2008).

Table 2. The four classes of PHA synthases (adapted from Rehm, 2003)

Class	Subunits	Occurrence	Substrates
I	PhaC ~60 - 65 kDa	*Cupriavidus necator* + most other bacteria except those listed below	$3HA_{SCL} - CoA$ $[\sim C_3 - C_5]$ Also: $4HA_{SCL} \cdot CoA$ $5HA_{SCL} - CoA, 3MA_{SCL} - CoA$
II	PhaC ~60 - 65 kDa	*Pseudomonas oleovorans* + all Pseudomonas sp. belonging to rRNA homology group I	$3HA_{MCL} - CoA$ $[\sim \geq C_5]$
III	PhaC PhaE ~40 kDa ~40 kDa	*Chromatium vinosum* *Thiocapsa pfenningii* +other sulphur purple bacteria + all cyanobacteria	$3HA_{SCL} - CoA$ $[\sim C_3 - C_5]$ $3HA_{MCL} - CoA$ $[\sim C_5 - C_8]$ Also: 4HA-CoA, 5HA-CoA
IV	PhaC PhaR ~40 kDa ~22 kDa	*Bacillus cereus SPV*	$3HA_{SCL} - CoA$ $[\sim C_3 - C_5]$

PHA Synthases in the Cell

Studies on *Zoogloea ramigera* and *Cupriavidus necator* demonstrate that PHA synthases occur in a soluble form in the cell cytoplasm, when cells are not accumulating PHA. Once PHA biosynthesis starts, it has been found that enzyme activity is associated with the granules. Gold labelled antibodies have been used to exhibit that PHA synthases are bound to the PHA granules and are localized to the granule surface (Gerngross *et al.*, 1993; Stuart *et al.*, 1995).

Ballard *et al.*, 1987 set out to study the relation between the number of granules in a cell and its effect on the level of polymer accumulation. It was reported that in *Cupriavidus necator* the number of granules per cell is fixed at the earliest stages of polymer accumulation. Owing to cell wall material constraints, it was established that once the cell accumulates 80% dry cell weight of polymer, the polymer production in *Cupriavidus necator* is stopped, even though the PHA synthase activity is high (Ballard *et al.*, 1987). The PHA granules are typically 0.2 to 0.5μm in diameter and possess a membrane coat of about 2nm in thickness. The coat is composed of lipid and protein. Dunlop and Robards, 1973, were able to isolate granules from *Bacillus cereus* and used freeze-etching to study the ultra-structure of thin sections. They found that the granules consisted of a central core with a diameter between 140 and 370nm which occupied less than 50% of the granule volume. This core was surrounded by an outer coat. The proposed structure of a PHA granule as it appears *in vivo* is shown in Figure 4.

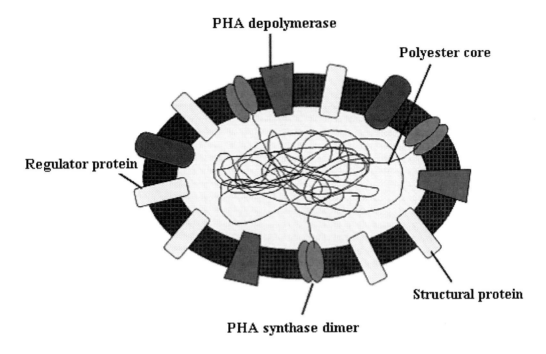

Figure 4. The proposed structure of a PHA granule as it appears *in vivo*.

Structural Features of PHA Synthases

The 88 PHA synthases from different bacteria have a very strong resemblance. There are six conserved blocks of amino acid sequence regions. However, the N-terminal region is highly variable and the enzyme activity is thought to be associated with the highly conserved C-terminal region. The fact that the conserved C-terminal region is hydrophobic in class I and class II synthases imply that this region could be the binding domain that attaches the synthase to the hydrophobic polyester core. Interestingly, class III and IV synthases do not possess a hydrophobic C-terminus. However, PhaE and PhaR respectively contain a hydrophobic C-terminus which might carry out a parallel function to that proposed for the C-terminus of the class I and class II PHA synthases (Muh *et al.*, 1999). Eight amino acid residues have been found to be conserved in all the 59 PHA synthases known (Rehm *et al.*, 2002; Schubert *et al.*, 1991). The highly variable nature of the N-terminus of PHA synthases indicated that this fragment was not necessary for a functionally active enzyme. However, Zheng *et al.*, 2006 found that in *Cupriavidus necator*, the polypeptide F79 – E88, in the N-terminus of the PHA synthase was crucial for the enzyme activity. This region was involved in protein-protein interactions between PHA synthase and other PHA proteins. Also, the polypeptide D70 – D78 was involved in PHA polymer chain transfer regulation. Hence, the amphiphilic α-helix assembled by the amino acid region, D70 – E88 of the N-terminal region played a significant role in maintaining the PHA synthase activity and regulating molecular weight and polydispersity of the accumulated polymer. This supported the earlier work carried out by Schubert *et al.*, 1991, who found that a recombinant *E. coli* strain lacking the N-terminus (approximately 100 amino acids) of the *phaC* gene lost the PHA polymerisation ability. However, the protein without the first 30 amino acid residues had identical PHA synthase function to that of the wild type. Hence it is evident that different regions of the N-terminus contribute differently to enzyme function (Schubert *et al.*, 1991).

Recently, it was reported that saturation mutation of the PHA synthase in *Aeromonas caviae*, led to changes in the molecular weight of the PHAs produced. *Aermonas caviae* produces a copolymer, P(3HB-co-3HHx) from vegetable oils as carbon source. Since the organism could accumulate only less than 30% dry cell weight of this polymer, the PHA synthase from *Aeromonas caviae* was introduced into a PHA negative strain of *Cupriavidus necator* to produce a recombinant strain. Mutation studies carried out on this strain revealed that the steric size of the amino acids at positions downstream of the active site histidine, affect PHA synthase activity and molecular weight of the polymer. A series of substitutions made at A505 revealed that this residue affected substrate specificity, copolymer composition and molecular weight of PHA. The production of such mutants with further modification will make it possible to synthesise tailor-made PHAs (Tsuge *et al.*, 2007).

To date, the tertiary structure of PHA synthases has not been determined. These enzymes consist of 49.7% variable-loop and 39.9% α-helical structure (Cuff *et al.*, 1998). All PHA synthases appear to contain a lipase box (GX[S/C]XG). However, the active site serine of lipases is substituted with a cysteine. Further analysis has proven that these enzymes have a α/β hydrolase domain at the C-terminal region. It is now established that PHA synthases belong to the α/β hydrolase superfamily and possess a catalytic triad comprising of a highly conserved cysteine from a 'lipase box', an aspartic acid and a histidine (Jia *et al.*, 2000; Rehm, 2003).

Threading models for classes I, II, III and IV PHA synthases have been published (Valappil *et al.*, 2007a).

Genetic Organisation of the PHA Biosynthetic Genes

Most of the genes involved in PHA metabolism are clustered together in the bacterial genomes (Figure 5).

In class I enzymes (prototype: *Cupriavidus necator*), the genes for the PHA synthase (*phaC*), β-ketothiolase (*phaA*) and NADP-dependent acetoacetyl CoA reductase (*phaB*) constitute the *phaCAB* operon (Peoples and Sinskey, 1989). In class II synthases (prototype: *Pseudomonas aeruginosa*) two different *phaC* genes are separated by the *phaZ* gene. *phaZ* codes for an intracellular PHA depolymerase. Furthermore, downstream of the synthase operon, the *phaD* gene is collinearly located along with the genes *phaI* and *phaF*, which are transcribed in the opposite direction. The function of the *phaD* gene is unknown and the latter genes code for structural and regulatory proteins (Rehm, 2003).

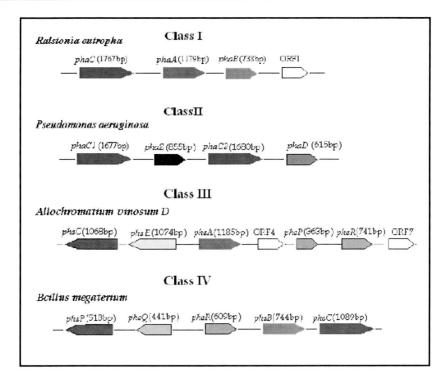

Figure 5. Genetic organisation of PHA synthase genes (Rehm, 2003).

In class III synthases (prototype: *Allochromatium vinosum*), both *phaC* and *phaE* are directly linked in their genomes and constitute a single operon (Rehm, 2003). Class IV synthases (prototype: *Bacillus megaterium*) have the *phaC* and *phaR* genes separated by the *phaB* gene. The *phaRBC* operon is divergently transcribed as a tricistronic operon and immediately upstream, the *phaP* and *phaQ* genes are transcribed in an opposite direction to the *phaRBC* operon, each from a separate promoter (McCool and Cannon, 1999).

The Mechanism of Action of PHA Synthases

The mechanism of action of PHA synthases is based on the reaction mechanism employed by the α/β hydrolases (Rehm, 2003) (Figure 6).

The proposed mechanism involves covalent catalysis by a cysteine-histidine pair and the non-covalent catalysis of an aspartate to activate the 3-hydroxyl of a second HBCoA (the coenzyme A form of hydroxybutyrate) to form an ester bond. The cysteine is activated for a nucleophilic attack by a histidine in the active site. Following this, the hydroxyl group of the second HBCoA is activated for nucleophilic attack by a second base catalyst to form a dimeric oxoester that is covalently or non-covalently bound to the synthase (Stubbe and Tian, 2003). The second base catalyst function has been attributed to the aspartate residue in the catalytic triad. Replacement of the aspartate residue with an asparagine, impaired the turnover of the substrate, 3-hydroxybutyryl-CoA and chain elongation was truncated (Rehm, 2003). The growing polymer chain exits through a tunnel accommodating an oligomeric HB (Stubbe and Tian, 2003).

In Class I PHA synthases, the catalytic triad comprises of Cys-319, Asp-480 and His-508. The highly conserved Trp-425 residue is thought to play an important role in protein-protein interaction by generating a hydrophobic surface for PhaC dimerisation. In Class III PHA synthases, the catalytic triad comprises of Cys-149, Asp-302 and His-331 residues. The putative active-site nucleophile, cysteine was located at the elbow of the strand–elbow–helix motif (Rehm, 2003).

Class II PHA synthases, on the other hand, represent a unique group with unique features not found in the other classes. Cys-296, His-480/His-453 and Asp-452 form the catalytic triad and the histidine residues work inter-changeably. Unlike class I or II, the replacement of the Cys-296 by a serine, produced a highly active enzyme (Rehm, 2003).

PHA synthases catalyse the formation of water-insoluble polyester which is located at the polyester–water interface, i.e. attached to the surface of PHA granules. The water-soluble substrate binds to the water-exposed regions of the PHA synthase, to enable the oriented synthesis of the growing polymer chain (Stubbe and Tian, 2003).

The interfacial activation of the PHA synthases is explained based on that of the *B. glumae* lipase, on which the PHA Class I synthase was modelled. This lipase structure, was obtained in the 'closed' conformation exhibiting the active site buried underneath a helical segment (α5), called a 'lid' or a 'flap'. In the PHA synthase model, the active site was also buried underneath this structurally conserved helical segment. During transition to the open conformation of the lipase, due to interfacial activation, the active site becomes accessible to the solvent and a hydrophobic surface is exposed by the movement of the lid. Rolling back of the lid from the active site occurs at an oil–water interface. The soluble PHA synthase turns into an amphipathic molecule upon availability of substrate (Rehm, 2003; Stubbe and Tian, 2003).

PHA Synthases in *Bacillus*

Class IV PHA synthases has been the latest addition to the PHA synthase family. This separate class of PHA synthases from *Bacilli* has generated a lot of interest. Though some preliminary work has been carried out on the structure of PHA inclusion bodies in *Bacilli*, reports of the characterization of the enzyme are few (Lundgren *et al.*, 1964, Ellar *et al.*, 1968, Griebel *et al.*, 1968, Dunlop and Robards, 1973).

In 1999, McCool and Cannon successfully cloned the *Bacillus megaterium* PHA gene cluster into *E. coli* and *P. putida* (McCool and Cannon, 1999). They found that both PhaC and PhaR are essential for PHA synthase activity, by using *in vivo* and *in vitro* methods. In the same report, they suggested the existence of two forms of PHA synthase: an active form in PHA accumulating cells and an inactive form in PHA non-accumulating cells. In the latter case, PhaC is more prone to degradation (McCool and Cannon, 2001).

Satoh *et al.*, 2002 have also found the occurrence of PhaC and PhaR in *Bacillus* INT005. They constructed two recombinant *E.coli* strains: one that contained the entire *phaRBC* operon and the other that carried the *phaC* gene on its own. PHA synthase activity could be detected only in the former (Satoh *et al.*, 2002). Law *et al.*, 2003 have cloned the *pha* genes from *B. megaterium* into *Bacillus subtilis* and have demonstrated that the presence of the *phaP* and *phaQ* genes, in addition to the *phaRBC* operon, are vital for PHA accumulation in cells (Law *et al.*, 2003). The *phaRBC* operon sequence from *Bacillus cereus* SPV has also been sequenced (Valappil *et al.*, 2008).

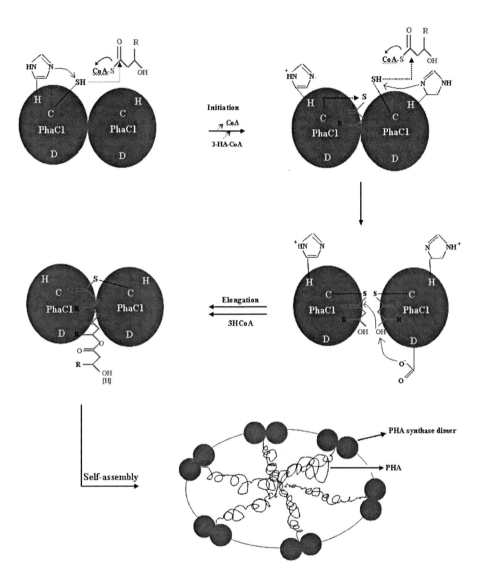

Figure 6. Proposed catalytic mechanism of PHA synthases (Rehm, 2003).

Class IV PHA synthases were recognised as a separate class in 2003. This fairly recent class of enzymes is poorly characterised. The complete characterisation of this group of enzymes will enable better understanding of their mechanism and in turn help in the development of better polymer production strategies.

PHA Production from Microorganisms

For almost 50 years, P(3HB) was the only PHA known until Wallen and Rohwedder identified a number of additional 3-hydroxy fatty acids in active sludge samples (Wallen and Rohwedder, 1974). The type of PHA produced depends on the carbon source supplied to the microbe. The carbon source chosen is given paramount importance because it contributes greatly to the cost of the production process. Hence it is essential that exceptionally good PHA producers are identified and the fermentation processes optimised.

Effect of Carbon Sources on PHA Production

Earlier studies indicated that with the increasing ratio of carbon to nitrogen in the growth medium of the bacteria, the rate of PHA accumulation increased (Macrae and Wilkinson, 1958). However, Suzuki *et al.* (1988) reported that a certain minimum concentration of a nitrogen source was necessary in the P(3HB) production phase. Feeding a small quantity of ammonia in a fed-batch culture of *Protomonas extorquens* resulted in a more rapid accumulation of P(3HB) than when ammonia was not supplied. Excess of ammonia feeding on the other hand caused degradation of the accumulated polymer (Suzuki *et al.*, 1988). In contrast, certain organisms such as *Alcaligenes latus* and *Paracoccus denitrificans* can accumulate PHA during active cell growth, without any nutrient limitation (Maekawa *et al.*, 1993, Kim *et al.*, 1997).

Of the many bacteria that produce P(3HB), *Cupriavidus necator* is perhaps the most intensively studied. *Cupriavidus necator* produces P(3HB) when grown on organic substrates such as ethanol and glucose. When grown on mixtures of 5-chloropentanoic acid and pentanoic acid, *Cupriavidus necator* can form terpolymers of 3-hydroxybutyrate, 3-hydroxyvalerate and 5-hydroxyvalerate (Doi *et al.*, 1987). *Delftia acidovorans* can also incorporate 3-hydroxyvalerate (3HV) monomers into polyhydroxyalkanoate (PHA) copolymers. When a mixture of sodium 3-hydroxybutyrate and sodium valerate is used as the carbon source for *Delftia acidovorans*, poly(3-hydroxybutyrate-*co*-3-hydroxyvalerate), P(3HB-*co*-3HV) containing 0–90 mol% of 3HV is obtained (Ching-Yee, L. and Kumar, S., 2007).

Most recently, Dai *et al.* (2008) have reported the production of PHA copolymers that comprise four monomeric units - 3-hydroxybutyrate (3HB), 3-hydroxyvalerate (3HV), 3-hydroxy-2-methylvalerate (3HMV) and 3-hydroxy-2-methylbutyrate (3HMB). The organisms they used included *Defluviicoccus vanus*-related glycogen accumulating organisms (DvGAOs) such as *Candidatus competibacter* phosphatis, under anaerobic conditions without any nutrient limitations. They were able to control the composition of the polymer by manipulating the ratio of propionate and acetate provided in the feed. By incorporating substantial amounts of 3HMV and 3HMB monomer units "defects" were introduced into the PHBV copolymer structure and this helped to lower the crystallinity of the polymer (Dai *et al.*, 2008).

When grown on only valeric acid, *Chromobacterium violaceum* synthesises a polymer of 100 mol% of 3HV unit. P(3HB-co-3HV) which contains 95 mol% of 3HV was produced by

Pseudomonas sp. HJ-2 when grown on valeric acid. When grown on 3-hydroxypivalic acid, *Rhodococcus ruber* NCIMB 40126 and *Nocardia corallina*, produced poly(3-hydroxypivalate) or copolymers of 3HB and 3-hydroxypivalate (Kim and Lenz, 2001).

Among the *Pseudomonas* sp., *Pseudomonas oleovorans* and *Pseudomonas putida* are the most thoroughly studied for MCL-PHA production. *Pseudomonas putida* KT2442 synthesises polymers with 3-hydroxydecanoate and minor constituents of 3-hydroxyhexanoate, 3-hydroxyoctanoate and 3-hydroxydodecanoate as monomeric units, when grown on the unrelated carbon source, glucose (Kim *et al.*, 2007). *Pseudomonas* sp. usually synthesise 3-hydroxyoctanoate as the main monomer when grown on carbon sources containing even number of carbon atoms and 3-hydroxynonanoate when grown on carbon sources containing odd number of carbon atoms (Kim and Lenz, 2001).

There have been many papers reporting a range of different PHAs produced by the genus *Bacillus* when grown on different carbon sources. Chen *et al.*, 1991 have reported that co-polymers of P(3HB-co-3HV) are accumulated when *Bacillus cereus* cultures are fed with odd chain n-alkanoic acids such as propionic, valeric or heptanoic acids (Chen *et al.*, 1991). Caballero *et al.*, 1995 have reported the biosynthesis and characterization of 3HB rich polymers consisting of 2-4 mol% of 3-hydroxycaproate (3HC) units and a terpolymer consisting of 3HC and 3-hydroxyoctanoate (3HO) units in *Bacillus cereus* when grown on caproate or octanoate (Caballero *et al.*, 1995). *Bacillus cereus* UW85 produces a tercopolymer of 3-hydroxybutyrate, 3-hydroxyvalerate and 6-hydroxyhexanoate units when grown on Є-caprolactone as the sole carbon source (Labuzek and Radecka, 2001). It was established that glucose, sucrose and fructose were the most suitable carbon sources for P(3HB) accumulation in *Bacillus mycoides* RLJ B-017 during optimisation studies. Among the nitrogen sources, beef extract and di-ammonium sulphate proved ideal for P(3HB) synthesis (Borah *et al.*, 2002). When fed with substrates such as butyrate, valerate, hexanoate, octanoate, decanoate and Є-caprolactone, *Bacillus* sp. INT005 produced a range of polymers such as P(3HB), P(3HB-co-3HV), P(3HB-co-3HHx), P(3HB-co-4HB-co-3HHx) and P(3HB-co-6HHx-co-3HHx) (Tajima *et al.*, 2003).

Bacillus cereus SPV is capable of using a wide range of carbon sources including glucose, fructose, sucrose, various fatty acids and gluconate for the production of PHAs. The media used for the polymer production, the Kannan and Rehacek medium (Kannan and Rehacek, 1970) was novel in the context of the genus *Bacillus*. It was observed that the PHA, once produced, was found to remain at a constant maximal concentration and did not degrade. This strain was able to synthesise various PHAs with 3-hydroxybutyrate (3HB), 3-hydroxyvalerate (3HV) and 4-hydroxybutyrate (4HB)-like monomer units from structurally unrelated carbon sources such as fructose, sucrose and gluconate. The PHAs isolated had molecular weights ranging between $0.4\text{-}0.8 \times 10^6$ and low polydispersity index values (M_W/M_N) ranging from 2.6 to 3.4 (Valappil *et al.*, 2007b).

Large-Scale Production of PHAs

The commercialisation of PHAs has been held back by the high production costs associated with them as compared to petroleum derived plastics. Hence, a lot of effort has been directed towards the advancement of efficient bacterial strains and economical fermentation processes. Based on the characteristics of their PHA accumulation, bacteria can be grouped into two categories. The first group of bacteria accumulate PHAs when their cell growth is hampered by the limitation of an essential nutrient such as nitrogen, potassium, oxygen or sulphur. The other group accumulates PHAs during their growth phase. *Cupriavidus necator* belongs to the first group whereas *Alcaligenes latus* and recombinant *E. coli* belong to the latter (Lee and Chang, 1995) (Lee, 1996). The first group requires expensive two-stage cultivation for PHA accumulation: the cells have to be grown to a high concentration under non-limiting conditions first; to be followed by a second phase of PHA accumulation. Alternatively, the second group just requires single stage fermentation (Grothe *et al.*, 1999). *Cupriavidus necator* has been used for the commercial production of P(3HB-co-3HV) by Zeneca BioProducts, UK and Monsanto, USA.

Many bacteria have been metabolically engineered to produce PHAs competently. For the first time, the PHA biosynthetic genes from *Cupriavidus necator* were introduced into *Escherichia coli* and a large amount of P(3HB) was synthesised (Schubert *et al.*, 1988). Poly(3HB) was accumulated to a level greater than 80g L^{-1} in a fed batch culture of recombinant *E. coli* in Luria Bertani medium containing an initial feed of 20g L^{-1} glucose (Kim *et al.*, 1992). It was reported that P(3HB) production by recombinant *E. coli* depended on the amount of intracellular acetyl-CoA and NADPH available (Lee *et al.*, 1996).

The fed-batch culture strategy is the recommended method to achieve high cell density and in turn, high productivity of a desired product. Optimisation studies carried out on *Cupriavidus necator* suggests that it is vital to maintain the concentration of the carbon source at an optimal value and to provide the nutrient limitation trigger of PHA accumulation at an optimal time point. Under ammonium limitation conditions, *Cupriavidus necator* DSM 545 accumulated up to 24g L^{-1} of P(3HB) in fed batch culture, using glucose as the carbon source (Ramsay *et al.*, 1990). *Azotobacter vinelandii* UWD accumulates P(3HB) during the exponential growth phase (Page and Knosp, 1989).

The pH-stat fed-batch culture of *Alcaligenes latus* on sucrose resulted in a P(3HB) concentration of 68.4g/L in 18 hours with a high productivity of 3.97g P(3HB) L^{-1} h^{-1} by using a high inoculum size of 13.7 g/L. However, the P(3HB) content achieved was only 50% dcw (Yamane *et al.*, 1996). Wang and Lee (1997) tried a range of limiting conditions to optimise the strategy of nutrient limitation in increasing P(3HB) content in *Alcaligenes latus*. After trying out the effect of limiting nitrogen, phosphorus, magnesium or sulphur in flask cultures, nitrogen limitation was decided upon as the best strategy since it allowed the enhancement of P(3HB) production (Wang and Lee, 1997).

Methylotrophic bacteria are also producers of PHAs and can use cheap methanol as a carbon source. Suzuki *et al.*, 1986 carried out an extensive study of the effects of physiological parameters such as temperature, pH and methanol concentration on the growth and accumulation of P(3HB) by *Protomonas extorquens* sp. Strain K. Using fed-batch culture, 136g/L of P(3HB) was produced in 175 hours. After optimisation of the medium composition, 149g/L of P(3HB) was produced in 170 hours (Suzuki *et al.*, 1988).

The fed-batch culture of recombinant *E. coli* requires a large amount of oxygen to maintain the dissolved oxygen level above 20% of air saturation. Since oxygen is a very expensive commodity and oxygen transfer is very poor in a large scale fermenter, P(3HB) production under insufficient oxygen supply was considered (Wang and Lee, 1997). Insufficient oxygen supply (a dissolved oxygen level of 1-3%) during the active growth phase of the cells, resulted in the inhibition of cell growth and there was no apparent enhancement of polymer production. However, cell growth and P(3HB) production remained unaffected when insufficient oxygen was supplied during the active P(3HB) synthesis phase. P(3HB) was actively accumulated and there was simultaneous increase of P(3HB) content. Cell concentration, P(3HB) concentration and P(3HB) content obtained were as high as 204.3g/L, 157.1g/L and 77% dry cell weight respectively, resulting in the P(3HB) productivity of 3.2 g/L/hour (Wang and Lee, 1997).

Bacillus sp. JMa5 accumulated P(3HB) at a concentration of 25 – 35% dcw when grown on molasses media during fermentation and the accumulation was growth associated. In addition, they observed that high ratios of carbon to nitrogen and carbon to phosphorus further promoted P(3HB) production. Due to low dissolved oxygen supply, sporulation occurred and as a result, P(3HB) contents were reduced.(Wu *et al.*, 2001).

Bacillus cereus SPV accumulates about 38% P(3HB) per cell dry weight when glucose is used as the main carbon source at shaken flask levels. When production was scaled up to 20L batch fermentations, about 29% P(3HB) per cell dry weight was accumulated in 48 hours. However, the cells were able to accumulate P(3HB) at a concentration of 38% dcw when a simple glucose feeding strategy was adopted. Acidic pH ensured that the accumulated P(3HB) did not degrade, thus emphasising the potential of *Bacillus cereus* SPV for commercial P(3HB) production. Different extraction procedures were tried out such as the soxhlet method, the chloroform method and the chloroform-hypochlorite dispersion method. It was found that the chloroform extraction method produced the highest crude yield of P(3HB), about 31% dry cell weight, though the purest form of P(3HB) was obtained from the soxhlet method, 99% (Valappil *et al.*, 2007c).

Detection, Isolation and Characterisation of PHAs

Traditional Methods

The traditional method of PHA detection in bacterial cells is by staining the cells with Sudan Black B. Nile Blue A stain, on the other hand has been recommended as a better staining tool by Ostle and Holt since they are specific for PHAs and do not stain glycogen and polyphosphate inclusion bodies. Nile Blue A is a water-soluble basic oxazine dye with greater affinity for PHA than Sudan Black. It gives a bright orange fluorescence at 460nm (Ostle and Holt, 1982).

Early efforts to isolate PHA from bacteria were accomplished by repeated centrifugation of DNase-treated cell extracts layered on glycerol causing the granules to collect at the surface (Merrick and Duodoroff, 1964). Some researchers have used both differential and density gradient centrifugation with glycerol to purify granules from *Bacillus megaterium* (Griebel *et al.*, 1968). PHAs can be readily extracted from microbial cells using chlorinated hydrocarbons. Chloroform refluxing has been commonly used. The resulting solution is concentrated and precipitated using cold methanol or ethanol (Brandl *et al.*, 1988). However, using solvents on a large-scale is expensive. Lafferty and Heinzle have used ethylene carbonate and propylene carbonate to extract P(3HB) from biomass (Anderson and Dawes, 1990).

In 1926, Lemoigne described the first analytical method for P(3HB) analysis. He extracted the polymer with hot alcohol and purified it using chloroform and diethyl ether. The polymer was then saponified and distilled, giving rise to a volatile and non-volatile residue. The volatile fraction was trapped, crystallized, and weighed accurately, and contained the dehydrated product of 3-hydroxybutyric acid: crotonic acid (trans-2-butenoic acid). The non-volatile fraction was also weighed accurately and contained primarily 3-hydroxybutyric acid. From the weight of both fractions the amount of P(3HB) in the starting material, *Bacillus* `M' could be calculated. This method though accurate, was time-consuming and required large amounts of biomass. Williamson and Wilkinson devised a novel spectrophotometric method in 1958 whereby *Bacillus* cells were treated with sodium hypochlorite. The turbid solution obtained was measured spectrophotometrically. The drawback was that the method had to be fine-tuned for every new bacterium. In 1961, Slepecky and Law described a more specific analytical method whereby the polymer could be treated with concentrated sulphuric acid to give rise to crotonic acid. The absorbance at 235nm was used as a measure of the concentration of the crotonic acid formed. However, all the methods for quantifying PHAs described above are either time-consuming or inaccurate.

GC and MS

A method developed by Braunegg *et al.* in 1978, enabled detection of the polymer at concentrations as low as 10μM. They subjected cells to direct acid or alkaline methanolysis, followed by gas chromatography (GC) of the methyl ester. However, an unfavourable partitioning coefficient of the 3-hydroxybutyric acid methyl ester after phase separation during sample cleanup resulted in partial recovery (Braunegg *et al.*, 1978). Riis *et al.* (1988) suggested improvements to the above method by recommending the use of propanol and hydrochloric acid, in order to minimise degradation of the polymer. This method is based on the derivatisation of 3-hydroxybutyric acid to a propyl ester. As the propyl ester is more lipophilic compared to the methyl ester, the partitioning coefficient is improved (Riis and Mai, 1988). Further, it was found that using a capillary instead of a packed column, removed solvent interference with PHA related peaks and a better resolution was obtained (Comeau *et al.*, 1988). Brandl *et al.*(1989) have described a method where they have determined the PHA content and quantity by GC after subjecting lyophilised cells to methanolysis for 140 minutes at 100^0C to produce methyl esters of the constituent 3-hydroxyalkanoic acids (Brandl *et al.*, 1989). Lageveen *et al.* (1988) and many others have modified the GC technique to analyse the MCL PHAs. They used 15% sulphuric acid in methanol as the transmethylation reagent.

The methyl esters were extracted into the chloroform phase after 140 minutes at 100^0C. In 1983, Findlay and White quantitated PHA samples from *Bacillus megaterium* by acid ethanolysis, followed by GC-MS (mass spectrometric) analysis of the resulting 3-hydroxyalkanoic acid ethyl esters (Findlay and White, 1983).

NMR

Nuclear magnetic resonance (NMR) is usually used to detect the presence of two or more distinctive polymers. The complexity of the carbonyl signals in the ^{13}C-NMR spectrum can be used to determine whether a given PHA sample consists of homopolymers or a copolymer. Doi *et al.* (1986) have used NMR to prove that the copolymer poly(3HB-co-3HV) was produced when *Alcaligenes eutrophus* was grown on acetate and propionate (Doi *et al.*, 1986). NMR has also been used in the past to determine the position of the hydroxyl group in the alkyl chain (Valentin *et al.*, 1991).

HPLC

Karr *et al.*, 1983 modified the Slepecky and Law method to analyse the crotonic acid produced from the polymer using ion-exchange high-performance liquid chromatography (HPLC) and UV detection. Samples containing quantities between 0.01 and 14μg of polymer from *Rhizobium japonicum* could be analysed. The GC method described by Braunegg *et al.* (1978) possibly underestimates the total amount of polymer since it involves the hydrolysis of the polymer using a 3% sulphuric acid solution, which could result in some degradation. Hence, the HPLC method offers an interesting alternative (de Rijk *et al.*, 2002).

Biodegradation

Perhaps one of the greatest advantages that PHAs possess over the other biodegradable polymers is their ability to degrade under both aerobic and anaerobic conditions. They can also be degraded by thermal means or by enzymatic hydrolysis. In a biological system, PHAs can be degraded using microbial depolymerases as well as by nonenzymatic and enzymatic hydrolysis in animal tissues (Gogolewski *et al.*, 1990).

The biodegradability of a polymer is governed primarily by its physical and chemical properties. It has been found that low molecular weight PHAs are more susceptible to biodegradation. The melting temperature is another important factor to be considered when studying biodegradation. As the melting point increases, the biodegradability decreases. With increasing melting temperature, the enzymatic degradability decreases. Tokiwa *et al.*, 1977 found that lipases cannot hydrolyse the optically active P(3HB) *in vitro*. This could be due to the high melting temperature of the latter (178°C) (Tokiwa and Suzuki, 1977).

Mochizuki *et al.*, 1997 have explained that biodegradation of solid polymers is influenced by chemical structure (especially functional groups and hydrophilicity-hydrophobicity balance) and highly ordered structures (mainly crystallinity, orientation and morphological properties) (Mochizuki and Hirami, 1997). Tokiwa *et al.*, 2004 have reaffirmed that crystallinity plays a very important role in biodegradability. They have also identified that highly ordered structures, i.e. highly crystalline materials have lower biodegradability (Nishida and Tokiwa, 1992).

Biodegradation in the Environment

The microbial population in a given environment and the temperature, contribute to biodegradability in the environment (Tokiwa and Calabia, 2004). Micro-organisms from the families of *Pseudonocardiaceae*, *Micromonosporaceae*, *Thermomonosporaceae*, *Streptosporangiaceae* and *Streptomycetaceae* predominantly degrade P(3HB) in the environment (Tokiwa and Calabia, 2004). Extracellular depolymerases from these microorganisms degrade polyhydroxyalkanoates present in the environment (Tokiwa and Calabia, 2004). The microorganisms present in the environment attack the polymers on the surface (Holmes, 1985). These microbes secrete extracellular enzymes that solubilise the polymer and these soluble products are then absorbed through their cell walls and utilised (Doi, 1990). The PHA depolymerase enzymes act on the polymer mainly by hydrophobic interactions. Degradation by these depolymerases initially produces oligomers. Some microbes produce an additional dimer hydrolase which further break down the oligomers into the corresponding monomer (Tanaka *et al.*, 1981). These extracellular depolymerases are quite well understood (Merrick and Doudoroff, 1964).

The rate of biodegradation of PHAs depend on environmental conditions like temperature, moisture, pH, nutrient supply and those related to the PHA materials themselves, such as, monomer composition, crystallinity, additives and surface area (Abe and Doi, 2002).

Biodegradation in Living Systems

The mechanism of biodegradability of the PHAs acquires utmost importance *in vivo* because the rate of degradation of the material should equal the regenerative rate of tissues, in order to be effectively used as scaffolds in tissue engineering applications. Literature on this aspect is limited and gives contradictory views. Studies by Miller and Williams suggest that P(3HB) and P(3HB-co-3HV) monofilaments implanted in animals, did not lose their mass and maintain their physical and mechanical properties over a period of 6-12 months (Miller and Williams, 1987). However, Duvernoy *et al.*, 1995 have reported that P(3HB) films lose their mass by 30-80% within a year.

In vivo PHAs are degraded by the enzymes present in blood and animal tissues in addition to non-enzymatic degradation. Atkins and Peacock (1996) studied the PHA-depolymerising activity of calf serum, pancreatin and synthetic gastric juice on P(3HB-co-3HV) microspheres with caprolactone. Weight loss was observed in the following order: bovine serum > pancreatin juice > synthetic gastric juice (Atkins and Peacock, 1996).

In vitro studies have been carried out to investigate the stability of the polymers in model systems that mimic *in vivo* conditions, using varying values of pH, temperature and medium salinity. The polymers used in these tests are of different compositions, molecular mass and crystallinity. The degradation is monitored by tracking the reduction in molecular mass and degree of crystallinity. Based on studies carried out, it is clear that PHA degradation is multiphasic. In the first few weeks, the amorphous regions are eroded by random scission. Over a few weeks, polymer chains are disrupted and the crystallinity of the polymer increases. Monomers, dimers and tetramers are formed due to which the molecular mass decreases. Later on, erosion processes are initiated leading to reduction in polymer mass. This can occur over 2-3 years and is dependent on environmental circumstances and the physicochemical properties of the PHA (Volova, 2004).

Applications

The number of research publications dealing with biosynthesis, fermentation, and characterisation of the PHA family of biopolymers has increased in the past two decades (US Congress, 1993). In 1990, about 25% of the plastics market (about 7.5 million tonnes) was consumed by the packaging industry. In 1993, it was predicted that the demand for biodegradable materials would increase within 3 years and a major portion of that demand would be from the packaging industry. The plastics market in the US is considerable with production amounting to about 170 million tons per year and expanding at around 4-5% per annum (Peoples and Snell, 2006). Over the last decade, the applications have increased both in variety and specialisation.

Medical Applications

Biodegradable polymers are finding significant applications in the medical field. PHAs, especially P(3HB), P(3HB-3HV), P(4HB), P(3HO) and P(3HB-3HHx) are frequently used in tissue engineering. They are widely used as bone plates, osteosynthetic materials and surgical sutures. They are useful in the slow release of drugs and hormones. However, the use of P(3HB) in such applications is restricted by its poor rate of biodegradation and high resistance to hydrolysis in sterile tissues (Steinbuchel and Fuchtenbusch, 1998). PHA fibres are especially sought after to make swabs and dressing materials for surgery (Babel *et al.*, 1990). PHAs are produced by large-scale fermentations rather than chemical synthesis and hence do not contain the left over undesirable metal catalysts used in chemical synthesis (Kim and Lenz, 2001). PHAs such as P(3HB), P(3HB-co-3HV), P(4HB), P(3HB-co-4HB), P(3HBco-3HHx) and P(3HHx-co-3HO) have been tested in animals. These tests have revealed that all these polymers were biocompatible in various host systems (Valappil *et al.*, 2006).

Sevastianov *et al.*, 2003 have examined the possible effects of P(3HB) and P(3HB-3HV) on the immune system and have shown that both these polymers do not affect the haemostasis system at the cell response level. However, they tend to trigger the coagulation system and the complement reaction. The bacterial lipopolysaccharides that come along with the polymer could be responsible for such an effect. Hence for medical purposes, the PHA production and purification strategy has to be stringent and should meet specific requirements (Sevastianov *et al.*, 2003). When P(3HB) or P(3HB-3HV) sutures were implanted intramuscularly in test animals for about a year, it was found that there was no acute vascular reaction at the site of implantation (Shishatskaya *et al.*, 2004). Commercially available PHAs are of industrial grade rather than medical grade and hence contain residual protein, surfactants and/or high levels of endotoxins, a potent pyrogen, and thus fall short of the standards required to meet regulatory approval. The extraction procedures employed for PHAs to be used in medical applications need to ensure highly pure end-products, absence of halogenated solvents, non-degradation of the polymer and high efficacy.

In order to obtain FDA approval, the endotoxin content in a particular PHA medical device should not exceed 20 US Pharmacopeia endotoxin units (EU) (Williams *et al.*, 1999). It is the lipopolysaccharides present in Gram-negative bacteria, used exclusively to produce PHAs that induce immunogenic reactions and hence are a serious risk. Gram-positive bacteria are potentially better for biomedical applications since they lack such endotoxins (Valappil *et al.*, 2006). Current FDA regulations state that medical products should not contain more than 0.5 EU/mL of endotoxin and in cases where the device will be used in contact with cerebrospinal fluid, the limit extends to 0.06 EU/ml (Valappil *et al.*, 2006).

The greatest contribution of PHAs to medicine has been in the cardiovascular area. Tepha specialises in manufacturing pericardial patches, artery augments, cardiological stents, vascular grafts, heart valves, implants and tablets, sutures, dressings, dusting powders, prodrugs and microparticulate carriers using PHAs (Williams and Martin, 2002).

When hydroxyapatite particles are incorporated into P(3HB), a bioactive and biodegradable composite is formed and this can be used in hard tissue regeneration (Doyle *et al.*, 1991). The combination of hydroxyapatite with P(3HB) or P(3HB-3HV) is particularly beneficial in bone tissue engineering because these composites possess mechanical strength similar to that of human bones (Galego *et al.*, 2000). P(3HB) and P(3HB-3HV) can be used as implant patches that are good scaffolds for tissue regeneration in low pressure systems (Malm *et al.*, 1994).

In their patent application (1991), Kwan and Steber have stressed the possible role of P(3HB-3HV) in veterinary medicine. P(3HB-3HV) capsules can be used as a biodegradable matrix containing medicine that can remain in the rumen of cattle delivering doses of medicine at controlled time intervals (Kwan and Steber, 1991).

Poly(4-hydroxybutyrate), P(4HB), is the new absorbable biomaterial used for medical purposes. Currently, P(4HB) is one of the promising materials available as opposed to other known absorbable thermoplastic polyester materials. P(4HB) is a strong, pliable thermoplastic and is more flexible than synthetic absorbable polymers such as polyglycolide (PGA) and poly-L-lactide (PLLA) and has high tensile strength. Production by fermentation gives rise to high molecular weight P(4HB) that is necessary for most industrial uses. Tepha markets P(4HB) for medical applications under the name of PHA4400. They produce P(4HB) using recombinant *Escherichia coli* K12 because it is a very well-understood and efficient micro-organism. Ethylene oxide is used to sterilise P(4HB). γ-irradiation at $25 - 50$ kGy is also used for this purpose. However, as the irradiation dose is increased, there is loss of molecular weight and increased polydispersity. P(4HB) is biocompatible and has been used as a cardiovascular patching material with great success. It has also been used to make heart valves and in vascular grafting. Sutures, ligaments, surgical meshes and pericardial substitutes made with P(4HB) are now being used (Martin and Williams, 2003). In 2007, the TephaFLEX absorbable suture, manufactured by Tepha received FDA approval for marketing in the US (http://www.tepha.com/publications/media.htm). The TephaFLEX material has biological and material properties that are suitable for implantable medical devices.

Biodegradable polymers such as homo- and co-polymers of lactate and glycolate are widely used in commercially available sustained release products for drug delivery. These products are administered parenterally and the drug is diffused over a 30 day period. However, lactate and glycolate co-polymers degrade by bulk hydrolysis. Hence, drug release cannot be fully controlled (Pouton and Akhtar, 1996). In the early 1990s, PHAs became candidates for use as drug carriers due to their biodegradability, biocompatibility and their degradation by surface erosion (Pouton and Akhtar, 1996). Korstako et al., have investigated the in vitro and in vivo release of 7-hydroxyethyltheophylline from compressed P(3HB) tablets containing the drug. Introduction of additives such as microcrystalline cellulose and lactose helped enhance drug release due to increased matrix porosity. Drug release depended on matrix porosity, copolymer composition and the molecular weight of the drug and was independent of the molecular weight of the polymer (Gould et al., 1987; (Korsatko et al., 1983). In 1994, Jones et al. described a P(3HB) matrix useful for long-term release of metoclopramide which when implanted subdermally in cattle was effective in treating fescue toxicosis caused by cattle feeding on endophyte-infected pastures (Jones et al., 1994). Bissery et al., 1984 compared the use of P(3HB) and PLA(poly-lactic acid) microspheres as carriers in drug targeting, in the release of lomustine (CCNU), an anti-cancer agent. They noticed that drug release from P(3HB) was rapid with release completed in 24 hours, whereas that from PLA microspheres was completed in 7 days (Bissery et al., 1984; Bissery et al., 1985). Akhtar et al., 1992 have reported that the crystallinity of the polymer is an important parameter that affects drug release (Akhtar et al., 1992). Sulbactam-cefoperazone antibiotic was integrated into rods made of poly(3-hydroxybutyrate-co-22 mol%-3-hydroxyvalerate) and studies were carried out to use these rods as antibiotic-loaded carriers to treat implant-related and chronic osteomyelitis. The rods were implanted into a rabit tibia that was artificially infected by Staphylococcus aureus. After 15 days, the infection subsided and within 30 days there was complete healing (Yagmurlu et al., 1999). The high melting points of P(3HB) and P(HB-HV) is sometimes a disadvantage because in turn, these polymers crystallise rapidly and make drug entrapment technically difficult. Medium chain PHAs with lower melting points could be a considerable advancement in drug delivery (Pouton and Akhtar, 1996).

PHAs have also been used to manufacture sutures, wound dressings, nerve conduits and carrier scaffolds. Sutures have to possess high in vivo tensile strength during its duration in the human body while the wound is healing and should get rapidly absorbed once that healing period is over (Singh and Maxwell, 2006). All studies carried out so far indicate that PHA sutures can be developed as future natural absorbable sutures (Valappil et al., 2006). An ideal nerve conduit that is used to bridge nerve gaps must be readily vascularised, biodegradable, should have low antigenicity, porous for oxygen diffusion and should avoid long-term compression (Valappil *et al.*, 2006). Non-woven P(3HB) sheet was used as a wrap to repair transacted superficial radial nerves and evaluated in cats for 12 months. The axonal regeneration could be compared to closure with an epineural suture with a nerve gap of 2-3mm. The inflammatory response was mild involving mainly macrophages and gradually decreased within a period of 6 months to form a fibrous capsule. This observation was interesting since it is known that macrophages are known to contribute regeneration-promoting factors at the injury site. In fact, the nerves reconstructed with P(3HB) had less adhesions of the conduit to surrounding tissues when compared to the sutured nerves. This suggested that the scar tissue will not disturb the regeneration process or damage gliding structures close to the nerve (Hazari *et al.*, 1999; Ljunberg *et al.*, 1999). P(4HB) nerve guide conduits used to bridge 10mm gaps in 30 male Spague-Dawley rats sciatic nerve, displayed axonal regeneration rate of 0.8mm daily and there was no evidence of wound infections, inflammation or anastomotic failures in a span of 20 days (Opitz *et al.*, 2004). A comprehensive review detailing the biomedical applications of polyhydroxyalkanoates is available (Valappil *et al.*, 2006).

Recent reports have shown that graft copolymers of methyl methacrylate and P(3HB) blocks have been synthesised and assessed as possible components in acrylic bone cements for use in orthopaedic applications. The copolymers were incorporated in a commercially available acrylic bone cement brand, Antibiotic Simplex® (AKZ). Acrylic bone cements are usually used for total joint replacement surgery. Copolymer content greater than 13.5 wt% in the cement resulted in the lack of processability due to the difference in the solubility of the polymers. When added in amounts less than 13.5 weight %, the volumetric porosity ranged between 13.5 to 16.9%. The ultimate compressive strength of the cement exceeded the minimum value. Other parameters such as setting time, doughing time, fracture toughness, degradability and biocompatibility of these cements containing the graft copolymer, are yet to be determined and studies are being carried out for the same (Nguyen and Marchessault, 2006).

Studies have been carried out to investigate the difference in biocompatibility between maleated poly(3-hydroxybutyrate-*co*-3-hydroxyhexanoate) and underivatised P(3HB-co-3HHx) using mouse fibroblast L929 and human microvascular endothelial cells (HMEC). P(3HB-co-3HHx) has strong mechanical properties and is biocompatible. However, maleated P(3HB-co-3HHx) showed increased thermostability and accelerated biodegradation. Moreover, L929 and HMEC proliferation rates on maleated P(3HB-co-3HHx) were increased by 120% and 260% respectively. The molecular weight of P(3HB-co-3HHx) was considerably lowered when maleated. Hence, maleated PHBHHx has great potential as a new biomaterial in tissue regeneration (Li *et al.*, 2008).

Tang *et al.*, (2008) carried out electrospinning of polyhydroxyalkanoate (PHA) copolymers of poly[(R)-3-hydroxybutyrate-co-5mol%-(R)-3-hydroxyhexanoate], poly[(R)-3-hydroxybutyrate-co-7mol%-4-hydroxybutyrate]   and   poly[(R)-3-hydroxybutyrate-co-97mol%-4-hydroxybutyrate] to fabricate scaffolds that have increased biocompatibility and bioabsorption. They reported that the content and type of the second monomer and the diameter of the fibre directly affected the bioabsorption. The mechanical studies on these scaffolds were comparable to that of human skin and hence it was proven that adequate biomechanical support was provided by the scaffold structure. It was also found that scaffolds with high 3-hydroxybutyrate content were found to be highly crystalline and hence displayed a slower absorption rate. Tissue response improved with the increase in 4-hydroxybutyrate content. Thus it is possible to prepare tailor-made tissue engineering scaffolds by manipulating the molar fraction of monomers in PHA copolymers (Tang *et al.*, 2008).

Industrial Applications

The use of PHAs in the industry has been impressive. PHA latex can be used to cover paper or cardboard to make water-resistant surfaces as opposed to the combination of cardboard with aluminium that is currently used and is non-biodegradable. This also works out be cost-effective since a very small amount of PHAs is required for this purpose (Lauzier *et al.*, 1993). PHAs can be used to make foils, films and diaphragms (Babel *et al.*, 1990).

Biomer, a German company owns the technology to produce P(3HB) from *Alcaligenes latus* on a large scale. The cells are grown in a mineral medium using sucrose as a carbon source. The strain used can accumulate up to 90% P(3HB) in dry cell weight. The polymer thus produced is used to make articles such as combs, pens and bullets. The polymer pellets are sold commercially for use in classical transformation processes. The product is very hard and can be used at temperatures ranging from -30^0C to 120^0C. This product degrades within two months in the environment (Chen, 2005).

Metabolix, a US based company, now markets among others, Metabolix PHA which is a blend of P(3HB) and poly(3-hydroxyoctanoate). This is an elastomer that has been approved by the FDA for production of food additives. Metabolix has created a recombinant *Escherichia coli* K12 strain for this purpose. These cells can accumulate up to 90% of PHA in dry cell weight in 24 hours (Clarinval and Halleux, 2005). In 2001, Metabolix was awarded a Department of Energy (DOE) grant of $2 million to develop PHA production directly in switchgrass. Switchgrass, a perennial plant that usually thrives on marginal land, is a leading candidate for bio-based production in North America because it can be grown on land of marginal use for other crops.

Industrial production of P(3HB-3HHx) by *Aeromonas hydrophila* has been accomplished by Tsinghua University (China) in collaboration with Guangdong Jiangmen Center for Biotech Develpoment (China), KAIST (Korea) and Procter and Gamble (USA) (Chen, 2001). The polymer produced is used to make flushables, nonwovens, binders, flexible packaging, thermoformed articles, synthetic paper and medical devices among others.

PHAs have found yet another potential usage as oil blotting materials as reported by Kumar *et al.* (2007). PHAs have the ability to rapidly absorb and retain oil. Moreover, they also have a natural oil-indicator property that displays changes in opacity after absorbing oil. These interesting properties could lead to PHAs being used in cosmetics and skin care industries. Currently oil-blotting materials are derived from plants but thermoplastics are found to be softer and more flexible when compared to the former. And it has been observed that the surface microstructure like porosity and smoothness affect the oil absorbing property of PHA films. PHAs could also be used as food wrappers to remove excess frying oil. It might also be possible that PHAs could be used to clean oil spills in the ocean and on beaches (Kumar *et al.*, 2007).

It is also possible to use PHAs to make the following articles due to their piezoelectric nature: pressure sensors for keyboards, stretch and acceleration measuring instruments, material testing, shock wave sensors, lighters, gas lighters, acoustics: microphone, ultrasonic detectors, sound pressure measuring instruments; oscillators: headphones, loudspeakers, for ultrasonic therapy and atomisation of liquids (Babel *et al.*, 1990). The gas barrier property of P(3HB-3HV) is useful for applications in food packaging and for making plastic beverage bottles. The same property can be exploited for making coated paper and films which can be used for coated paper milk cartons. P(3HB) or its copolymers can be used to make the non-woven cover stock and the plastic film moisture barriers in nappies and sanitary towels along with some speciality paramedical film applications in hospitals (Hocking and Marchessault, 1994).

Polystyrene waste amounts to about 14 million tonnes per year in the US alone. A research group at the University College Dublin, have devised a new method to convert polystyrene waste into biodegradable PHAs. Firstly, polystyrene is converted to styrene oil by pyrolysis, in a fluidised bed reactor. The styrene oil is composed of 82.8% (w/w) of styrene and low levels of other aromatic compounds. When provided as the sole carbon source, *Pseudomonas putida* CA-3 converted the styrene oil to medium chain length PHA. Some researchers have warned that since pyrolysis is an 'energy-demanding' process and can generate hazardous wastes, the whole process is not without its demerits. On the other hand, this process could be the optimum way to use up waste polystyrene cost-effectively and efficiently (Booth, 2006; Ward *et al.*, 2006).

Agricultural Applications

PHAs have been used as mulch films for agricultural purposes (Hocking and Marchessault, 1994). In recent years, Procter and Gamble have produced Nodax™ that can be used to manufacture biodegradable agricultural film. Nodax™ is a copolymer containing mainly 3(HB) and small quantities of MCL monomers. Nodax™ can degrade anaerobically and hence can be used as a coating for urea fertilizers to be used in rice fields or for herbicides and insecticides.

One of the specialised applications of P(3HB-3HV) in agriculture is in the controlled release of insecticides. Insecticides could be integrated into P(3HB-3HV) pellets and sown along with the farmer's crops. The insecticide would be released at a rate related to the level of pest activity since the bacteria breaking down the polymer would be affected by the same environmental conditions as that of the soil pests (Holmes, 1985).

Another use of PHAs in agriculture is in bacterial inoculants used to enhance nitrogen fixation in plants. The bacterial culture used in inoculant preparations for agricultural purposes need to withstand stressful environments. The bacterial cells have to be stored for long periods, and endure desiccation and hot conditions. Inoculants have to possess the ability to sustain high survival rates within the carrier. Therefore research in this area has focussed on the addition of elements like nutrients or other synthetic products that can increase the quality of carriers leading to prolonged survival (Lopez *et al.*, 1998). From studies carried on *Azospirillum brasilense* inoculants, it was observed that while carriers may vary, the plant growth promotion outcome was more constant with *A. brasilense* inoculants containing high amounts of intracellular PHA (Fallik and Okon, 1996). This was confirmed by field experiments in Mexico with maize and wheat. Better consistency was achieved in increasing crop yield by using peat inoculants prepared with PHA-rich *Azospirillum* cells (Dobbelaere *et al.*, 2001). Hence, intracellular PHA is of paramount significance for improving the shelf life, efficiency and reliability of commercial inoculants (Kadouri *et al.*, 2003; Kadouri *et al.*, 2005).

Nanocomposites

A nanocomposite is "a hybrid material consisting of a polymer matrix reinforced with a fibre, platelet, or particle having one dimension in the nanometre (nm) scale (10^{-9}m)" (Pandey *et al.*, 2005). There are different kinds of nanocomposites and among them layered silicate nanocomposites are well-known since these are economical with high aspect ratio and exceptional barrier properties (Pandey *et al.*, 2005).

Several PHAs were incarcerated in layered silicates to determine whether by doing so the material properties of the former could be improved. When P(3HB) was used as the host matrix for octadecylammonium modified montmorillonite (MMT) and fluoromica, it was found that the rate of degradation of P(3HB) was higher in MMT during nanocomposite preparations than in the fluoromicas. It is still unclear as to how the fluoromicas help protect P(3HB). The presence of clay particles might have decreased the degradation rates in these nanocomposites. In addition, the occurrence of Al Lewis acid sites in MMT was thought to be the reason for the higher degradation rate of this composite. The Al Lewis acid sites catalyse the ester linkage hydrolysis (Pandey *et al.*, 2005).

Poly(3-hydroxyoctanoate) latex films were prepared and found to possess outstanding thermoplastic properties. When these latex films were used as the host matrix for nanocomposite materials along with a colloidal solution of hydrolysed starch or cellulose whiskers as fillers, "high-performance" materials were produced. However, polymer-filler interactions, aspect ratio and geometrical constraints are parameters that require further optimisation for successful reinforcement of PHA matrices (Pandey *et al.*, 2005).

Nanocomposites are the new, promising generation of materials. The material properties of the polymers improve significantly after reinforcing with suitable filler materials. However, polymers have to be modified significantly to exhibit better matrix properties. It is hoped that nanocomposites will enable PHAs compete more effectively with petroleum based plastics (Pandey *et al.*, 2005).

Chiral Hydroxyalkanoates

Pure chiral compounds are finding increasing applications in pharmaceutical industries. With their natural effectiveness, enantiomerically pure drugs are preferred to their chemically synthesised counterparts. Chiral drugs ensure greater wellbeing and increased effectiveness to patients at low dosages (Angelo, 1996). PHA synthase is the key enzyme for PHA biosynthesis and can be a source for a rich chiral pool (Steinbuchel and Valentin, 1995).

Tasaki et al., 1999, tried the use of ketone bodies including a 3HB dimer and trimer residue mixture as a novel nutrition source. The mixture, when injected into rats was found to be completely converted into monomers. This could enable the potential use of 3HB as an energy substrate in injured patients since 3HB has high penetration rates and rapid diffusion rate in peripheral tissues. 3HB has also been used as the starting material for producing carbapenem antibiotics and macrolides (Tasaki et al., 1999). Dendrimers with biodegradability, monodispersity and large number of surface-functional moieties can be synthesised using 3HB monomers and used as in vivo drug carriers (Seebach et al., 1996). Optically pure monomers have also been used in the production of sex hormones and S-citronellol, a fragrance (Holmes, 1985).

Hydroxyalkanoate monomers can also be used to synthesise novel chiral polyesters. Researchers have been successful in synthesising two new 3-hydroxyalkanoates: 3-hydroxy-3-cyclopropylpropionate and 3-hydroxy-4-chlorobutyrate and their CoA thioester derivatives. Using the polymerase enzyme of E. shaposhnikovii, which can produce PHA as an intracellular reserve; these were polymerised in aqueous solutions. The two new polymers were crystalline in nature and both possessed one chiral centre each (Kamachi et al., 2001). Novel β- and γ-peptides have also been produced by replacing amino acids with 3-hydroxybutyrate residues in peptides and by replacing the chain-bound oxygens in 3HB or 4HB with NH (Park et al., 2001). Among their useful characteristics are their resistance to peptidases and environmental microbial degradation and their longevity in mammalian serum (Seebach et al., 2001). Some of them also possess useful antibacterial, antiproliferative and haemolytic properties (Chen, 2005).

PHA Blends and Composites

Blending PHAs with high or low molecular weight molecules help improve their material properties. It also helps to reduce production costs. Savenkova et al., 2000, investigated and compared the effect of plasticisers such as PEG, Laprol, DBS and PIB on P(3HB)-based films and their biodegradation in soil, so as to choose an appropriate matrix for pesticides. They found that Laprol was a good, cheap and non-toxic plasticiser and had the maximum effect on the rate of degradation of the composite at a concentration of 30% w/w. As the concentration of Laprol increased, the rate of biodegradation decreased significantly. The average life-time of a P(3HB)-Lap film was 6-8 months. The principal microorganisms that attack P(3HB) in soil were also identified, namely: Penicillium, Cephalosporium, Paecilomyces and Trichoderma (Savenkova et al., 2000).

PHA blends with natural fibres have been tested to check the effect of these fibres on the mechanical properties of PHAs. When P(3HB-3HV) films were reinforced with pineapple leaf fibres (30% w/w), the tensile strength of the films increased by 100%. Avella et al., 2000, reinforced P(3HB) with heat exploded wheat straw and hemp fibres and the composites thus obtained can be used in agricultural mulching and transplantation (Avella et al., 2000).

PHA blends with polycaprolactone (PCL) are reported to be tough with excellent mechanical properties. The polymers are not miscible but the blends exhibited improvement in heat stability and impact strength (Urakami et al., 2000).

Recent reports suggest the possibility of combining mcl-PHAs with different rubbers such as natural rubber, nitrile rubber and butadiene rubber at room temperature to give rise to a different type of polymer with a melting point of 90^0C. Addition of natural rubber increased the thermal decomposition temperature of the blend. Blends with rubber also affected the crystallisation and melting properties of the mcl-PHAs which in turn improves the processing problems faced with PHAs. This also leads to the synthesis of biodegradable materials whose rate of biodegradation can be controlled by adjusting the rubber:PHA ratio (Bhatt et al., 2007).

Researchers have been able to design composites of PHAs in permutation with inorganic phases. This helps to enhance the mechanical properties of the former, as well as influencing the degradation rate and bioactivity. The inorganic phases normally used are hydroxyapatite, bioactive glass, and glass-ceramic fillers or coatings. A comprehensive review detailing the development of such composites, their properties and applications in tissue engineering is available (Misra et al., 2006).

P(3HB)/Bioglass® composites including multiwalled carbon nanotubes (MWCNTs) have been reported. Bioglass® type 45S5 has been used in combination with a variety of biodegradable polymers to form composites successfully. It has excellent properties as a biomaterial and elicits the formation of a thin layer of hydroxyapatite when in contact with biological fluids in vitro and in vivo. CNTs on the other hand, are electrically conducting materials that are chemically stable and will enable the monitoring of scaffolds in situ while implanted in the body. Thus it will be possible to gain important information about the scaffold's interaction with the surrounding biological environment. In this study, it was seen that it was possible to incorporate MWCNTs into the P(3HB)/Bioglass® base. Tests of these composites in Simulated Body Fluid (SBF) resulted in the formation of the hydroxyapatite layer on the composite surface. Thus it should be possible to use these composites in bone tissue engineering applications (Misra et al., 2007).

Other Possible Routes of Production of PHAs

Cyanobacteria

An interesting area of research dwells on the prospect of using cyanobacteria as PHA producers. Cyanobacteria indigenously accumulate PHA by oxygenic photosynthesis (Asada et al., 1999). It is known that cyanobacteria accumulate mainly P(3HB) under photoautotrophic or mixotrophic growth conditions with acetate. However, PHA accumulation per cell is very low (6% dry cell weight) and this is a limiting factor in considering cyanobacteria as prospective commercial producers of the polymer (Stal, 1992). Synechococcus MA19, a unicellular thermophile can however, accumulate upto 30% dcw of P(3HB) under photoautrophic and nitrogen deficient conditions (Miyake et al., 1996). After optimisation, it was observed that the strain could accumulate up to 55% dcw of P(3HB)

(Nishioka *et al.*, 2001). Asada *et al.*, 1999 have also constructed a genetically engineered cyanobacteria that accumulates P(3HB) using *Synechococcus sp.* PCC7942. The PHA synthetic genes from *Alcaligenes eutrophus* have been transformed into *Synechococcus sp.* PCC7942 since it does not indigenously accumulate P(3HB). However, production of PHA in a large-scale from cyanobacteria is still impractical due to lack of an established mass cultivation system. It is therefore imperative that further research be carried out for the development of better large-scale production systems involving cyanobacteria and other photosynthetic microbes Asada *et al.*, 1999).

PHA Production in Transgenic Plants

Currently, PHAs for commercial applications are being produced by microbial fermentations. However, continuing efforts are being made to devise cost-effective means such as transgenic plants to produce PHAs. Transgenic plants have always been utilised for producing genetically modified (GM) crops for food purposes and have always attracted suspicion and ethical debates. Producing PHAs in plants, on the other hand, will help connect the low cost/ high-volume sustainable production capacity of crops with the vast amount of developing polymer industries and should not promote much controversy. This will enable PHAs to possess both superior cost and performance factors (Snell and Peoples, 2002). Many research groups have reported their attempts to produce PHAs in plant systems. The only raw materials required will be carbon dioxide for carbon and sunlight as the energy source. The overall costs would make PHA production economical. Transgenic plants containing the PHA synthase genes have been created. The transgenic plants were stunted but accumulated about 15% dcw of P(3HB) in the leaf expression systems (Valentin *et al.*, 1999). Co-polymers of 3HB and 3HV were produced in *Arabidopsis* and *Brassica rapa*. A molecular mass of 10^6 was attained which is excellent for commercial applications. Nawrath and group have attempted the production of PHAs in plastids whereas Poirier and group targeted PHA production in seeds (Nawrath *et al.*, 1994). *Arabidopsis* plants accumulated P(3HB) up to 14% of the leaf dry weight (Valentin *et al.*, 1999; Nawrath *et al.*, 1994). Transgenic PHA producing plants have also been produced using tobacco (Wang *et al.*, 2005), cotton (John and Keller, 1996) and flax (Wrobel *et al.*, 2004) systems. However, plant systems have other disadvantages. It is difficult to produce copolymers in plant systems since the production is under the control of endogenous metabolic precursors in the plant. Similarly, the recovery of the polymer from plant tissues is tricky and expensive. Monsanto had previously produced poly(3HB) and poly(3HB-co-HV) in *Arabidopsis* and *Brassica rapa* and these polymers resembled the bacterial polymers (Slater *et al.*, 1999).

Researchers in Japan have also recently devised a transgenic *Arabidopsis thaliana*, harbouring an engineered PHA synthase gene from *Pseudomonas* sp. 61-3, *fabH* gene (codes for 3-ketoacyl-ACP synthase III) from *E. coli* and a *phaAB* gene (codes for a ketothiolase and acetoacetyl-CoA reductase) from *Cupriavidus necator* and the enzymes were targeted to the plastids. A polymer consisting of 3-hydroxybutyrate unit and a small portion of 3-HA units (C_5-C_{14}) was produced. The *fabH* gene aided a 2-fold increase in PHA content on an average but the maximum PHA amount remained the same and any further increase led to stunted growth. However, PHA production in plastids countered that. The same group also carried out seed-specific PHA production in rice and tuber-specific production in potato (Matsumoto *et al.*, 2006). Purnell *et al.* are actively investigating the accumulation of PHA in sugarcane. PHA biosynthetic genes from *Cupriavidus necator* were expressed in the plastids. Although the maximum polymer accumulated in leaves was only 0.26%, the characteristic stunted growth associated with transgenic plants producing PHAs was not observed (Purnell *et al.*, 2006). These set of experiments have definitely enhanced hope in the direction of cheaper PHA production.

Metabolix has now successfully transferred the PHA metabolic pathway into switchgrass (*Panicum virgatum*). Switchgrass grows quickly, converting solar energy into chemical energy. Switchgrass also absorbs carbon dioxide from the atmosphere as it grows, thus reducing the increasing concentration of this gas in the atmosphere. They have developed a detailed biorefinery cost and engineering analysis using switchgrass, which is being used to promote large-scale PHA manufacturing (Peoples, 2004). Hence, altered plant phenotypes, low productivity and transgenic stability are problems that have to be resolved before transgenic plants become the chosen mode of PHA production (Snell and Peoples, 2002). A summary of transgenic plants producing polyhydroxyalkanoates is given below (Table 3).

Table 3. Summary of PHA production in transgenic plants

Plant species	Subcellular compartment WHERE pha WAS PRODUCED	Tissue	PHA produced	PHA yield (%dry cell weight)
Arabidopsis thaliana	Plastid	Shoot	P(3HB)	14 – 40
	Plastid	Shoot	P(3HB-co-3HV)	1.6
	Peroxisome	Whole plant	PHA$_{MCL}$	0.6
Alfalfa	Plastid	Shoot	P(3HB)	0.2
Corn	Plastid	Shoot	P(3HB)	6
Cotton	Cytoplasm	Fibre	P(3HB)	0.3
	Plastid	Fibre	P(3HB)	0.05
Maize	Peroxisome	Cell suspension	P(3HB)	2
Potato	Plastid	Shoot	P(3HB)	0.02
Rapeseed	Cytoplasm	Shoot	P(3HB)	0.1
	Plastid	Seed	P(3HB)	8
	Plastid	Seed	P(3HB-co-3HV)	2.3
Tobacco	Cytoplasm	Shoot	P(3HB)	0.01
	Plastid	Shoot	P(3HB)	0.04

Other Areas

There have been reports about the accumulation of polyhydroxyalkanoates viz., P(3HB) and P(3HB-co-3HV), when moderate halophiles such as *Halomonas boliviensis* are grown on sugars, propionic, butyric, valeric, hexanoic, heptanoic and octanoic acids as sole carbon sources. These microbes accumulate about 50-88% dcw of PHAs. Batch production of polymer using sucrose as sole carbon source resulted in polymer volumetric productivity of 0.4 g/L/h and and cell concentrations 14 g/L respectively (Quillaguaman, 2006).

Industrially Available PHAs and Their Applications

Nodax™

Nodax™ comprises polymer containing monomer units of 3-hydroxybutyrate and a comparatively small quantity of medium chain length monomers with side groups of at least three carbon units or more. The MCL-units used include 3-hydroxyhexanoate, 3-hydroxyoctanoate and 3-hydroxydecanoate. The members of the Nodax™ family are grouped into grades depending on the average molecular weight, the MCL-monomer in the copolymer and the length of the side group chain of the MCL-monomer (Noda *et al.*, 2005).

The advantage of incorporating MCL-PHA monomers into P(3HB) is reflected in the lower crystallinity and melting temperature which in turn provides the toughness and ductility vital for commercial applications. It was reported that the flexibility of the copolymer is determined by the length of the side group chain of the MCL-PHA (Noda *et al.*, 2005).

The polymer has been developed by Procter and Gamble and promises anaerobic and aerobic degradability, hydrolytic stability and elastic and mechanical properties to suit specific needs. Nodax™ is available as foams, fibres or nonwovens, films and latex among others. This polymer can be used to make flushables that can degrade in septic systems and this would include hygienic wipes and tampon applicators. They can also be used to manufacture medical surgical garments, upholstery, carpet, packaging, compostable bags and lids or tubs for thermoformed articles (Noda *et al.*, 2005).

DegraPol

DegraPol is a block-copolyester urethane and is chemically synthesised from P(3HB)-diol and α,ω-dihydroxy-poly(Є-caprolactone-block-diethylene-glycol-block-Є-caprolactone). Saad *et al.*, 1999 have used α,ω-dihydroxy-oligo[(R-3-hydroxybutyrate-co-R-3-hydroxyvalerate)-*block*-ethylene glycol] as the diol in all their studies. It is biocompatible both *in vitro* and *in vivo*. Studies carried out showed that DegraPol foams on degradation were well phagocytosed by macrophages and osteoblasts. DegraPol finds exciting applications in bone healing methods like autologous osteoblast or chondrocyte transplantation (Saad *et al.*, 1999).

BIOPOL®

BIOPOL® was initially manufactured by ICI and is now produced by Metabolix. As mentioned earlier, BIOPOL® is a co-polymer of poly(3-hydroxybutyrate-co-3-hydroxyvalerate). The copolymer is a thermoplastic and has a melting point in the range of 140°–180°C. With an increase in the amount of 3HV in the copolymer, the crystallinity decreases and the polymer become more elastic. BIOPOL® can be used to coat paper and paperboards. In addition to being suitable for injection, blow moulding and film production, BIOPOL® has antistatic properties that can be exploited for electric and electronic packaging. Companies like Fluka and Toray are now engaged in developing BIOPOL® for medical applications (Clarinval and Halleux, 2005). BIOPOL® monofilaments have been used to make fishing nets and ropes. BIOPOL® fibres have been used to make ropes and nets for crab cages. The nets exhibited good strength and biodegradability in the sea. BIOPOL® coated with polyvinyl alcohol is a good matrix for growing seaweed. When BIOPOL® is reinforced with PCL; they develop anti-algal properties and can be used as nets for seafood cultivation (Asrar and Gruys, 2002). The sole distributor for BIOPOL® rigid packaging, Berlin Packaging, has been enjoying tremendous success. Brocato International, Baton Rouge, La., introduced their hair care products in BIOPOL® bottles in 1992. BIOPOL® has been used to produce shampoo bottles, motor oil bottles and disposable razors (Clarinval and Halleux, 2005).

Economics and Legal Aspects

Bioplastics account for about 50,000 tonnes of the 50 million tonnes European polymer market. Many big supermarket chains such as Sainsburys, Tesco and Ikea are now switching to biodegradable shopping bags (Carmichael, 2006).

Initially, the cost of producing BIOPOL® from *Cupriavidus necator* using glucose and propionic acid was as high as US$16 per kilogram (Lee, 1996). Companies like Procter and Gamble worked towards reducing the costs of PHAs to make them more competitive against the established products. The cost of Nodax™ was around US$2.20 per kilogram (Keenan *et al.*, 2005). Procter and Gamble have however stopped the production of Nodax™ in 2006 (Poirier *et al.*, 2006). PHAs are finding increasing number of applications in the biomedical field since in medical applications the high price is tolerated *in lieu* of the high quality of the PHA products. Metabolix and ADM (Archer Daniels Midland Company) have announced that they will build the first commercial plant to develop large-scale production of PHAs.

The plant is expected to be functional in 2008 and will be situated in North America with the capacity to produce 50,000 tonnes of polymer per year. The plant will produce PHA from fully biological fermentation processes that convert agricultural raw materials (e.g. corn sugar) into a multipurpose range of plastics. Governments are now becoming more aware of the importance of biodegradable materials. Specific organisations have been set up for the standardisation of biodegradable materials (Table 4).

Table 4. Overview of normalisation institutes

Name (short)	Name (long)	Geographical spread
ASTM	American Society for Testing and Materials	USA/ Canada
CEN	Comité Européen de Normalisation (European Committee for Standardisation)	EU and EFTA countries and Czech Republic (EFTA = Iceland, Norway, Switzerland)
DIN	Deutsches Instut für Normung eV	Germany
ISO	International Organisation for Standardisation	Worldwide
JIS	Japanese Institute for Standardisation	Japan
OECD	Organisation for Economic Cooperation and Development	OECD countries

Details of these tests and standardisation procedures are described in Wilde, 2005. According to the European norm EN 13432 a compostable material should possess biodegradability (i.e. converted into carbon dioxide under microbial action), disintegrability (i.e. fragmentation and loss of visibility in the final compost), and absence of negative effects on composting process and low levels of heavy metals. The European standard EN 13432 has been adopted by EU bioplastics manufacturers and is symbolised by a 'seedling logo' (Wilde, 2005).

UK, Germany, Poland and the Netherlands have placed the 'seedling' logo on all products that display biodegradability. Based on new legislation, all shopping bags in France will have to be biodegradable by 2010 (Carmichael, 2006).

Conclusion

The demand for plastics in the world seems to be increasing by the day. The global market for biodegradable polymers exceeds 114 million pounds and is expected to rise at an average annual growth rate (AAGR) of 12.6% to 206 million pounds in 2010. A graphical representation of the rising manufacturing capacity for biopolymers in the world is shown below (Figure 7) (Carmichael, 2006).

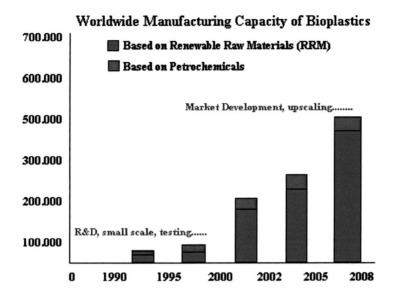

Figure 7. Worldwide manufacturing capacity of bioplastics (adapted from the European Bioplastics website, http://www.european-bioplastics.org/index.php?id=141)

PHAs have been commercial for well over 20 years, but face a variety of roadblocks such as high prices and lack of an industrial infrastructure to deal with many of these materials. The idea that novel polymers with physical and chemical properties more suited to a particular application can be synthesised merely by controlling the composition of the carbon source has aroused great industrial interest. However, the high cost of producing PHAs is threatening the rapid progress in their commercial application. Metabolix aims to bring down the cost of PHA to approximately $1/lb (Pandey *et al.*, 2005). Another main obstacle for biodegradable polymers is the deficiency of proper disposal facilities. There are no means to separate these polymers from other waste. Hence strategies have to be set in place to ensure that the public is educated and aware of proper waste disposal methods (Bhattacharya *et al.*, 2005).

PHAs thus have a wide variety of applications and chief among them are the medical applications which currently seem to be the most economically practical area. However, new niches are being carved out for their application and with the extensive research being carried out in this area; PHAs will soon emerge as the new class of promising biodegradable materials.

Acknowledgement

S.E. Philip was financially supported by a scholarship from the University of Westminster, London, UK.

References

Abe H. and Y. Doi, Molecular and material design of biodegradable poly(hydroxyalkanoate)s, in *Biopolymers* 3b, Polyesters II, Y. Doi and A. Steinbuchel, Editors. 2002, Weinheim Wiley-VCH. p. 105 - 132.

Akhtar, S., C.W. Pouton, and L.J. Notarianni, Crystallization behaviour and drug release from bacterial polyhydroxyalkanoates. *Polymer Bulletin*, 1992. 33: p. 117 - 126.

Anderson, A.J. and E.A. Dawes, Occurrence, metabolism, metabolic role and industrial uses of bacterial polyhydroxyalkanoates. *Microbiology Reviews*, 1990. 54: p. 450 - 472.

Angelo, D.P., Chiral chemistry is still evolving, driven by techniques and business demands. *Genetic Engineering News*, 1996. 6: p. 15.

Asasda, Y., et al., Photosynthetic accumulation of poly-(hydroxybutyrate) by cyanobacteria - the metabolism and potential for CO_2 recycling. *International Journal of Biological Macromolecules*, 1999. 25: p. 37 - 42.

Asrar, J. and K.J. Gruys, Biodegradable polymer (BIOPOL®), in *Biopolymers- Polyesters III: Applications and commercial products* Wiley - VCH, Y. Doi and A. Steinbuchel, Editors. 2002, Wiley - VCH.

Atkins, T.W. and S.J. Peacock, The incorporation and release of bovine serum albumin from poly-hydroxybutyrate-hydroxyvalerate microcapsules. *Journal of Microincapsulation*, 1996. 13: p. 709 - 717.

Avella, M., et al., Poly(3-hydroxybutyrate-co-3-hydroxyvalerate) and wheat straw fibre composites: thermal, mechanical properties and biodegradation behaviour. *Journal of Materials Science*, 2000. 35: p. 829 - 836.

Babel, W., V. Riis, and E. Hainich, Mikrobelle thermoplaste: biosynthese, eigenschaften und anwendung. *Plaste Und Kautschuk*, 1990. 37: p. 109 - 115.

Ballard, D. G., Holmes, P. A., Senior, P. J., Formation of polymers of ß-hydroxybutyric acid in bacterial cells and a comparison of the morphology of growth with the formation of polyethylene in the solid state, in *Recent advances in mechanistic and synthetic aspects of polymerization, Fontanille*, M. and Guyot, A., Editors. 1987, Reidel (Kluwer) Publishing Co. 215: 293 – 314.

Bhatt, R., Shah, D., Patel, K.C., Trivedi, U. PHA–rubber blends: Synthesis, characterization and biodegradation. *Bioresource Technology*. Article in press.

Bhattacharya, M., et al., Material properties of biodegradable polymers, in *Biodegradable polymers for industrial applications*, R. Smith, Editor. 2005, CRC: FL, USA. p. 336 - 356.

Bissery, M.C., F. Valeriote, and C. Thies, In vitro and in vivo evaluations of CCNU-loaded microspheres prepared from polylactide and poly(ß-hydroxybutyrate), in *Microspheres and Drug Therapy, Pharmaceutical, Immunological and Medical Apsects*, S.S. Davis, et al., Editors. 1984, Elsevier: Amsterdam. p. 217 - 227.

Bissery, M.C., F. Valeriote, and C. Thies, Fate and effect of CCNU-loaded microspheres made of poly(DL-lactide) or poly(ß-hydroxybutyrate) in mice. *Proceedings of the International Symposium on Controlled Release and Bioactive Materials*, 1985. 12: p. 69 - 70.

Bohlmann, G.M., Polyhydroxyalkanoate production in crops. Feedstocks for the future: renewables for the production of chemicals and materials *ACS SYMPOSIUM SERIES*, 2006. 921: p. 253 - 270.

Borah, B., Thakur, P. S., Nigam, J. N. The influence of nutrintional and environmental conditions on the accumulation of poly-ß-hydroxybutyrate in Bacillus mycoides RLJ B-017. *Journal of Applied Microbiology*, 2002. 92: 776 – 783.

Brandl, H., Knee, Jr., Fuller, R. C., Gross, R. A., Lenz, R. W. Ability of the phototrophic bacterium Rhodospirillum rubrum to produce various poly(ß-hydroxyalkanoates): potential sources of biodegradable polyesters. *International Journal of Biological Macromolecules*, 1989. 11: 49 – 55.

Braunegg, G., Sustainable poly(hydroxyalkanoate) (PHA) production, in Degradable polymers: Principles and Applications, G. Scott, Editor. 2003, Kluwer Academic.

Booth, B., Polystyrene to biodegradable PHA plastics. *Environmental Science and Technology*, 2006. 40(7): p. 2074 - 2075.

Caballero, K. P., Karel, S.F., Register, R. A. Biosynthesis and characterization of hydroxybutyrate-hydroxycaproate copolymers. *International Journal of Biological Macromolecules*, 1995. 17: 86 – 92.

Carmichael, H., Compost-ready packaging. *Chemistry and Industry*, 2006. 16: p. 20 - 22.

Chen, G.Q., et al., Industrial production of poly(hydroxybutyrate-co-hydroxyhexanoate). *Applied Microbiology and Biotechnology*, 2001. 57: p. 50 - 55.

Chen, G.Q., Polyhydroxyalkanoates, in *Biodegradable polymers for industrial applications*, R. Smith, Editor. 2005, CRC: FL, USA. p. 32 - 56.

Chen, G. Q., Konig, K. H., Lafferty, R. M. Occurrence of poly-(-)-3-hydroxyalkanoates in the genus Bacillus. *FEMS Microbiology Letters*, 1991. 84: 173 – 176.

Ching-Yee, L., Kumar, S. Biosynthesis and native granule characteristics of poly(3-hydroxybutyrate-co-3-hydroxyvalerate) in Delftia acidovorans. *International Journal of Biological Macromolecules*, 2007. 40: p. 466 - 471.

Clarinval, A.-M. and J. Halleux, Classification of biodegradable polymers, in *Biodegradable polymers for industrial applications*, R. Smith, Editor. 2005, CRC: Fl, USA. p. 3 - 56.

Comeau, Y., Hall, K., Oldham, W. K. Determination of poly-ß-hydroxybutyrate and poly-ß-hydroxyvalerate in activated sludge by gas-liquid chromatography. *Applied and Environmental Microbiology*, 1988. 54: 2325 – 2327.

Congress, U.S., *Biopolymers: Making Materials Nature's Way -Background Paper*. 1993, Office of Technology Assessment: Washington D. C.

Cuff, J. A., Clamp, M. E., Siddiqui, A. S., Finlay, M., Barton, G. J. JPred: a consensus secondary structure prediction server. *Bioinformatics*, 1998. 14: 892 - 893

Dai Y, Lambert L, Yuan Z, Keller J. Characterisation of polyhydroxyalkanoate copolymers with controllable four-monomer composition. *Journal of Biotechnology*, 2008. 134(1-2): p. 137 – 145

Davis, R., Chandrashekar, A., Shamala, T.R. Role of (R)-specific enoyl coenzyme A hydratases of Pseudomonas sp in the production of polyhydroxyalkanoates. *Antonie van Leeuwenhoek*, 2008. 93: p.285–296.

Dobbelaere, S., et al., Responses of agronomically important crops to inoculation with Azospirillum. *Aust. J. Plant. Physiol.*, 2001. 28: p. 871 - 879.

Doi, Y., Kunioka, M., Nakamura, M., Soga, K. Biosynthesis of polyesters by Alcaligenes eutrophus: incorporation of ^{13}C-labeled acetate and propionate. *Journal of the Chemical Society, Chemical Communications*, 1986. 23: 1696 – 1697.

Doi, Y., Tamaki, A., Kunioka, M., Soga, K. Biosynthesis of terpolyesters of 3-hydroxybutyrate, 3-hydroxyvalerate and 5-hydroxyvalerate in Alcaligenes eutrophus from 5-chloropentanoic and pentanoic acids. Die Makromolekulare Chemie, *Rapid Communications*, 1987. 8: 631 – 635.

Doi, Y., *Microbial Polyesters*. 1990, New York: VCH Publishers.

Doi, Y., S. Kitamura, and H. Abe, Microbial synthesis and characterization of poly(3-hydroxybutyrate-co-3-hydroxyhexanoate). *Macromolecules*, 1995. 28: p. 4822 - 4828.

Doyle, C., E.T. Tanner, and W. Bonfield, In vitro and in vivo evaluation of polyhydroxybutyrate and of polyhydroxybutyrate reinforced with hydroxyapatite. *Biomaterials*, 1991. 12: p. 841 - 847.

Dunlop, W. F. and Robards, A. W. Ultrastructural study of poly-ß-hydroxybutyrate granules from Bacillus cereus. *Journal of Bacteriology*, 1973. 114: 1271 – 1280.

Duvernoy, O., et al., A biodegradable patch used as a pericardial substitute after cardiac surgery: 6- and 24-month evaluation with CT. *Thoracic and Cardiovascular Surgery*, 1995. 43(5): p. 271 - 274.

Ellar, D., Lundgren, D. G., Okamura, K., Marchessault, R. H. Morphology of poly-ß-hydroxybutyrate granules. *Journal of Molecular Biology*, 1968. 35: 489 – 502.

Fallik, F. and Y. Okon, Inoculants of Azospirillum brasilense: Biomass production, survival and growth promotion of Setaria italica and Zea mays. *Soil Biol. Biochem.*, 1996. 28: p. 123 - 126.

Findlay, R. H., White, D. C. Polymeric ß-hydroxyalkanoates from environmental samples and Bacillus megaterium. *Applied and Environmental Microbiology*, 1983. 45: 71 – 78.

Fukui, T. and Y. Doi, Cloning and analysis of the poly(3-hydroxybutyrate-co-3-hydroxyhexanoate) biosynthesis genes of Aeromonas caviae. *Journal of Bacteriology*, 1997. 179(15): p. 4821- 4830.

Fukui, T., N. Shiomi, and Y. Doi, Expression and characterization of (R)-specific enoyl coenzyme A hydratase involved in polyhydroxyalkanoate biosynthesis by Aeromonas caviae. *Journal of Bacteriology*, 1998. 180(3): p. 667 - 673.

Gagnon, K.D., et al., Crystallization behaviour and its influence on the mechanical properties of a thermoplastic elastomer produced by Pseudomonas oleovorans. *Macromolecules*, 1992. 25(14): p. 3723 - 3728.

Galego, N., et al., Characterization and application of poly(ß-hydroxyalkanoates) family as composite biomaterials. *Polymer Test*, 2000. 19: p. 485 - 492.

Gerngross, T. U., Reilly, P., Stubbe, J., Sinskey, A. J., Peoples, O. P. Immunocytochemical analysis of poly-beta-hydroxybutyrate (PHB) synthase in Alcaligenes eutrophus H16: localization of the synthase enzyme at the surface of PHB granules. *Journal of Bacteriology*, 1993. 175: 5289 – 5293.

Gogolewski, S., et al., Tissue response and in vivo degradation of selected polyhydroxyacids: polylactides (PLA), poly(3-hydroxybutyrate) (PHB), and poly(3-hydroxybutyrate-co-3-hydroxyvalerate) (PHB/PHV). *Biomaterials*, 1990. 11: p. 679 - 685.

Gould, P.L., S.J. Holland, and B.J. Tighe, Polymers for biodegradable medical devices IV.Hydroxybutyrate-hydroxyvalerate copolymers as non-disintegrating matrices for controlled release oral dosage forms. *International Journal of Pharmaceuticals*, 1987. 38: p. 231 - 237.

Griebel, R., Smith, Z., Merrick, J. M. Metabolism of poly-ß-hydroxybutyrate. I. Purification, composition and properties of native poly-ß-hydroxybutyrate granules from Bacillus megaterium. *Biochemistry*, 1968. 7:3676 – 3681.

Grothe, E., Moo-Young, M., Chisti, Y. Fermentation optimisation for the production of poly(ß-hydroxybutyric acid) microbial thermoplastic. *Enzyme and Microbial Technology*, 1999. 25: 132 – 141.

Hazari, A., R.G. Johansson, and B.K. Junemo, A new resorbable wrap around implant as an alternative nrve repair technique. *Journal of Hand Surgery*, 1999. 24: p. 291 - 295.

Hocking, P.J. and R.H. Marchessault, Biopolyesters, in *Chemistry and Technology of Biodegradable Polymers*, G.J.L. Griffin, Editor. 1994, Blackie Academic and Professional: London.

Holmes, P.A., Application of PHB: a microbially produced biodegradable thermoplastic. *Physics in Technology*, 1985. 16: p. 32 - 36.

Holmes, P.A., Biologically produced R(3)-hydroxyalkanoate polymers and copolymers. *Development in crystalline polymers*, ed. D.C. Bassett. 1988, London: Elsevier. 1 - 65.

Howells, E.R., Opportunities in biotechnology for the chemical industry. *Chemistry and Industry*, 1982. 8: p. 508 - 511.

Huijberts, G.N., et al., Pseudomonas putida KT2442 cultivated on glucose accumulates poly(3-hydroxyalkanoates) consisting of saturated and unsaturated monomers. *Applied and Environmental Microbiology*, 1992. 58(2): p. 536 – 544.

Jendrossek, D., Microbial degradation of biopolyesters. *Advances in biochemical engineering and biotechnology*, 2001. 71: p. 294 - 325.

Jia, Y., Kappock, J., Frick, T., Sinskey, A. J., Stubbe, J. Lipase provide a new mechanistic model for polyhydroxybutyrate (PHB) synthases: characterization of the functional residues in Chromatium vinosum PHB synthase. *Biochemistry,* 2000. 39: 3927 – 3936.

John, M. and G. Keller, Metabolic pathway engineering in cotton: Biosynthesis of polyhydroxybutyrate in fiber cells. *Proc. Natl. Acad. Sci. USA*, 1996. 93: p. 12768 - 12773.

Jones, R.D., J.C. Price, and J.M. Bowen, In vitro and in vivo release of metoclopramide from a subdermal diffusion matrix with potential in preventing fescue toxicosis in cattle. *Journal of Controlled Release*, 1994. 30: p. 35 - 44.

Kadouri, D., E. Jurkevitch, and Y. Okon, Involvement of the reserve material poly-beta-hydroxybutyrate (PHB) in Azospirillim brasilense in stress endurance and root colonization. *Applied Environ. Microbiol.*, 2003. 69: p. 3244 - 3250.

Kadouri, D., E. Jurkevitch, and Y. Okon, Ecological and agricultural significance of bacterial polyhydroxyalkanoates. *Critical Reviews in Microbiology*, 2005. 31: p. 55 - 67.

Kamachi, M., et al., Enzymatic polymerization and characterization of new poly(3-hydroxyalkanoate)s by a bacterial polymerase. *Macromolecules*, 2001. 34: p. 6889 - 6894.

Keenan, T.M., S.W. Tanenbaum, and J.P. Nakas, Biodegradable polymers from renewable forest resources, in *Biodegradable polymers for industrial applications*, R. Smith, Editor. 2005, CRC: FL, USA. p. 219 - 250.

Kim, B. K., Yoon, S. C., Nam, J. D., Lenz, R. W. Effect of C/N ratio on the production of poly(3-hydroxyalkanoates) by the methylotroph Paracoccus denitrificans. *Journal of Microbiology and Biotechnology*, 1997. 7: 391 – 396.

Kim, Y.B. and R.W. Lenz, Polyesters from microorganisms. *Advances in Biochemical Engineering/ Biotechnology*, 2001. 71: p. 51 - 79.

King, P.P., Biotechnology: an industrial view. *Journal of Chemical Technology and Biotechnology*, 1982. 32: p. 2 - 8.

Korsatko, W., et al., Poly-D-(-)3hydroxybutyric acid (polyHBA) a biodegradable carrier for long term medication dosage. I. Development of parenteral matrix tablets for long-term administration of pharmaceuticals. *Pharmaceutical Industry*, 1983. 45: p. 525 - 527.

Korsatko, W., et al., Poly-D-(-)3hydroxybutyric acid (polyHBA) a biodegradable carrier for long term medication dosage.II. The biodegradation in animals and in vitro-in vivo correlation with the liberation of pharmaceuticals from parenteral matrix tablets. *Pharmaceutical Industry*, 1983. 45: p. 1004 - 1007.

Kumar S., Ching-Yee L., Lay-Koon G., Tadahisa I., Mizuo M. The Oil-Absorbing Property of Polyhydroxyalkanoate Films and its Practical Application: A Refreshing New Outlook for an Old Degrading Material. *Macromolecular Bioscience*, 2007. 7: p. 1199–1205

Kwan, L. and W. Steber, European Patent Application. 1991.

Labuzek, S., Radecka, I. Biosynthesis of PHB tercopolymer by Bacillus cereus UW85. *Journal of Applied Microbiology*, 2001. 90: 353 – 357.

Lageveen, R.G., et al., Formation of polyesters by Pseudomonas oleovorans: effect of substrates on formation and composition of poly-(R)-3-hydroxyalkanoates and poly-(R)-3-hydroxyalkenoates. *Applied and Environmental Microbiology*, 1988. 54: p. 2924 - 2932.

Lauzier, C.A., et al., Film formation and paper coating with poly(ß-hydroxyalkanoate), a biodegradable latex. *Tappi Journal*, 1993. 76(5): p. 71 - 77.

Law, K. H., Cheng, Y. C., Leung, Y. C., Lo, W. H., Chua, H., Yu, H. F. Construction of recombinant Bacillus subtilis strains for polyhydroxyalkanoate synthesis. *Biochemical Engineering Journal*, 2003. 16: 203 – 208.

Lee, S.Y., Plastic bacteria? Progress and prospects for polyhydroxyalkanoate production in bacteria. *Trends in Biotechnology*, 1996. 14: p. 431 - 438.

Lee, S. Y., Chang, H. N. Production of poly(hydroxyalkanoic acid). *Advances in Biochemical Engineering and Biotechnology*, 1995. 52: 27 – 58.

Li, X., Sun, J., Chen, S., Chen, G. In vitro investigation of maleated poly (3-hydroxybutyrate-co-3-hydroxyhexanoate) for its biocompatibility to mouse fibroblast L929 and human microvascular endothelial cells. *Journal of Biomedical Materials Research Part A*, 2008. Epub ahead of print.

Ljungberg, C., et al., Neuronal survival using a resorbable synthetic conduit as an alternative to primary nerve repair. *Microsurgery,* 1999. 19: p. 250 - 264.

Lopez, N.I., J.A. Ruiz, and B.S. Mendez, Survival of poly-3-hydroxybutyrate-producing bacteria in soil microcosms. W. J. *Microbiol. Biotechnol.*, 1998. 14: p. 681 - 684.

Lundgren, D. G., Pfister, R. M., Merrick, J. M. Structure of poly-ß-hydroxybutyric acid granules. *Journal of General Microbiology*, 1964.34: 441 – 446.

Macrae, R.M. and J.F. Wilkinson, Poly-beta-hydroxybutyrate metabolism in washed suspensions of Bacillus cereus and Bacillus megaterium. *Journal of General Microbiology,* 1958. 19(1): p. 210 - 222.

Madison, L. and G. Huisman, Metabolic engineering of poly(3-hydroxyalkanoates) : from DNA to plastic. *Microbiology and Molecular Biology Reviews*, 1999. 63(1): p. 21 - 53.

Maekawa, B., Kayama, N., Doi, Y. Purification and properties of 3-ketothiolase from Alcaligenes latus. *Biotechnology Letters*, 1993. 15: 691 - 696.

Malm, T., et al., Enlargement of the right-ventricular outflow tract and the pulmonary-artery with a new biodegradable patch in transannular position. *European Surgical Research*, 1994. 26: p. 298 - 308.

Martin, D.P. and S.F. Williams, Medical applications of poly-4-hydroxybutyrate: a strong flexible absorbable biomaterial. *Biochemical Engineering Journal*, 2003. 16: p. 97 - 105.

Matsumoto, K., et al. Production of polyhydroxyalkanoate (PHA) copolymer in transgenic plants and development of plant tissue-specific PHA production systems. in *International Symposium on Biological Polyesters - Conference proceedings*. 2006. Minneapolis, Minnesota, USA.

McCool, G. J., Cannon, M. C. Polyhydroxyalkanoate inclusion body-associated proteins and coding region in Bacillus megaterium. *Journal of Bacteriology*, 1999. 181: 585 - 592.

McCool, G. J., Cannon, M. C. PhaC and PhaR are required for polyhydroxyalkanoic acid synthase activity in Bacillus megaterium. *Journal of Bacteriology*, 2001. 183: 4235 - 4243.

McChalicher, C.W.A., Srienc, F. Investigating the structure-property relationship of bacterial PHA block copolymers. *Journal of Biotechnology*, 2007. 132: p. 296 - 302.

Merrick, J.M. and M. Doudoroff, Depolymerisation of poly-ß-hydroxybutyrate by intracellular enzyme system. *Journal of bacteriology*, 1964. 88: p. 60 - 71.

Miller, N.D. and D.F. Williams, On the biodegradation of poly-beta-hydroxybutyrate (PHB) homopolymer and poly-beta-hydroxybutyrate-hydroxyvalerate copolymers. *Biomaterials*, 1987. 8: p. 129 - 137.

Misra, S.K., et al., Polyhydroxyalkanoate (PHA)/ inorganic phase composites for tissue engineering applications. *Biomacromolecules*, 2006. 7(8): p. 2249 - 2258.

Misra, S.K., Watts, P.C.P., Valappil, S.P., Silva, S.R.P., Roy, I., Boccaccini, A.R. Poly(3-hydroxybutyrate)/Bioglass® composite films containing carbon nanotubes. *Nanotechnology*, 2007. 18: 1 - 7.

Miyake, M., M. Erata, and Y. Asada, A thermophilic cyanobacterium, Synechococcus sp. MA19, capable of accumulating poly-beta-hydroxybutyrate. *Journal of Fermentation and Bioengineering,* 1996. 82: p. 512 - 514.

Mochizuki, M. and M. Hirami, Structural effects on the biodegradation of aliphatic polyesters. *Polymers for advanced technologies*, 1997. 8(4): p. 203 - 209.

Mochizuki, M., Properties and application of aliphatic polyester products, in *Biopolymers-Polyesters III: Applications and commercial products*, Y. Doi and A. Steinbuchel, Editors. 2002, Wiley - VCH. p. 1 - 23.

Muh, U., Sinskey, A. J., Kirby, D. P., Lane, W. S. and Stubbe, J. PHA synthase from Chromatium vinosum: cysteine 149 is involved in covalent catalysis. *Biochemistry,* 1999. 38: 826 - 837.

Nawrath, C., Y. Poirier, and C. Somerville, Targeting of the polyhydroxybutyrate biosynthetic pathway to the plastids of Arabidopsis thaliana results in high levels of polymer accumulation. *Proceedings of the National Academy of Sciences of USA.,* 1994. 91: p. 12760 – 12764.

Nguyen, S. and R.II. Marchessault, Graft copolymers of methyl methacrylate and poly([R]-3-hydroxybutyrate) macromonomers as candidates for inclusion in acrylic bone cement formulations: Compression testing. *Journal of Biomedical Materials Research - Part B - Applied Biomaterials,* 2006. 77(1): p. 5 - 12.

Nishida, H. and Y. Tokiwa, Distribution of poly(ß-hydroxybutyrate) and poly(^{a}-caprolactone) degrading microorganisms and microbial degradation behaviour on plastic surfaces. *Polymeric Materials Science and Engineering,* 1992. 67: p. 137 - 138.

Nishioka, M., et al., Production of the poly-beta-hydroxyalkanoate by thermophilic cyanobacterium, Synechococcus sp. MA19, under phosphate-limited condition. *Biotechnology Letters,* 2001. 23: p. 1095 - 1099.

Noda, I., et al., Preparation and properties of a novel class of polyhydroxyalkanoate copolymers. *Biomacromolecules,* 2005. 6: p. 580 - 586.

Opitz, F., et al., Tissue Engineering of Ovine Aortic Blood Vessel Substitutes Using Applied Shear Stress and Enzymatically Derived Vascular Smooth Muscle Cells. *Annals of Biomedical Engineering,* 2004. 32(2): p. 212 - 222.

Ostle, A. G., Holt, J. G. Nile Blue A as a fluorescent stain for poly-ß-hydroxybutyrate. *Applied and Environmental Microbiology,* 1982. 44:238 - 241.

Page, W. J., Knosp, O. Hyperproduction of poly-ß-hydroxybutyrate during exponential growth of Azotobacter vinelandii UWD. *Applied and Environmental Microbiology,* 1989. 55: 1334 - 1339.

Pandey, J.K., et al., Recent advances in biodegradable nanocomposites. *Journal of Nanoscience and Nanotechnology,* 2005. 5: p. 497 - 526.

Park, S.H., S.H. Lee, and S.Y. Lee, Preparation of optically active beta-amino acids from microbial polyester polyhydroxyalkanoates. *Journal of Chemical Research, Synopses,* 2001. 11: p. 498 - 499.

Pederson, E.N., W.J. McChalicher, and F. Srienc, Bacterial synthesis of PHA block copolymers. *Biomacromolecules,* 2006. 7: p. 1904 - 1911.

Peoples, O. P., Sinskey, A. J. Poly-ß-hydroxybutyrate (PHB) biosynthesis in Alcaligenes eutrophus H16: Identification and characterisation of the PHB polymerase gene (phbC). *The journal of Biological Chemistry,* 1989. 264: 15298 - 15303.

Peoples, O.P. PHA Production in Plants: Enabling a Sustainable Plastics Biorefinery. in *4th International Crop Science Congress.* 2004. Brisbane, Australia.

Peoples, O.P. and K.D. Snell. Progress on plant based PHA production systems. in *International Symposium on Biological Polyesters.* 2006. Minneapolis, Minnesota, US.

Poirier, Y., B. Orts, and J. Beilen. Foundation paper for the Biopolymers Flagship. in *EPOBIO Workshop: Products from plants - the Biorefinery Future.* 2006. Wageningen International Conference Centre, The Netherlands.

Pouton, C.W. and S. Akhtar, Biosynthetic polyhydroxyalkanoates and their potential in drug delivery. *Advanced Drug Delivery Reviews*, 1996. 18: p. 133 - 162.

Purnell, M.P., et al. Spatio-temporal characterisation of polyhydroxybutyrate accumulation in sugarcane. in *International Symposium on Biological Polyesters-Conference Proceedings*. 2006. Minneapolis, Minnesota, USA.

Quillaguaman, J., et al. PHA production from renewable sources by the moderate halophile Halomonas boliviensis. in *International Symposium on Biological Polyester - Conference Proceedings*. 2006. Minneapolis, Minnesota, USA.

Ramsay, B. A., Lomaliza, K., Chavarie, C., Dube, B., Bataille, P., Ramsay, J. A. Production of poly-(ß-hydroxybutyric-co-ß-hydroxyvaleric) acids. *Applied and Environmental Microbiology*, 1990. 56: 2093 - 2098.

Rehm, B.A., N. Kruger, and A. Steinbuchel, A new metabolic link between fatty acid de novo synthesis and polyhydroxyalkanoic acid synthesis. The phaG gene from Pseudomonas putida KT2440 encodes a 3-hydroxy-acyl carrier protein-coenzyme a transferase. *Journal of Biological Chemistry*, 1998. 273: p. 24044 - 24051.

Rehm, B. A., Antonio, R. V., Spiekermann, P., Amara, A. and Steinbuchel, A. Molecular characterization of the poly(3-hydroxybutyrate) (PHB) synthase from Ralstonia eutropha: in vitro evolution, site-specific mutagenesis and development of a PHB synthase protein model. *Biochimica et Biophysica Acta*, 2002. 1594: 178 - 190.

Rehm, B.A., Polyester synthases: natural catalysts for plastics. *Biochemical Journal*, 2003. 376: 15 - 33.

Riis, V., Mai, W. Gas chromatographic determination of poly-beta-hydroxybutyric acid in microbial biomass after hydrochloric acid propanolysis. *Journal of Chromatography*, 1988. 445: 285 - 289.

Saad, B., et al., New versatile, elastomeric, degradable polymeric materials for medicine. *International Journal of Biological Macromolecules*, 1999. 25: p. 293 - 301.

Satoh, Y., Minamoto, N., Tajima, K., Munekata, M. Polyhydroxyalkanoate synthase from Bacillus sp. INT005 is composed of phaC and phaR. *Journal of Bioscience and Bioengineering*, 2002. 94: 343 - 350.

Savenkova, L., et al., Mechanical properties and biodegradation characteristics of PHB-based films. *Process Biochemistry*, 2000. 35: p. 573 - 579.

Schubert, P., Kruger, N., Steinbuchel, A. Molecular analysis of the Alcaligenes eutrophus poly(3-hydroxybutyrate) biosynthetic operon: identification of the N-terminus of the poly(3-hydroxybutyrate) synthase and identification of the promoter. *Journal of Bacteriology*, 1991. 173: 168 - 175.

Seebach, D., et al., Synthesis and enzymatic degradation of dendrimers from (R)-3-hydroxybutanoic acid and trimesic acid. *Angewandte Chemie International Edition*, 1996. 35: p. 2795 - 2797.

Seebach, D., et al., From the biopolymer PHB to biological investigations of unnatural beta- and gamma-peptides. *Chimia*, 2001. 55: p. 345 - 353.

Senior, P.J. and E.A. Dawes, Poly-ß-hydroxybutyrate biosynthesis and the regulation of glucose metabolism in Azotobacter beijerinckii. *Biochemical Journal*, 1971. 125: p. 55 - 66.

Senior, P.J. and E.A. Dawes, The regulation of poly-ß-hydroxybutyrate metabolism in Azotobacter beijerinckii. *Biochemical Journal*, 1973. 134(1): p. 225 - 238.

Sevastianov, V.I., et al., Production of purified polyhydroxyalkanoates (PHAs) for applications in contact with blood. *Journal of Biomaterial Science Polymer* Edn., 2003. 14(10): p. 1029 - 1042.

Shishatskaya, E.I., et al., Tissue response to the implantation of biodegradable polyhydroxyalkanoate sutures. *Journal of material sci - Mater M*, 2004. 15: p. 719 - 728.

Singh, S. and D. Maxwell, Tools of the trade. *Best Practice and Research in Clinical Obstetrics and Gynaecology*, 2006. 20: p. 41 - 59.

Slater, S., et al., Metabolic engineering of Arabidopsis and Brassica for poly(3-hydroxybutyrate-co-3-hydroxyvalerate) copolymer production. *Nature Biotechnology*, 1999. 17: p. 1011 – 1016.

Snell, K.D. and O.P. Peoples, Polyhydroxyalkanoate polymers and their production in transgenic plants. *Metabolic Engineering*, 2002. 4: p. 29 - 40.

Stal, L.J., Polyhydroxyalkanoate in cyanobacteria: an overview. *FEMS Microbiology Reviews*, 1992. 103: p. 169 - 180.

Steinbuchel, A. and H.E. Valentin, Diversity of bacterial polhydroxyalkanoic acids. *FEMS Microbiology Letters*, 1995. 128: p. 219 - 228.

Steinbuchel, A. and B. Fuchtenbusch, Bacterial and other biological systems for polyester production. *Tibtech*, 1998. 16: p. 419 - 427.

Steinbuchel, A. and Hein, S., Biochemical and molecular basis of microbial synthesis of polyhydroxyalkanoates in microorganisms. *Advances in Biochemical Engineering and Biotechnology*, 2001. 71: 81 - 123.

Stuart, E. S., Lenz, R. W., Fuller, R. C. The ordered macromolecular surface or the polyester inclusion in Pseudomonas oleovorans. *Canadian Journal of Microbiology*, 1995. 41:84 - 93.

Stubbe, J., Tian, J. Polyhydroxyalkanoate (PHA) homeostasis: the role of the PHA synthase. *Nat. Prod. Rep.* ,2003. 20: 445 - 457.

Suzuki, T., Deguchi, H., Yamane, T., Shimizu, S., Gekko, K. Control of molecular weight of poly-p-hydroxybutyric acid produced in fed-batch culture of Protomonas extorquens. *Applied Microbiology and Biotechnology*, 1988. 27: 487 - 491.

Taguchi, K., et al., Metabolic pathways and engineering of PHA biosynthesis, in *Biopolymers: Polyesters, Y. Doi and A. Steinbuchel*, Editors. 2002, Wiley-VCH. p. 217 - 372.

Tajima, K., Igari, T., Nishimura, D., Nakamura, M., Satoh, Y., Munekata, M. Isolation and characterization of Bacillus sp. INT005 accumulating polyhydroxyalkanoate (PHA) from gas field soil. *Journal of Bioscience and Bioengineering*, 2003. 95: 77 - 81.

Tanaka, Y., et al., Purification and properties of D(-)-3-hydroxybutyrate - dimer hydrolase from Zoogloea ramigera I-16-M. *European Journal of Biochemistry*, 1981. 118: p. 177 - 182.

Tang, H.Y., Daisuke, I., Atsushi, M., Sunao, M., Tetsuji, Y., Kumar, S., Razip, S., Masahiro, F., Mizuo, M., Tadahisa, I. Scaffolds from electrospun polyhydroxyalkanoate copolymers: Fabrication, characterization, bioabsorption and tissue response. *Biomaterials*, 2008. 29: p. 1307 - 1317.

Tasaki, O., et al., The dimer and trimer of 3-hydroxybutyrate oligomer as a precursor of ketone bodies for nutritional care. *Journal of Parenteral and Enteral Nutrition*, 1999. 23: p. 321 - 325.

Tokiwa, Y. and T. Suzuki, Hydrolysis of polyesters by lipases. Nature, 1977. 270: p. 76 - 78.

Tokiwa, Y. and B.P. Calabia, Degradation of microbial polyesters. *Biotechnology letters*, 2004. 26: p. 1181 - 1189.

Tsuge, T., et al., Molecular cloning of two (R)-specific enoyl-CoA hydratase genes from Pseudomonas aeruginosa and their use for polyhydroxyalkanoate synthesis. *FEMS Microbiology Letters*, 2000. 184(2): p. 193 - 198.

Tsuge, T., Watanabe, S., Sato, S., Hiraishi, T., Abe, H., Doi, Y. and Taguchi, S. Variation in Copolymer Composition and Molecular Weight of Polyhydroxyalkanoate Generated by Saturation Mutagenesis of Aeromonas caviae PHA Synthase, *Macromolecular Bioscience*, 2007. 7: p. 846 - 854.

Urakami, T., et al., Development of biodegradable plastic-poly-beta-hydroxybutyrate/polycaprolactone blend polymer. Kobunshi Ronbunshu, 2000. 57: p. 263 - 270.

Valappil, S., et al., Biomedical applications of polyhydroxyalkanoates (PHAs), an overview of animal testing and in vivo responses. *Expert Review - Medical Devices*, 2006. 3(6): p. 853 - 868.

Valappil, S., C. Bucke, and I. Roy, Polyhydroxyalkanoates in Gram-positive bacteria: insights from the genera Bacillus and Streptomyces. *Antonie Van Leeuwenhoek*, 2007(a). 91:1 - 17.

Valappil, S., Peiris, D., Langley, G.J., Herniman, J.M., Boccaccini, A.R., Bucke, C., Roy, I. Polyhydroxyalkanoate (PHA) biosynthesis from Structurally unrelated carbon sources by a newly characterized Bacillus spp. *Journal of Biotechnology*, 2007(b). 127: p. 475–487.

Valappil, S., Misra, S.K., Boccaccini, A.R., Keshavarz, T., Bucke, C., Roy, I. Large-scale production and efficient recovery of PHB with desirable material properties, from the newly characterised Bacillus cereus SPV. *Journal of Biotechnology*, 2007(c). 132: 251 - 258.

Valappil, S., Rai, R., Bucke, C., Roy, I. Polyhydroxyalkanoate biosynthesis in Bacillus cereus SPV under varied limiting conditions and an insight into the biosynthetic genes involved. *Journal of Applied Microbiology*, 2008. Epub ahead of print.

Valentin, H.E., et al., PHA production, from bacteria to plants. *International Journal of Biological Macromolecules*, 1999. 25: p. 303 -306.

Volova, T.G., Biodegradation of polyhydroxyalkanoates. *Polyhydroxyalkanoates - plastic materials of the 21st century: production, properties, applications.* 2004, New York: Nova Science Publishers, Inc.

Wallen, L. L., Rohwedder, W. K. Poly-ß-hydroxyalkanoate from activated sludge. *Environmental Science and Technology*, 1974. 8: 576 - 579.

Wang, Y.H., et al., Synthesis of medium-chain-length polyhydroxyalkanoates in tobacco via chloroplast genetic engineering. Chinese Science Bulletin, 2005. 50(11): p. 1113 - 1120.

Ward, P.G., et al., A two step chemo-biotechnological conversion of polystyrene to a biodegradable thermoplastic. *Environmental Science and Technology*, 2006. 40(7): p. 2433 - 2437.

Weiner, R.M., Biopolymers from marine prokaryotes. *Tibtech*, 1997. 15: p. 390 - 427.

Weiczorek, R., Pries, A., Steinbuchel, A. and Mayer, F., Analysis of a 24-kilodalton protein associated with the polyhydroxyalkanoic acid granules in Alcaligenes eutrophus. *Journal of Bacteriology*, 1995. 177: 2425 - 2435.

Wilde, B.D., International and National Norms on Biodegradability and Certification Procedures, in *Handbook of Biodegradable Polymers*, C. Bastioli, Editor. 2005, RAPRA. p. 145 - 175.

Williams, S.F., et al., PHA applications: addressing the price performance issue I. Tissue engineering. *International Journal of Biological Macromolecules*, 1999. 25: p. 111 - 121.

Williams, S.F. and D.P. Martin, Applications of PHAs in medicine and pharmacy, in *Biopolymers: Polyesters 3*, Y. Doi and A. Steinbuchel, Editors. 2002, Wiley - VCH: Weinheim. p. 91 - 128.

Wrobel, M., J. Zebrowski, and J. Szopa, Polyhydroxybutyrate synthesis in transgenic flax. *Journal of Biotechnology*, 2004. 107(1): p. 41 - 54.

Wu, Q., Huang, H., Hu, G. H., Chen, J., Ho, K. P., Chen, G. Q. Production of poly-3-hydroxybutyrate by Bacillus sp. JMa5 cultivated in molasses media. *Antonie van Leeuwenhoek*, 2001. 80: 111 - 118.

Yagmurlu, M.F., et al., Sulbactam-cefoperazone polyhydroxybutyrate-co-hydroxyvalerate (PHBV) local antibiotic delivery system: in vivo effectiveness and biocompatibility in the treatment of implant-related experimental osteomyelitis. *Journal of Biomedical Materials Research*, 1999. 46: p. 494 - 503.

Zheng, Z., Li, M., Xue, X., Tian, H., Li, Z., Chen, G. Q. Mutation on N-terminus of polyhydroxybutyrate synthase of Ralstonia eutropha enhanced PHB accumulation. *Applied Microbiology and Biotechnology*, 2006. 72: 896 - 905.

In: Biotechnology: Research, Technology and Applications   ISBN 978-1-60456-901-8
Editor: Felix W. Richter   © 2008 Nova Science Publishers, Inc.

Chapter 2

Polyhydroxyalkanoates: In Nature, in the Laboratory and in Industry

Chih-Ching Chien
Graduate School of Biotechnology and Bioengineering
Yuan Ze University, 135 Yuan Tung Road
Chung-Li, 320 Taiwan

Abstract

With the increasing price of crude oil and the concerns about global warming, petrochemical based plastics now faces two major challenges: their non-degradability in the environment and diminishing supply of resource. Bio-based polymers present several advantages over conventional plastics; first, they originate from sustainable renewable resources and secondly, they are completely biodegradable. Polyhydroxyalkanoates (PHAs) is the most well-known biologically-based polyester. They accumulate naturally as intracellular inclusions in microorganisms grown under imbalanced nutritional conditions. PHAs can contain different and repeating units of carbon 'groups' and this structural diversity confer different physical properties. Thus, they can serve as a reliable substitute to synthetic plastics.

Although more than 90 genera of bacteria have been identified for their ability to accumulate a wide spectrum of PHA, none could serve as a strain suitable for industrial production of PHA. This is partially due to their lack of genetic tractability resulting in an inability for strain improvement. Recent efforts have been focused on reducing the cost for PHAs production in order to make the price of bio-polymers competitive against petrochemical based plastics. Strategies include strain improvement by recombinant DNA technology, high cell-density fermentation and the utilization of inexpensive carbon sources for the production of PHA. The launch of biobased plastics Mirel™ by Metabolix Inc. in 2007 clearly demonstrates that PHAs is starting to be accepted as an alternative to petro-based plastics. With the completion of the genome sequence of the well-known PHA-producing bacterium *Ralstonia eutropha* H16 and the advanced metabolic engineering technology available, opportunities abound for the improvement of PHA quality. This contribution describes the microorganisms (both indigenous and recombinant strains) that are capable of accumulating PHA in their natural habitats, in the laboratory and the feasibility of large scale industrial production.

Introduction

Plastics play multiple roles in all facets of our lives; they are present in almost all containers or devices used in the household, hospital and various industries. Plastics are polymers of repeating chemical units and depending on the chemical units and the degree of polymerization, they can attain various degrees of robustness or malleability depending on their intended use. However, most plastics share the dubious distinction of being synthesized from a non-renewable source: fossil fuel.

Arguably, one of the more popular environmental and social issues in the first decade of the 21st century concern global warming or rather, an attempt to stem its effects. It is generally accepted that the unregulated exploitation of fossil fuel and its use can enhance the rate of global warming resulting in a variety of environmental, epidemiological and social consequences. Concurrently, the demand for crude oil has not subsided as evidenced by increasing prices of crude oil. Both of these issues serve as a strike against the continued reliance on petrochemical plastics.

Parallel to the concern on global warming, an increased effort to search for alternative renewable resources exists. In the area of plastic technology, techniques employing alternative polymeric 'plastics' to replace petrochemical plastics have been tried. Among it is the use of biodegradable biological polymers. Yet, the glaring shortcomings of this technology continue to be its cost-effectiveness.

Biopolymers produced by microorganisms, are completely biodegradable and originate from a sustainable renewable resource. It has potential for broad biotechnological applications due to their biocompatibility and presents a considerable advantage with respect to other synthetic materials. Microbial polymers such as xanthan gum (synthesized by *Xanthomonas campestris* as extracellular polymeric substances) and dextrans have been produced in bulk and have numerous industrial and biotechnological applications [Sutherland 1982; 1998]. Biopolymers in the forms of lipid inclusions can also be found ubiquitously as intracellular inclusions such as polyhydroxyalkanoates (PHA) and triacylglycerols (TAG) in microorganisms (Figure 1) [Alvarez et al. 1996; Steinbüchel and Füchtenbusch 1998; Alvarez and Steinbüchel 2002; Chien et al. 2007]. The lipophilic inclusions PHAs and TAG are anticipated to have wide applicability in the areas of bio-plastic and bio-diesel technology and serve as attractive alternatives to petrochemical products.

Microorganisms are capable of forming a variety of intracellular inclusions depending on the growth conditions. The accumulation of PHAs occurs in some bacteria during growth conditions with an excess of carbon source but limiting in other nutrients (e.g. phosphate, nitrogen, etc). Some bacteria can accumulate PHAs up to 90% of the cell dry weight. This property can be harnessed for industrial production of PHAs. PHAs as polyesters of hydroxyalkanoates (poly-3-hydroxybutyrate (PHB)) were first reported by Maurice Lemoigen of the Pasteur Institute in *Bacillus megaterium* [Lemoigne, 1926].

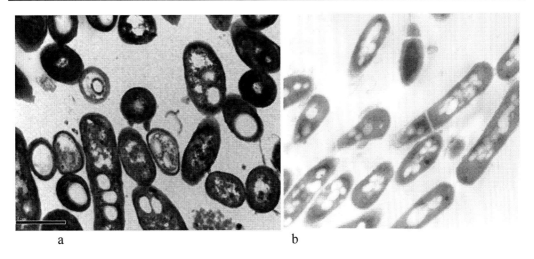

a   b

Figure 1. Granules of accumulated PHAs polymers were observed by transmission electron microscopy in an environmental *Bacillus* isolate (a) and TAGs in a *Rhodococcus opacus* (b).

Since then, about 150 different constituents have been observed in PHAs isolated from microorganisms belonging to more than 90 genera; however, few of them could be exploited for commercial production [Lee and Choi 2001; Luengo et al. 2003; Valappil et al. 2007]. The general metabolic pathway for the synthesis of PHAs by bacteria from different carbon sources is shown in Figure 2.

PHAs are classified according to the number of carbons of each repeating unit in the polymers (Figure 3) as short- (3-5 C-atoms) chain-length PHAs (scl-PHAs) and medium- (6 or more C-atoms) chain-length PHAs (mcl-PHAs) [Steinbüchel and Valentin 1995; Sudesh et al 2000; Kim and Lenz 2001]. The hydroxyl-substituted carbon atom (the chiral center) of the acid monomers can vary from 1 (3-hydroxyalkanoic acid) to 4 (6-hydroxyalkanoic acid) and are in the R configuration because of the stereospecificity of PHA synthase (the polymerizing enzyme of PHAs) [Anderson and Dawes 1990; Steinbüchel and Valentin 1995; Lee 1996a; Sudesh et al 2000].

The side chains (the R group) can be varied by means of substrates used or the type of growth condition. They can be aliphatic or aromatic, saturated or unsaturated, linear or branched, and halogenated or epoxidized [Steinbüchel and Valentin 1995; Madison and Huisman 1999; Pötter and Steinbüchel 2006].

The discovery of polythioesters (PTEs) synthesized by PHAs-accumulating bacteria showed an enormous diversity of biopolymers and can be produced by the PHA-synthesizing enzymes [Lütke-Eversloh et al 2001a, b; Lütke-Eversloh et al 2002a, b; Lütke-Eversloh and Steinbüchel 2003; Lütke-Eversloh and Steinbüchel 2004].

The diversity of the side chain provides a vast variation of the physicochemical properties of PHAs and therefore makes them an excellent biodegradable plastic compounds with a broad range of applications [Guerrero and Berlanga 2007].

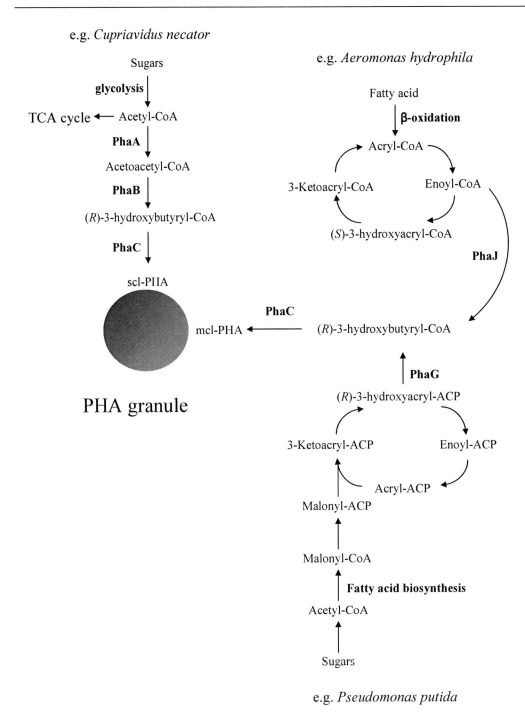

Figure 2. General metabolic pathways of microbial synthesis of PHAs using different carbon sources. [Sudesh et al. 2000; Philip et al. 2007].

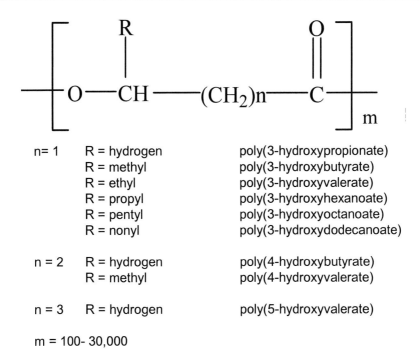

n= 1	R = hydrogen	poly(3-hydroxypropionate)
	R = methyl	poly(3-hydroxybutyrate)
	R = ethyl	poly(3-hydroxyvalerate)
	R = propyl	poly(3-hydroxyhexanoate)
	R = pentyl	poly(3-hydroxyoctanoate)
	R = nonyl	poly(3-hydroxydodecanoate)
n = 2	R = hydrogen	poly(4-hydroxybutyrate)
	R = methyl	poly(4-hydroxyvalerate)
n = 3	R = hydrogen	poly(5-hydroxyvalerate)

m = 100- 30,000

Figure 3. Basic structure of PHAs. [Lee 1996b; Philip et al. 2007].

Who's Who in PHA Production in Microbial World

Among the various types of lipid inclusions, PHAs can be found in members of most bacterial and archaeal genera while triacylglycerols (TAGs) and wax esters (WEs) exists in only a few [Alvarez and Steinbüchel 2002]. Unlike TAGs and WEs, synthesis of PHAs as a storage compound in nature occurs only in bacteria. However, yeasts such as *Saccharomyces cerevisiae* and *Pichia pastoris* have been used as surrogates for polyhydroxyalkanoate production in the peroxisome [Poirier et al. 2001; 2002; Zhang et al. 2006]. PHAs are not deposited as lipid bodies in eukaryotes in nature [Wältermann and Steinbüchel 2005].

The amounts of PHAs accumulated in microorganisms in their natural niches can vary greatly ranging from less than 1% to as high as 80% of their cellular dry weight [Kim and Lenz 2001] but only those with the ability to accumulate very high contents of PHAs are ideal candidates for further strain improvement for industry. Therefore, despite the prevalence of PHAs in bacteria in nature able to synthesize PHA, only few species have actually been intensively investigated for their application in industry [Kung et al. 2007]. Up to now, Gram-negative bacteria are currently main strains that being intensively developed for commercial source of PHAs-production [Pötter and Steinbüchel 2006; Valappil et al. 2007]. Perhaps the most intensively studied PHA-producing strain is the facultative chemolithoautotrophic hydrogen oxidizing bacterium *Ralstonia eutropha* (formerly known as *Alcaligenes eutrophus* and now *Cupriavidus necator*), which synthesize PHB; the most frequently occurring type of PHAs [Schlegel et al. 1961; Vandamme and Coenye 2004].

Other Gram-negative bacteria that have been intensively investigated for industrial production of PHAs include *Pseudomonas putida*, *Pseudomonas oleovorans*, *Aeromonas hydrophila* and recombinant strains of *Escherichia coli* [Fuchtenbusch and Steinbüchel 1999; Kim et al. 2000; Li et al. 2007; Ouyang et al. 2007; Valappil et al. 2007]. Recently, more attention has been paid to other microorganisms including Gram-positive bacteria, some halophilic bacteria and Archaea for the ability of diverse PHAs synthesis [Hezayen et al. 2002; Chen et al. 2006; Huang et al. 2006; Valappil et al. 2006; Han et al. 2007; Quillaguamán ct al. 2008]. Some important microbial strains for PHA synthesis are discussed in this section.

Cupriavidus necator (= *Alcaligenes eutrophus* = *Ralstonia eutropha* = *Wautersia eutropha*)

Although *Cupriavidus necator* is not the first microorganism that has been discovered for the ability to accumulate intracellular PHB, it can be considered as the most well-known PHB-producing bacterium. Biosynthesis of PHB in *C. necator* involves a three-step reaction catalyzed by three enzymes [Steinbüchel and Schlegel 1991; Zinn et al. 2001; Reddy et al. 2003]. The first reaction catalyzed by a β-ketothiolase (PhaA) condenses two molecules of acetyl-CoA into acetoacetyl-CoA with the release of one CoA [People and Sinskey 1989a]. The resulting acetoacetyl-CoA is than reduced to *R*-(3)-hydroxybutyryl-CoA by an NADPH-dependent acetoacetyl-CoA reductase (PhaB) [Haywood et al. 1988; People and Sinskey 1989a]. This reaction occurs stereospecifically and all resulting 3-hydroxybutyryl-CoA synthesized are in the *R*-configuration at position 3. In the last step of PHB biosynthesis the monomers of *R*-(3)-hydroxybutyryl-CoA is polymerized into the PHB polymer by the PHB synthase (PhaC) [People and Sinskey 1989b]. *C. necator* is also the first microorganism whose PHB biosynthesis genes (The PHA operon *pha*CAB) were cloned and heterologously expressed in *E. coli* [Schubert et al. 1988; Slater et al. 1988]. With the increasing interest in PHA biosynthesis, additional genes were discovered to be involved in PHA metabolism. The completion of the genome sequence of the bioplastic-producing "Knallgas" bacterium *R. eutropha* (*C. necator*) H16 also exhibited 37 additional *pha*A isologs and 15 *pha*B isologs [Pollmann et al. 2006]. The existence of a second gene for PHB synthase (*pha*C2) was also discoverd. Other genes that are involved in the PHB metabolism were genes encoding for phasins (*pha*P1, *pha*P2, *pha*P3 and *pha*P4), PHB depolymerases (*pha*Z1and *pha*Z2 on Chromosome 1, *pha*Z3, *pha*Z5, *pha*Z6and *pha*Z7 on chromosome 2, and *pha*Z4 on pHG1) and PHB-oligomer hydrolases (*pha*Y1 and *pha*Y2) [Pollmann et al. 2006].

C. necator generally only accumulates PHB in nature. However, in laboratory, different strategies have been used to encourage PHA copolymer production by this model bacterium. By supplying acetate and propionate as the carbon sources, *C. necator* H16 was able to accumulate poly(3-hydroxybutyrate-*co*-3-hydroxyvalerate [P(3HB-*co*-3HV)]. Additionally, the amount of 3HV can be varied between 0 and 45% by adjusting different ratios of acetate and propionate added in the medium [Doi et al. 1987a]. Butyrate, valerate and pentanoic acid were also used for the synthesis of P(3HB-*co*-3HV) by *C. necator* [Doi et al. 1987b; 1988b]. An unusual copolyester poly(3-hydroxybutyrate-*co*-4-hydroxybutyrate) [P(3HB-*co*-4HB)] was also synthesized by *C. necator* with special feeding strategies [Doi et al. 1988a; Kunioka et al. 1988; 1989; Song and Kim 2005]. Production of P(3HB-*co*-4HB) by *C. necator* with flask and fed-batch cultures have also been evaluated [Kim et al. 2005]. Polythioesters can also be accumulated in *C. necator* when sulfur containing precursors were provided [Lütke-Eversloh and Steinbüchel 2004].

Many fermentation studies have been done on *C. necator* in order to obtain higher cell densities and larger quantity of PHAs. In order to achieve high cell densities of more than 100 g/L, a fed-batch mode is the preferred operation strategy. The cell densities of *C. necator* were as high as 164 g/L in 50 h with 121 g/L of PHB produced when nitrogen was limited at a biomass of 70 g/L in a fed-batch culture [Kim et al. 1994]. Fed-batch cultures of *C. necator* with phosphate-limitation have also been examined and the cell densities obtained were over 200 g/L with the PHB productivity levels over 3.0 g/L/h [Ryu et al. 1997; 1999; Shang et al. 2003]. Phosphate concentration apparently played a key role in accumulation of PHB as well as in cell densities in high cell density fed-batch culture of *C. necator* as shown in an unstructured model [Shang et al. 2007]. Continuous culture techniques have also been used for production of PHB and [P(3HB-*co*-3HV)] using pH regulation and/or glucose concentration [Gostomski and Bungay 1996; Chung et al. 1997; Henderson and Jones 1997; Yu et al. 2005]. Other fermentation strategies for PHB production by *C. necator* include use of repeated batch cultivation, airlift bioreactor and dispersed bioreactor have also been discussed have also been discussed [Tavares et al 2004; Khanna and Srivastava 2005; Patnaik 2006;2007].

Pseudomonas

Instead of accumulating PHB, *Pseudomonas* strains synthesize mcl-PHAs (C_6 to C_{14}). The most intensively investigated microorganisms of this group are *P. oleovorans* and *P. putida*, and both are able to synthesize a wide variety of mcl-PHAs with different organic carbon sources including alkanes, alkenes and carboxylic acids [Lim and Lenz 2001]. When hydrocarbons serve as the substrate for *Pseudomonas*, mcl-PHA is produced via the fatty acid oxidation pathway [Lageveen et al 1988; Huisman et al. 1991; Madison and Huisman 1999]. However, sugars such as glucose can also be used for PHAs formation by *Pseudomonas* spp. such as *P. putida*; the pathway involves the fatty acid *de novo* biosynthesis [Huijberts et al. 1992; Rehm et al. 1998]. The enzymes and related genes that inter-connect the major metabolic pathways and PHAs synthesis include 3-hydroxy acyl-acyl carrier protein: CoA transferase (PhaG), enoyl-CoA hydratase (PhaJ), 3-ketoacyl-CoA reductase (FabG) and epimerase have been well studied and characterized [Rehm et al. 1998; Ren et al. 2000; Fiedler et al. 2000; 2002; Tsuge et al. 2000; 2003; Davis et al. 2008].

Alternative to the homopolymers of scl-PHAs (PHB and PHV) that are often accumulated in microorganisms, the mcl-PHAs synthesized by *Pseudomonas* are copolymers [Du and Yu 2002; Nishikawa et al. 2002]. For example, *P. oleovorans* synthesizes copolymer of 3-hydroxynonanoate and 3-hydroxyheptanoate from nonanoic acid and copolymer of 3-hydroxyoctanoate and 3-hydroxyhexanoate from octanoic acid [Huisman et al. 1989]. Some *Pseudomonas* spp. have also demonstrated the ability to incorporate both scl- and mcl-PHA monomers [Steinbüchel and Wiese 1992; Abe et al. 1994; Lee et al. 1995; Kato et al. 1996; Chung et al. 1999; Ashby et al. 2002; Sujatha et al. 2007]. Homopolymers of PHB synthesized by some *Pseudomonas* strains has also been reported [Chung et al, 1999; Yao et al. 1999]. Production of PHAs by fed-batch fermentation and continuous culture has also been investigated in order to increase the yield of mcl-PHA from *Pseudomonas* [Preusting et al. 1993; Sun et al. 2006; 2007].

Aeromonas

Strains of *Aeromonas caviae* and *Aeromonas hydrophila* are able to accumulate poly(3-hydroxybutyrate-*co*-3-hydrohexanoate) (PHBHHx) which are composed of both scl-PHA monomer 3-hydroxybutyrate (3HB) and mcl-PHA monomer 3-hydrohexanoate (3HHx) [Doi et al. 1995; Fukui and Doi 1997; Lee et al. 2000; Lu et al. 2004a]. The copolymer was reported to have similar mechanical properties to low density polyethylene with a high degree of flexibility compared to that pf PHB [Shimamura et al. 1994; Doi et al. 1995; Feng et al. 2002]. The precursors for biosynthesis of PHBHHx by *Aeromonas* are supplied through the fatty acid ß-oxidation pathway [Fukui and Doi 1997; Fukui et al. 1998]. The type I PHA synthase (PhaC) was responsible for the biosynthesis of PHBHHx from lauric acid in *A. hydrophila* [Rehm and Steinbüchel 1999; Lu et al. 2004a]. *A. hydrophila* CGMCC 0911 produced only traces of PHA when glucose or gluconate were used as carbon source [Lu et al. 2004a]. A second PHA synthase (type II PHA synthase) was identified in wild type of *A. hydrophila* which was responsible for the synthesis of mcl-PHA in *A. hydrophila* when PhaC was inactivated [Hu and You 2007]. The precursor for type II PHA synthase was provided by the transacylase (PhaG) and PHA was synthesized via the fatty acid de novo pathway [Hoffmann et al. 2000]. The presence of type II PHA synthase and PhaG gene in *A. hydrophila* indicates that the bacterium may have more than one PHA biosynthesis systems [Hu and You 2007]. Besides PHBHHX and mcl-PHA, a terpolyester consisting of 3-HB, 4-hydroxybutyrate (4-HB) and 3HHx [P(3HB-*co*-4HB-*co*-3HHx)] can be produced by a recombinant *A. hydrophila* 4AK4 harboring genes encoding phasing (*pha*P), PHA synthase (*pha*C) and (*R*)-specific enoyl-CoA hydratase (*pha*J) from *A. caviae* [Xie and Chen 2008].

Large scale production of PHBHHX by *Aeromonas* has been examined by two-stage fed-batch fermentation [Lee et al. 2000; Chen et al. 2001]. Industrial-scale production of PHA by *A. hydrophila* was also evaluated by using glucose and lauric acid as carbon sources in a two-stage fermentation strategy. Glucose was used as the carbon source for the growth phase while lauric acid for the PHA synthesis phase with limited nitrogen or phosphorus [Chen et al., 2001]. A final 50 g l^{-1} of cell mass was obtained with PHA productivity of 0.54 g l^{-1} h^{-1} in this approach (Chen et al., 2001). Enhancement of the production of PHBHHx by *Aeromonas* has also been investigated by several metabolic engineering strategies such as in vitro evolution of PHA synthase [Kichise et al. 2002], over-expression of the phasing gene [Tian et al. 2005] and introduction of the *vgb* gene (encoding *Vitreoscilla* hemoglobin) into *Aeromonas* [Qiu et al. 2006]. Production of PHBHHx by cloning PHA synthesis genes into recombinant *Escherichia coli* has also been studied and will be discussed below.

Gram-Positive Bacteria

Up to now, almost all the endeavors on PHAs production on the industrial scale were focused on Gram-negative bacteria (e.g. *C. necator*, *Pseudomonas* spp, *Aeromonas* spp and recombinant *E. coli*), however, many Gram-positive bacteria are also able to accumulate significant amount of PHAs in nature. Some examples include *Bacillus* spp. [Slepecky and Law 1961; Chen et al. 1991; McCool and Cannon 1999; Tajima et al. 2003; Kalia et al. 2007; Valappil et al. 2007]; *Streptomyces* spp. [Manna et al. 1999; Ramachander and Rawal 2005; Kalia et al. 2007 ; Valappil et al. 2007]; *Rhodococcus* spp. [Haywood et al. 1991; Anderson et al. 1995]; *Corynebacterium* spp. [Haywood et al. 1991; Jo et al. 2007]; *Nocardia* spp. [Valentin and Dennis 1996; Hall et al. 1998] and *Staphylococcus* spp. [Szewczyk and Mikucki 1989; Wong et al. 2000; Kalia et al. 2007]. Polymers produced by Gram-positive bacteria have the advantages of lacking lipopolysaccharide (LPS). This is especially vital since LPS is a strong antigen [Valappil et al. 2007]. However, the several drawbacks of using Gram-positive bacteria for PHA production include a relatively long generation time and the recalcitrance of cells to lysis (due to the thick bacterial cell wall). Nevertheless, some of these Gram-positive bacteria are able to synthesize unusual type of PHAs and the genes relayed to PHA synthesis from these Gram-positive bacteria are still valuable for heterologous expression in other recombinant microorganisms.

Recombinant Microorganisms

It is generally accepted that that a microorganism used for industrial production is most likely a genetically engineered strain. Recombinant *C. necator* over-expressing *phbCAB* have been demonstrated to produce slightly higher amounts of PHB by cell dry weight compared to that the wild type strain [Park et al. 1995]. Production of P(3HB-*co*-3HV) and P(3HB-*co*-4HB) were also achieved by introducing *phbCAB*, *phbAB* and *phbC* genes into *C. necator* and the molar fraction of 3HV and 4HB were increased, respectively [Lee et al. 1997]. A PHB-negative mutant strain (PHB-4) of *C. necator* harboring a low-substrate-specific PHA synthase from *Pseudomonas stutzeri* was used for producing PHA copolymers consisting of scl-PHA and mcl-PHA [Luo et al. 2006]. Mutants of PHA synthase have also been introduced into this PHB-negative strain to generate recombinant strains of *C. necator* able to synthesize PHA from renewable carbon sources [Taguchi et al. 2003]. A *C. necator* strain harboring the PHA-biosynthesis genes from *A. caviae* was able to synthesize terpolymers of 3HB, 3HV and (*R*)-3-hydroxyheptanoate (3HHp) [Fukui et al. 1997]. Metabolic engineering of *Pseudomonas* spp. has also been investigated for PHA production [Huisman et al. 1992; Kraak et al. 1997]. The relationship of PHA synthesis and expression level of PHA synthase PhaC1 was evaluated by the recombinant *Pseudomonas* strains [Kraak et al. 1997]. The genetically engineered *Pseudomonas* strains able to utilize different substrate such as CO_2 or triacylglycerols for the biosynthesis of PHA have also been reported [Yagi et al. 1996; Solaiman et al. 2001]. By transforming the *phbCAB* operon of *C. necator* into *Pseudomonas putida*, a blending of mcl-PHA and PHB could be synthesized using octanoate and gluconate as carbon sources [Shin et al. 2002]. Heterologous expression of PHA biosynthesis genes of

Pseudomonas sp. in PHA-negative mutants of *Pseudomonas putida* resulted in the accumulation of PHA consisting of scl-PHA (3HB) and mcl-PHA (3HHx, 3HO, 3HD and 3HDD) [Matsusaki et al. 1998]. In addition, PHBHHx could also be produced from gluconate, hexanoate or octanoate by recombinant *P. putida* harboring PHA biosynthesis genes from *A. caviae* [Fukui and Doi 1997]. Another natural PHA-producer that has been intensively investigated by metabolic engineering is *Aeromonas* [Qiu et al. 2006; Hu et al. 2007; Xie and Chen 2008]. The content and concentration of PHBHHx could be increased up to 70 wt% and 4.0 g l^{-1}, respectively by co-expression of *phbAB* with *vgb* genes in *A. hydrophila* [Qiu et al. 2006]. Site-directed mutagenesis and saturation mutagenesis have also been used to improve PHA production in *Aeromonas* spp. [Hu et al. 2007; Tsuge et al. 2007]. Terpolyesters consisting of 3HB, 4HB and 3HHx [P(3HB-*co*-4HB-*co*-3HHx)] was synthesized by recombinant *A. hydrophila* harboring PHA synthesis genes *phaPCJ* [Xie and Chen 2008].

Although metabolic engineering of natural PHA-producers have been carried out by many groups, recombinant *E. coli* will still plays a key role for the commercial production of PHAs. This is due to our deep understanding of the biology of this organism as well as the vast array of molecular tools available for genetic manipulation. *E. coli* is considered to be a suitable workhorse because of the following advantages: (1) A well established high cell density fermentation process; (2) It is able to utilize a broad spectrum of substrates including some inexpensive carbon sources; (3) It is relatively easy to lyse the cell for eventual purification of PHA; (4) It does not have intracellular depolymerases which may degrade the PHAs and (5) it is amenable to a variety of genetic strategies and metabolic engineering for strain improvement [Fidler and Dennis 1992; Hahn et al. 1995; Li et al. 2007]. Recombinant *E. coli* harboring PHA biosynthesis genes is able to accumulate large amount of PHB, which could reach 80-90% of dry cell weight at the end of fermentation [Kim et al. 1992]. One of the challenges during the fermentation was the high demand for oxygen, and one strategy is to introduce the *Vitreoscilla* hemoglobin gene (*vgb*) into recombinant E. coli [Yu et al. 2002].

Recombinant *E. coli* has also been engineered for the production of a variety of scl-PHA copolymers such as P(3HB-*co*-4HB) and P(3HB-*co*3HV) [Slater et al. 1992; Valentin and Dennis 1997]. Copolymers of scl-PHA and mcl-PHA can also be synthesized by recombinant *E. coli* harboring PHA synthases with broad substrate specificity from *A. caviae*, *A. hydrophila* and *Pseudomonas* spp. [Park et al. 2001; Lu et al. 2004b; Nomura et al. 2004]. Providing *phbAB* genes of *C. necator* and *phaC* gene of *Pseudomonas* sp. 61-3 in *E. coli fadA* and/or *fadB* mutant could result in the biosynthesis of copolymers of scl-PHA and mcl-PHA with monomers consisting of C4, C6, C8 and C10 [Park and Lee 2004]. In addition, co-expression of *fabG* (encoding a 3-ketoacyl-ACP reductase) and *phaC* in *E. coli* could also led to the biosynthesis of PHA consisting of C4, C6, C8 and C10 monomers from unrelated carbon source [Nomura et al. 2005;2008]. Metabolic engineering of *E. coli* in order to produce mcl-PHAs has been also achieved by expression PHA synthase gene *phaC1* from *P. aeruginosa* in *E. coli fadB* mutant LS1298 [Langenbach et al. 1997]. Stable mcl-PHAs producing recombinant *E. coli* was constructed by cloning the *phaC1* on chromosome for scale-up production of chiral mcl-PHAs [Prieto et al. 1999]. One of the major costs for the PHB production is the carbon sources of the raw material [Choi and Lee 1999]. Therefore, *E. coli* have been genetically engineered to utilize inexpensive carbon sources such as whey-based media for the production of PHB [Wong and Lee 1998; Kim et al. 2000]. Other low cost substrates have also been evaluated for the production of PHAs by recombinant *E. coli* in order to reduce the high production cost of PHAs [Fonseca and Antonio 2007].

Purification, Properties and Applications of PHA

Industrially produced PHAs need to be extracted and purified from cells after fermentation. The conventional method for the recovery of PHAs consists of PHAs extraction with warm chloroform after the removal of other cellular lipophilic components by reflux in hot methanol [Kessler et al. 2001; Suriyamongkol et al. 2007]. Highly purified PHAs can be obtained by this method; however, the use of the highly toxic solvent chloroform is unfavorable in most of practices. Additionally, chloroform is not only environmentally unfriendly, but also impractical for use in scale up extractions [Byrom 1987]. Thus, many alternative protocols have been introduced for extraction PHAs from bacterial. These approaches include recovery of PHAs by enzymatic digestion [de Koning et al. 1997; Yasotha et al. 2006]; digestion with sodium hypochlorite followed by the treatment of cells with a surfactant [Tamer et al. 1998; Dong and Sun 2000] and supercritical fluid method [Hejazi et al. 2003].

Intracellular PHAs are amorphous whereas PHAs are partially crystalline polymers with a degree of crystallinity in the range of 55-80% in the extracellular state [Sudesh et al. 2000]. Depending on the composition of monomer units, PHAs can exhibit a wide variety of physicochemical properties ranging from hard crystalline to elastic. Changing the composition of PHA homopolymers or copolymers and heteropolymers can result in very different physical properties of the polymer for different applications. These features make PHAs highly competitive against the petrochemical-based plastics. The thermal and mechanical properties of PHB, PHBV, polypropylene (PP), polyethylene-terepthalate (PET) and low-density polyethylene (LDPE) are listed and compared in Table 1.

Table 1. Comparison of polymer properties of PHB, PHBV, PP, PET and LDPE [de Koning and Lemstra 1992; Sudesh et al. 2000]

Polymer[a]	PHB	PHBV	PP	PET	LDPE
Tm (°C)[b]	179	145	176	262	130
Tensile modulus (GPa)	3.5	1.2	1.7	2.2	0.2
Tensile strength (MPa)	40	32	38	56	10
Elongation to break (%)	5	50	400	100	620
Tg (°C)[c]	4	-1	-10	69	-30

[a] PHBV: P(3HB-*co*-20 mol%-3HV); PP; polypropylene; PET: polyethylene-terepthalate; LDPE:low-density polyethylene.
[b] Tm: melting temperature.
[c] Tg: glass-transition temperature.

Because of their thermoplastic and/or elastomeric, nontoxic, biocompatible and biodegradable properties, PHAs can be used in many facets of applications. PHAs (PHB and PHBV) were used initially for the manufacture of bottles, fibers, films, latex and some products of agricultural or packaging materials [Steinbüchel and Füchtenbusch 1998; Angelova and Hunkeler 1999; Luengo et al. 2003]. Also, new plastics can be obtained by blending of PHAs with other polymers to improve the mechanical properties [Zinn et al. 2001; Bhatt et al. 2007]. Because of their biocompatible and biodegradable, PHAs have the potential to be employed for medical applications, tissue engineering and as a drug carrier [Zinn et al. 2001; Chen and Wu 2005; Williams 2005; Valappil et al. 2006]. However, new properties of different bio-polyesters will be investigated exponentially by means of genetic engineering and material science technology. Therefore, the application and use of PHAs will not be restricted to the areas described above and PHAs are biopolymers with an extremely promising future.

Conclusion

Applied microbiological practices have been around since the dawn of human civilization. Today, microbes are still bringing the new frontiers of biotechnology to us in the 21st century. As microbiology and biotechnology move into the sustainable environmental sector and renewable energy sector, the prospects of these industries are bright. PHAs have drawn attention to the industrial world during the last major petroleum crisis of the 1970s. A British company, ICI/Zeneca Bioproducts was motivated and investigated in this industry. Biopol™ (co-polymer of PHB and PHV) was the trade name of the products by ICI/Zeneca Bioproducts and later by Monsanto until 1995. Proctor and Gamble also developed a family of biodegradable polyhydroxyalkanoate (PHA) copolymers consists of scl-and mcl-monomers and was marketed under the trade name Nodax™. There are many more industrial factories that aimed for the production of microbial plastics. Metabolix, Inc., founded in 1992, is developing a proprietary platform technology for the co-production of bioplastics and biomass for biofuels and commercializes the production of Mirel™ bioplastics in 2007 (http://www.metabolix.com/). Meredian, Inc. acquired PHA technology from Procter and Gamble Co. in October 2007 and aimed to use the technology to manufacture over 600 million pounds of biopolymers annually using renewable resources (http://www.meredianpha.com/). Many industrial entities in Europe and China are also involved in the developments and investments of PHAs technology. In March 2008, DSM Venturing, the corporate venturing unit of Royal DSM, announced that it has participated in an investment in Tianjin Green Bio-Science Co., Ltd (China) to build what it claims will be China's largest manufacturing plant for PHA, with an annual production capacity of 10,000 tons. These all attest to the bright and promising future of PHAs as commercially viable products for the alternatives of conventional plastics.

When facing the problems of increasing price of crude oil and the concerns about environmental pollutions caused from conventional plastics, where should scientists look for the solutions? As stated many years ago by Baas Becking "everything is everywhere; but the milieu selects… in nature and in the laboratory" [Baas Becking 1934], and by Jackson W.

Foster "Never underestimate the power of the microbes" [Foster 1964], we can be confident to say that microbial industrial and biotechnology will offer a possible and promising solution.

Acknowledgements

This work was partially supported by grant (contract number NSC96-2221-E-155-023) from National Science Council, Taiwan. The author thanks Dr. Thomas Lie (University of Washington, Seattle, USA) for critical editing assistance. I am also indebt to the biopolymer research team (Professors Y. M. Sun, H. S. Wu, H. K. Hong S. S. Kung and Y. H. Wei) of Yuan Ze University, Taiwan.

References

Abe, H., Doi, Y., Fukushima, T., and Eya, H. (1994). Biosynthesis from gluconate of a random copolyester consisting of 3-hydroxybutyrate and medium-chain-length 3-hydroxyalkanoates by *Pseudomonas* sp. 61-3. *Int. J. Biol. Macromol.,16*(3),115-119.

Alvarez, H.M., Mayer, F., Fabritius, D., and Steinbüchel, A. (1996). Formation of intracytoplasmic lipid inclusions by *Rhodococcus opacus* strain PD630. *Arch. Microbiol.,165*(6), 377-386.

Alvarez, H.M., and Steinbüchel, A. (2002). Triacylglycerols in prokaryotic microorganisms. *Appl. Microbiol. Biotechnol., 60*(4), 367-376.

Anderson, A.J., and Dawes, E.A. (1990). Occurrence, metabolism, metabolic role, and industrial uses of bacterial polyhydroxyalkanoates. *Microbiol. Rev., 54*(4), 450-472.

Anderson, A.J., Williams, D.R., Dawes, E.A., and Ewing, D.F. (1995). Biosynthesis of poly(3-hydroxybutyvate-co-3-hydroxyvalerate) in *Rhodococcus rubber. Can. J. Microbiol.,41*(suppl.1),4-13.

Angelova, N., and Hunkeler, D. (1999). Rationalizing the design of polymeric biomaterials. *Trends Biotechnol.,17*(10),409-21.

Ashby, R.D., Solaiman, D.K., and Foglia, T.A. (2002). The synthesis of short- and medium-chain-length poly(hydroxyalkanoate) mixtures from glucose- or alkanoic acid-grown *Pseudomonas oleovorans*. *J. Ind Microbiol. Biotechnol.,28*(3),147-153.

Baas-Becking, L.G. M., (1934). Geobiologie of inleiding tot de milieukunde. W.P. van Stockum and Zoon N. V., The Hague, The Netherlands. *Cited from*: Leadbetter, E.R. (1996). Prokaryotic diversity: from ecophysiology and habitat. Pp. 14-24. *In:* Manual of Environmental Microbiology. ASM press.

Bhatt, R., Shah, D., Patel, K.C., and Trivedi, U.. (2008). PHA-rubber blends: Synthesis, characterization and biodegradation. *Bioresour Technol, 99(11),*4615-4620.

Byrom, D. (1987). Polymer synthesis by microorganisms: technology and economics. *Trends biotechnol.,5*(9),246-250.

Chen, C.W., Don, T.M., Yen, H.F. (2006). Enzymatic extruded starch as a carbon source for the production of poly(3-hydroxybutyrate-co-3-hydroxyvalerate) by *Haloferax mediterranei*. *Process Biochem., 41*,2289-2296.

Chen, G.Q., König, K.H., and Lafferty, R.M. (1991). Occurrence of poly-D(-)-3-hydroxyalkanoates in the genus *Bacillus*. *FEMS Microbiol Lett,68*(2),173-176.

Chen, G.Q., and Wu, Q. (2005). The application of polyhydroxyalkanoates as tissue engineering materials. *Biomaterials,26*(33),6565-6578.

Chen, G.Q., Zhang, G., Park, S.J., and Lee, S.Y. (2001). Industrial scale production of poly(3-hydroxybutyrate-co-3-hydroxyhexanoate). *Appl. Microbiol. Biotechnol,57*(1-2),50-55.

Chien, C.C., Chen, C.C., Choi, M.H., Kung, S.S., and Wei, Y.H. (2007). Production of poly-beta-hydroxybutyrate (PHB) by Vibrio spp. isolated from marine environment. *J. Biotechnol, 132*(3), 259-263.

Chung, Y.J., Cha, H.J., Yeo, J.S., and Yoo, Y.J.E. (1997). Production of poly(3-hydroxybutyric-co-3-hydroxyvaleric)acid using propionic acid by pH regulation. *J. Fermen Bioeng,83*(5),492-495.

Chung, C.W., Kim, Y.S., Kim, Y.B., Bae, K.S., and Rhee, Y.-H. (1999). Isolation of a *Pseudomonas* sp. strain exhibiting unusual behavior of poly(3-hydroxyalkanoates) biosynthesis and characterization of synthesized polyesters. *J. Microbiol Biotechno.,9*(6), 847-853.

Davis, R., Chandrashekar, A., and Shamala, T.R. (2008). Role of (R)-specific enoyl coenzyme A hydratases of *Pseudomonas* sp. in the production of polyhydroxyalkanoates. *Antonie Van Leeuwenhoek,93*(3),285-296.

de Koning, G.J.M., and Lemstra, P.J. (1992). The amorphous state of bacterial poly[(R)-3-hydroxyalkanoate] *in vivo*. *Polymer,33*(15),3292-3294.

de Koning, G.J.M., and Witholt, B. (1997). A process for the recovery of poly(hydroxyalkanoates) from pseudomonads part 1: Solubilization. *Bioprocess Eng.,17*(1), 7-13.

Doi, Y., Kitamura, S., and Abe, H. (1995). Microbial synthesis and characterization of poly (3-hydroxybutyrate-co-3-hydroxyhexanoate). *Macromolecules,28*(28),4822-4828.

Doi, Y., Kunioka, M., Nakamura, Y., and Soga K. (1987a). Biosynthesis of copolyesters in *Alcaligenes eutrophus* H16 from ^{13}C- labeled acetate and propionate. *Macromolecules, 20* (12), 2988-2991.

Doi, Y., Tamaki, A., Kunioka, M., and Soga, K. (1987b). Biosynthesis of an unusual copolyester (10 mol % 3-hydroxybutyrate and 90 mol % 3-hydroxyvalerate units) in *Alcaligenes eutrophus* from pentanoic acid. *J. Chem. Soc., Chem. Commun.*, (21), 1635-1636.

Doi, Y., Kunioka, M., Nakamura, Y., and Soga, K. (1988a). Nuclear magnetic resonance studies on unusual bacterial copolyesters of 3-hydroxybutyrate and 4-hydroxybutyrate . *Macromolecules,21*(9),2722-2727.

Doi, Y., Tamaki, A., Kunioka, M., and Soga, K. (1988b). Production of copolyesters of 3-hydroxybutyrate and 3-hydroxyvalerate by *Alcaligenes eutrophus* from butyric and pentanoic acids. *Appl. Microbiol. Biotechnol.,28* (4-5), 330-334.

Dong, Z., and Sun, X. (2000). A new method of recovering polyhydroxyalkanoate from *Azotobacter chroococcum. Chin. Sci. Bull.,45*(3), 252-256.

Du, G., and Yu, J. (2002). Metabolic analysis on fatty acid utilization by *Pseudomonas oleovorans*: mcl-poly(3-hydroxyalkanoates) synthesis versus β-oxidation. *Process Biochem.,38*(3),325-332.

Feng, L., Watanabe, T., Wang, Y., Kichise, T., Fukuchi, T., Chen, G.Q., Doi, Y., and Inoue, Y. (2002). Studies on comonomer compositional distribution of bacterial poly(3-hydroxybutyrate-co-3-hydroxyhexanoate)s and thermal characteristics of their factions. *Biomacromolecules,3*(5),1071-1077.

Fidler, S., and Dennis, D. (1992). Polyhydroxyalkanoate production in recombinant *Escherichia coli. FEMS Microbiol Rev,9*(2-4),231-235.

Fiedler, S., Steinbüchel, A., Rehm, B.H. (2000). PhaG-mediated synthesis of Poly(3-hydroxyalkanoates) consisting of medium-chain-length constituents from nonrelated carbon sources in recombinant *Pseudomonas fragi. Appl. Environ Microbiol.,66*(5),2117-2124.

Fiedler, S., Steinbüchel, A., and Rehm, B.H. (2002). The role of the fatty acid beta-oxidation multienzyme complex from *Pseudomonas oleovorans* in polyhydroxyalkanoate biosynthesis: molecular characterization of the *fad*BA operon from *P. oleovorans* and of the enoyl-CoA hydratase genes *pha*J from *P. oleovorans* and *Pseudomonas putida. Arch. Microbiol.,178*(2),149-60.

Fonseca, G.G., and Antonio, R.V. (2007). Polyhydroxyalkanoates production by recombinant *Escherichia coli* using low cost substrate. *Am. J. Food Technol.,2*(1), 12-20.

Foster, J. W. (1964). In M. P. Starr (Ed.), *Global Impact of Applied Microbiology* (p. 61). Nw York, Wiley. *Cited from*: Demain, A. L. (1983). New applications of microbial products. *Science, 219*(4585),709-714.

Füchtenbusch, B., and Steinbüchel, A. (1999). Biosynthesis of polyhydroxyalkanoates from low-rank coal liquefaction products by *Pseudomonas oleovorans* and *Rhodococcus rubber. Appl. Microbiol. Biotechnol., 52*(1), 91-95.

Fukui, T., and Doi, Y. (1997). Cloning and analysis of the poly(3-hydroxybutyrate-co-3-hydroxyhexanoate) biosynthesis genes of *Aeromonas caviae. J. Bacteriol,179*(15),4821-4830.

Fukui, T., Kichise, T., Yoshida, Y., and Doi, Y. (1997). Biosynthesis of poly(3-hydroxybutyrate-co-3-hydroxyvalerate-co-3-hydroxy-heptanoate) terpolymers by recombinant *Alcaligenes eutrophus. Biotechnol Lett.,19*(11),1093-1097.

Fukui, T., Shiomi, N., and Doi, Y. (1998). Expression and characterization of (R)-specific enoyl coenzyme A hydratase involved in polyhydroxyalkanoate biosynthesis by *Aeromonas caviae. J. Bacteriol.,180*(3),667-673.

Gostomski, P.A., and Bungay, H.R. (1996). Effect of glucose NH^{4+} levels on poly(β-hydroxybutyrate) production growth in a continuous culture of *Alcaligenes eutrophus. Biotechnol Prog.,12*(2),234-239.

Guerrero, R., and Berlanga, M. (2007). The hidden side of the prokaryotic cell: rediscovering the microbial world. *Int. Microbiol., 10*(3), 157-168.

Hahn, S.K., Chang, Y.K., and Lee, S.Y. (1995). Recovery and characterization of poly(3-hydroxybutyric acid) synthesized in *Alcaligenes eutrophus* and recombinant *Escherichia coli. Appl. Environ. Microbiol.,61*(1),34-39.

Hall, B., Baldwin, J., Rhie, H.G., and Dennis, D. (1998). Cloning of the *Nocardia corallina* polyhydroxyalkanoate synthase gene and production of poly-(3-hydroxybutyrate-co-3-hydroxyhexanoate) and poly-(3- hydroxyvalerate-co-3-hydroxyheptanoate). *Can. J. Microbiol.,44* (7),687-691.

Han, J., Lu, Q., Zhou, L., Zhou, J., and Xiang, H. (2007). Molecular characterization of the *pha*EC$_{Hm}$ genes, required for biosynthesis of poly(3-hydroxybutyrate) in the extremely halophilic archaeon *Haloarcula marismortui. Appl. Environ. Microbiol., 73*(19),6058-65.

Haywood, G.W., Anderson, A.J., Chu, L., Dawes EA. (1988). Characterization of two 3-ketothiolases in the polyhydroxyalkanoate synthesizing organism *Alcaligenes eutrophus. FEMS Microbiol Lett, 52,* 91-96.

Haywood, G.W., Anderson, A.J., Williams, D.R., Dawes, E.A., and Ewing, D.F. (1991). Accumulation of a poly(hydroxyalkanoate) copolymer containing primarily 3-hydroxyvalerate from simple carbohydrate substrates by *Rhodococcus* sp. NCIMB 40126. *Int. J. Biol. Macromol.,13*(2),83-88.

Hejazi, P., Vasheghani-Farahani, E., and Yamini, Y. (2003). Supercritical Fluid Disruption of *Ralstonia eutropha* for Poly(β-hydroxybutyrate) Recovery. *Biotechnol. Prog.,19*(5),1519-1523.

Henderson, R.A., and Jones, C.W. (1997). Physiology of poly-3-hydroxybutyrate (PHB) production by *Alcaligenes eutrophus* growing in continuous culture. *Microbiol,143*(7),2361-2371.

Hezayen, F.F., Steinbüchel, A., Rehm, B.H. (2002). Biochemical and enzymological properties of the polyhydroxybutyrate synthase from the extremely halophilic archaeon strain 56. *Arch. Biochem. Biophys., 15,*403(2):284-291.

Hoffmann, N., Steinbüchel, A., and Rehm, B.H.A. (2000). Homologous functional expression of cryptic *pha*G from *Pseudomonas oleovorans* establishes the transacylase-mediated polyhydroxyalkanoate biosynthetic pathway. *Appl. Microbiol. Biotechnol,54*(5),665-670.

Hu, F., and You, S. (2007). Inactivation of type I polyhydroxyalkanoate synthase in *Aeromonas hydrophila* resulted in discovery of another potential PHA synthase. *J. Ind. Microbiol Biotechnol,34*(3),255-260.

Hu, F., Cao, Y., Xiao, F., Zhang, J., and Li, H. (2007). Site-directed mutagenesis of *Aeromonas hydrophila* enoyl Coenzyme A hydratase enhancing 3-hydroxyhexanoate fractions of poly(3-hydroxybutyrate-co-3- hydroxyhexanoate). *Curr Microbiol,55*(1),20-24.

Huang, T.Y., Duan, K.J., Huang, S.Y., and Chen, C.W. (2006). Production of polyhydroxyalkanoates from inexpensive extruded rice bran and starch by *Haloferax mediterranei. J. Ind Microbiol. Biotechnol., 33,*701-706.

Huijberts, G.N., Eggink, G., de Waard, P., Huisman, G.W., and Witholt, B. (1992). *Pseudomonas putida* KT2442 cultivated on glucose accumulates poly(3-hydroxyalkanoates) consisting of saturated and unsaturated monomers. *Appl. Environ. Microbiol.,58*(2),536-544.

Huisman, G.W., de Leeuw, O., Eggink, G., and Witholt, B. (1989). Synthesis of poly-3-hydroxyalkanoates is a common feature of fluorescent pseudomonads. *Appl. Environ. Microbiol.,55*(8),1949-1954.

Huisman, G.W., Wonink, E., de Koning, G., Preusting, H., and Witholt, B. (1992). Synthesis of poly(3-hydroxyalkanoates) by mutant and recombinant *Pseudomonas* strains. *Appl. Microbiol Biotechnol,38*(1), 1-5.

Jo, S.-J., Matsumoto, K., Leong, C.R., Ooi, T., and Taguchi, S. (2007). Improvement of poly(3-hydroxybutyrate) [P(3HB)] production in *Corynebacterium glutamicum* by codon optimization, point mutation andgene dosage of P(3HB) biosynthetic genes. *J. Biosci. Bioeng.,104*(6), 457-463.

Kalia, V.C., Lal, S., and Cheema, S. (2007). Insight in to the phylogeny of polyhydroxyalkanoate biosynthesis: horizontal gene transfer. *Gene,389*(1),19-26.

Kato, M., Bao, H.J., Kang, C.-K., Fukui, T., and Doi, Y. (1996). Production of a novel copolyester of 3-hydroxybutyric acid and medium-chain-length 3-hydroxyalkanoic acids by *Pseudomonas* sp. 61-3 from sugars. *Appl. Microbiol. Biotechnol.,45*(3),363-370.

Kessler, B., Weusthuis, R., Witholt, B., Eggink, G. (2001). Production of microbial polyesters: fermentation and downstream processes. *Adv. Biochem. Eng. Biotechnol.,71*,159-182.

Khanna, S., and Srivastava, A.K. (2005). Repeated batch cultivation of *Ralstonia eutropha* for poly (β-hydroxybutyrate) production. *Biotechnol. Lett.,27*(18),1401-1403.

Kichise, T., Taguchi, S., and Doi, Y. (2002). Enhanced accumulation and changed monomer composition in polyhydroxyalkanoate (PHA) copolyester by in vitro evolution of *Aeromonas caviae* PHA synthase. *Appl. Environ. Microbiol.,68*(5),2411-2419.

Kim, B.S., Lee, S.C., Lee, S.Y., Chang, H.N., Chang, Y.K., and Woo, S.I. (1994). Production of poly(3-hydroxybutyric acid) by fed-batch culture of *Alcaligenes eutrophus* with glucose concentration control. *Biotechnol. Bioeng,43*(9),892-898.

Kim, B.S., Lee, S.Y., and Chang, H.N. (1992). Production of poly-β-hydroxybutyrate by fed-batch culture of recombinant *Escherichia coli*. *Biotechnol Lett,14*(9),811-816.

Kim, B.-S., O'Neill, B.K., and Lee, S.-Y. (2000). Increased poly(3-hydroxybutyrate) accumulation in recombinant *Escherichia coli* from whey by agitation speed control. *J. Microbiol Biotechnol,10*(5),628-631.

Kim, D.Y., Kim, Y.B., and Rhee, Y.H. (2000). Evaluation of various carbon substrates for the biosynthesis of polyhydroxyalkanoates bearing functional groups by *Pseudomonas putida*. *Int J Biol Macromol, 28*(1), 23-29.

Kim, J.S., Lee, B.H., and Kim, B.S. (2005). Production of poly(3-hydroxybutyrate-co-4-hydroxybutyrate) by *Ralstonia eutropha*. *Biochem. Eng. J.,23*(2),169-174.

Kim, Y.B., and Lenz, R.W. (2001). Polyesters from microorganisms. *Adv. Biochem. Eng. Biotechnol., 71,* 51-79.

Kung, S.S., Chuang, Y.C., Chen, C.H., and Chien, C.C. (2007). Isolation of polyhydroxyalkanoates-producing bacteria using a combination of phenotypic and genotypic approach. *Lett. Appl. Microbio., 44*(4), 364-371.

Kunioka, M., Kawaguchi, Y., and Doi, Y. (1989). Production of biodegradable copolyesters of 3-hydroxybutyrate and 4-hydroxybutyrate by *Alcaligenes eutrophus*. *Appl. Microbiol. Biotechnol.,30*,569-573.

Kunioka, M., Nakamura, Y., and Doi, Y. (1988). New bacterial copolyesters produced in *Alcaligenes eutrophus* from organic acids. *Polym. Commun.,29*(6),174-176.

Kraak, M.N., Smits, T.H.M., Kessler, B., and Witholt, B. (1997). Polymerase C1 levels and poly(R-3-hydroxyalkanoate) synthesis in wild- type and recombinant *Pseudomonas* strains. *J. Bacteriol.,179*(16),4985-4991.

Lageveen, R.G., Huisman, G.W., Preusting, H., Ketelaar, P., Eggink, G., and Witholt, B. (1988). Formation of Polyesters by Pseudomonas oleovorans: Effect of Substrates on Formation and Composition of Poly-(R)-3-Hydroxyalkanoates and Poly-(R)-3-Hydroxyalkenoates. *Appl. Environ. Microbiol.,54*(12),2924-2932.

Langenbach, S., Rehm, B.H., and Steinbüchel, A. (1997). Functional expression of the PHA synthase gene *phaC1* from *Pseudomonas aeruginosa* in *Escherichia coli* results in poly(3-hydroxyalkanoate) synthesis. *FEMS Microbiol. Lett.,150*(2),303-309.

Lee, E.Y., Jendrossek, D., Schirmer, A., Choi, C.Y., and Steinbüchel, A. (1995). Biosynthesis of copolyesters consisting of 3-hydroxybutyric acid and medium-chain-length 3-hydroxyalkanoic acids from 1,3-butanediol or from 3-hydroxybutyrate by *Pseudomonas* sp. A33. *Appl. Microbiol Biotechnol.,42*(6), 901-909.

Lee, S.H., Oh, D.H., Ahn, W.S., Lee, Y., Choi, J., and Lee, S.Y. (2000). Production of poly(3-hydroxybutyrate-co-3-hydroxyhexanoate) by high-cell-density cultivation of *Aeromonas hydrophila. Biotechnol. Bioeng,67*(2),240-244.

Lee, S.Y. (1996a). Plastic bacteria? Progress and prospects for polyhydroxyalkanoate production in bacteria. *Trends Biotechnol., 14,* 431-438.

Lee, S.Y. (1996b). Bacterial polyhydroxyalkanoates. *Biotechnol. Bioeng,49*(1),1-14.

Lee, S.Y., and Choi, J.I. (2001). Production of microbial polyester by fermentation of recombinant microorganisms. *Adv. Biochem. Eng. Biotechnol., 71,*183-207.

Lee, S.Y., Wong, H.H., Choi, J., Lee, S.H., Lee, S.C., and Han, C.S. (2000). Production of medium-chain-length polyhydroxyalkanoates by high-cell-density cultivation of *Pseudomonas putida* under phosphorus limitation. *Biotechnol. Bioeng., 68*(4), 466-470.

Lee, Y.-H., Park, J.-S., and Huh, T.-L. (1997). Enhanced biosynthesis of P(3HB-3HV) and P(3HB-4HB) by amplification of the cloned PHB biosynthesis genes in *Alcaligenes eutrophus. Biotechnol. Lett.,19*(8),771-774.

Lemoigne, M. (1926). Produits de deshydration et de polymerization de lácide β-oxybutyrique. *Bull Soc Chim Biol, 8,* 770-782.

Li, R., Zhang, H., and Qi, Q. (2007). The production of polyhydroxyalkanoates in recombinant *Escherichia coli. Bioresour Technol., 98*(12), 2313-2320.

Lu, X.Y., Wu, Q., and Chen, G.Q. (2004a). Production of poly(3-hydroxybutyrate- co-3-hydroxyhexanoate) with flexible 3-hydroxyhexanoate content in *Aeromonas hydrophila* CGMCC 0911. *Appl Microbiol Biotechnol,64*(1),41-45.

Lu, X.Y., Wu, Q., Zhang, W.J., Zhang, G., and Chen, G.Q. (2004b). Molecular cloning of polyhydroxyalkanoate synthesis operon from *Aeromonas hydrophila* and its expression in *Escherichia coli. Biotechnol Prog.,20*(5),1332-1336.

Luengo, J.M., García, B., Sandoval, A., Naharro, G., and Olivera, E.R. (2003). Bioplastics from microorganisms. *Curr. Opin. Microbiol., 6*(3), 251-260.

Luo, R., Chen, J., Zhang, L., and Chen, G. (2006). Polyhydroxyalkanoate copolyesters produced by *Ralstonia eutropha* PHB-4 harboring a low-substrate-specificity PHA synthase PhaC2Ps from *Pseudomonas stutzeri* 1317. *Biochem. Eng. J.,32*(3),218-225.

Lütke-Eversloh, T., Bergander, K., Luftmann, H., and Steinbüchel, A. (2001a). Identification of a new class of biopolymer: bacterial synthesis of a sulfur-containing polymer with thioester linkages. *Microbiology, 147,* 11-9.

Lütke-Eversloh, T., Bergander, K., Luftmann, H., and Steinbüchel, A. (2001b). Biosynthesis of poly(3-hydroxybutyrate-co-3-mercaptobutyrate) as a sulfur analogue to poly(3-hydroxybutyrate) (PHB). *Biomacromolecules, 2*(3), 1061-1065.

Lütke-Eversloh, T., Fischer, A., Remminghorst, U., Kawada, J., Marchessault, R.H., Bögershausen, A., Kalwei, M., Eckert, H., Reichelt, R., Liu, S.J., and Steinbüchel, A. (2002a). Biosynthesis of novel thermoplastic polythioesters by engineered *Escherichia coli. Nat. Mater., 1*(4), 236-240.

Lütke-Eversloh, T., Kawada, J., Marchessault, R.H., and Steinbüchel, A. (2002b). Characterization of microbial polythioesters: physical properties of novel copolymers synthesized by *Ralstonia eutropha. Biomacromolecules, 3*(1), 159-166.

Lütke-Eversloh, T., and Steinbüchel, A. (2003). Novel precursor substrates for polythioesters (PTE) and limits of PTE biosynthesis in *Ralstonia eutropha. FEMS Microbiol Lett., 221*(2), 191-196.

Lütke-Eversloh, T., and Steinbüchel, A. (2004). Microbial polythioesters. *Macromol. Biosci., 4*(3), 166-174.

Madison, L.L., and Huisman, G.W. (1999) Metabolic engineering of poly(3-hydroxyalkanoates): from DNA to plastic. *Microbiol Mol Biol Rev, 63*(1), 21-53.

Manna, A., Banerjee, R., and Paul, A.K. (1999). Accumulation of poly (3-hydroxybutyric acid) by some soil *Streptomyces*. Curr Microbiol,39(3),153-158.

Matsusaki, H., Manji, S., Taguchi, K., Kato, M., Fukui, T., and Doi, Y. (1998). Cloning and molecular analysis of the poly(3-hydroxybutyrate) and poly(3-hydroxybutyrate-co-3-hydroxyalkanoate) biosynthesis genes in *Pseudomonas* sp. strain 61-3. *J. Bacteriol.,180*(24),6459-6467.

McCool, G.J., and Cannon, M.C. (1999). Polyhydroxyalkanoate inclusion body-associated proteins and coding region in *Bacillus megaterium. J. Bacteriol.,181*(2),585-592.

Nomura, C.T., Taguchi, K., Taguchi, S., and Doi, Y. (2004). Coexpression of genetically engineered 3-ketoacyl-ACP synthase III (*fabH*) and polyhydroxyalkanoate synthase (*phaC*) genes leads to short-chain-length-medium-chain-length polyhydroxyalkanoate copolymer production from glucose in *Escherichia coli* JM109. *Appl. Environ. Microbiol,70*(2),999-1007.

Nomura, C.T., Taguchi, K., Gan, Z., Kuwabara, K., Tanaka, T., Takase, K., and Doi, Y. (2005). Expression of 3-ketoacyl-acyl carrier protein reductase (*fabG*) genes enhances production of polyhydroxyalkanoate copolymer from glucose in recombinant *Escherichia coli* JM109. *Appl. Environ. Microbiol.,71*(8),4297-4306.

Nomura, C.T., Tanaka, T., Eguen, T.E., Appah, A.S., Matsumoto, K., Taguchi, S., Ortiz, C.L., and Doi, Y. (2008). FabG mediates polyhydroxyalkanoate production from both related and nonrelated carbon sources in recombinant *Escherichia coli* LS5218. *Biotechnol. Prog.,24*(2),342-351.

Nishikawa, T., Ogawa, K., Kohda, R., Zhixiong, W., Miyasaka, H., Umeda, F., Maeda, I., Kawase, M.,and Yagi, K. (2002). Cloning and molecular analysis of poly(3-hydroxyalkanoate) biosynthesis genes in *Pseudomonas aureofaciens*. *Curr. Microbiol.,44*(2),132-135.

Ouyang, S.P., Luo, R.C., Chen, S.S., Liu, Q., Chung, A., Wu, Q., and Chen, G.Q. (2007). Production of polyhydroxyalkanoates with high 3-hydroxydodecanoate monomer content by fadB and fadA knockout mutant of *Pseudomonas putida* KT2442. *Biomacromolecules, 8*(8), 2504-2511.

Park, J.-S., Park, H.-C., Huh, T.-L., and Lee, Y.-H. (1995). Production of poly-β-hydroxybutyrate by *Alcaligenes eutrophus* transformants harbouring cloned *phbCAB* genes. *Biotechnol. Lett.,17*(7),735-740.

Park, S.J., Ahn, W.S., Green, P.R., and Lee, S.Y. (2001). Biosynthesis of poly(3-hydroxybutyrate-*co*-3-hydroxyvalerate-*co*-3-hydroxyhexanoate) by metabolically engineered *Escherichia coli* strains. Biotechnol Bioeng,74(1),81-86.

Park, S.J., and Lee, S.Y. (2004). Biosynthesis of poly(3-hydroxybutyrate- co-3-hydroxyalkanoates) by metabolically engineered *Escherichia coli* strains. *Appl. Biochem. Biotechnol.,113-116*,335-346.

Patnaik, P.R. (2006). Dispersion optimization to enhance PHB production in fed-batch cultures of *Ralstonia eutropha*. *Bioresour. Technol.,97*(16),1994-2001.

Patnaik, P.R. (2007). Analysis of the effect of flow interruptions on fed-batch fermentation for PHB production by *Ralstonia eutropha* in finitely dispersed bioreactors. *Chem. Eng. Commun.,194*(5),603-617.

Peoples, O.P., and Sinskey, A.J. (1989a). Poly-beta-hydroxybutyrate biosynthesis in *Alcaligenes eutrophus* H16. Characterization of the genes encoding beta-ketothiolase and acetoacetyl-CoA reductase. *J. Biol. Chem.,264*(26),15293-15297.

Peoples, O.P., and Sinskey, A.J. (1989b). Poly-beta-hydroxybutyrate biosynthesis in *Alcaligenes eutrophus* H16. Identification and characterization of the PHB polymerase gene (phbC). *J. Biol. Chem.,264*(26),15298-15303.

Philip, S., Keshavarz, T., and Roy, I. (2007). Polyhydroxyalkanoates: Biodegradable polymers with a range of applications. *J. Chem. Technol. Biotechnol.,82*(3), 233-247.

Poirier, Y., Erard, N., and Petétot, J.M. (2001). Synthesis of polyhydroxyalkanoate in the peroxisome of *Saccharomyces cerevisiae* by using intermediates of fatty acid beta-oxidation. *Appl Environ Microbiol, 67*(11), 5254-5260.

Poirier, Y., Erard, N., and Petétot, J.M. (2002). Synthesis of polyhydroxyalkanoate in the peroxisome of *Pichia pastoris*. *FEMS Microbiol Lett., 207*(1), 97-102.

Pohlmann, A., Fricke, W.F., Reinecke, F., Kusian, B., Liesegang, H., Cramm, R., Eitinger, T., Ewering, C., Pötter, M., Schwartz, E., Strittmatter, A., Voss, I., Gottschalk, G., Steinbüchel, A., Friedrich, B., and Bowien, B. (2006). Genome sequence of the bioplastic-producing "Knallgas" bacterium *Ralstonia eutropha* H16. *Nat Biotechnol,24*(10),1257-1262.

Pötter, M., and Steinbüchel, A. (2006). Biogenesis and structure of polyhydroxyalkanoate granules. In: J.M. Shively (Ed). *Microbiol Monogr: Inclusions in Prokaryotes* (110-136). Springer-Verlag Berlin Heidelberg.

Preusting, H., Hazenberg, W., and Witholt, B. (1993). Continuous production of poly(3-hydroxyalkanoates) by *Pseudomonas oleovorans* in a high cell density, two-liquid-phase chemostat. *Enzyme Microb. Technol.,15*,311-316.

Prieto, M.A., Kellerhals, M.B., Bozzato, G.B., Radnovic, D., Witholt, B., and Kessler, B. (1999). Engineering of stable recombinant bacteria for production of chiral medium-chain-length poly-3-hydroxyalkanoates. *Appl. Environ. Microbiol.,65*(8),3265-3271.

Qiu, Y.-Z., Han, J., and Chen, G.-Q. (2006). Metabolic engineering of *Aeromonas hydrophila* for the enhanced production of poly(3-hydroxybutyrate-co-3-hydroxyhexanoate). *Appl. Microbiol. Biotechno.,69*(5),537-542.

Quillaguamán, J., Doan-Van, T., Guzmán, H., Guzmán, D., Martín, J., Everest, A., and Hatti-Kaul, R. (2008). Poly(3-hydroxybutyrate) production by *Halomonas boliviensis* in fed-batch culture. *Appl. Microbiol. Biotechnol.,78*(2),227-232.

Ramachander, T.V.N., and Rawal, S.K. (2005). PHB synthase from *Streptomyces aureofaciens* NRRL 2209. *FEMS Microbiol. Lett.,242*(1),13-18.

Reddy, C.S., Ghai, R., Rashmi., and Kalia, V.C. (2003). Polyhydroxyalkanoates: an overview. *Bioresour Technol., 87*(2),137-146.

Rehm, B.H., Krüger, N., and Steinbüchel, A. (1998). A new metabolic link between fatty acid *de novo* synthesis and polyhydroxyalkanoic acid synthesis. The PHAG gene from *Pseudomonas putida* KT2440 encodes a 3-hydroxyacyl-acyl carrier protein-coenzyme a transferase. *J. Biol. Chem.,273*(37),24044-24051.

Rehm, B.H.A., and Steinbüchel, A. (1999). Biochemical and genetic analysis of PHA synthases and other proteins required for PHA synthesis. *Int. J. Biol. Macromol.,25*(1-3), 3-19.

Ren, Q., Sierro, N., Witholt, B., and Kessler, B. (2000). FabG, an NADPH-dependent 3-ketoacyl reductase of *Pseudomonas aeruginosa*, provides precursors for medium-chain-length poly-3-hydroxyalkanoate biosynthesis in *Escherichia coli. J. Bacteriol,182*(10),2978-2981.

Ryu, H.W., Cho, K.S., Kim, B.S., Chang, Y.K., Chang, H.N., and Shim, H.J. (1999). Mass production of poly(3-hydroxybutyrate) by fed-batch cultures of *Ralstonia eutropha* with nitrogen and phosphate limitation. *J. Microbiol. Biotechnol.,9*(6), 751-756.

Ryu, H.W., Hahn, S.K., Chang, Y.K., and Chang, H.N. (1997). Production of poly(3-hydroxybutyrate) by high cell density fed-batch culture of *Alcaligenes eutrophus* with phosphate limitation. *Biotechnol Bioeng,55* (1),28-32.

Schlegel, H.G., Gottschalk, G., and Von Bartha, R. (1961). Formation and utilization of poly-β-hydroxybutyric acid by Knallgas bacteria (Hydrogenomonas). *Nature, 191*, 463-465.

Schubert, P., Steinbüchel, A., and Schlegel, H.G. (1988). Cloning of the Alcaligenes eutrophus genes for synthesis of poly-beta-hydroxybutyric acid (PHB) and synthesis of PHB in *Escherichia coli. J Bacteriol, 170*(12),5837-5847.

Shang, L., Fan, D.D., Kim, M.I., Choi, J. and Chang, H.N. (2007). Modeling of poly(3-hydroxybutyrate) production by high cell density fed-batch culture of *Ralstonia eutropha. Biotechnol. Bioprocess Eng.,12*(4),417-423.

Shang, L., Jiang, M., and Chang, H.N. (2003). Poly(3-hydroxybutyrate) synthesis in fed-batch culture of *Ralstonia eutropha* with phosphate limitation under different glucose concentrations. *Biotechnol. Lett.,5*(17),1415-1419.

Shimamura, E., Kasuya, K., Kobayashi, G., Shiotani, T., Shima, Y., and Doi, Y. (1994). Physical Properties and Biodegradability of Microbial Poly(3-hydroxybutyrate-co-3-hydroxyhexanoate). *Macromolecules,27*(3),878-880.

Shin, H.-D., Lee, J.-N., and Lee, Y.-H. (2002). In vivo blending of medium chain length polyhydroxy-alkanoates and polyhydroxybutyrate using recombinant *Pseudomonas putida* harboring *phbCAB* operon of *Ralstonia eutropha*. *Biotechnol. Lett.,24*(20),1729-1735.

Slater, S., Gallaher, T., and Dennis, D. (1992). Production of poly-(3-hydroxybutyrate-*co*-3-hydroxyvalerate) in a recombinant *Escherichia coli* strain. *Appl. Environ. Microbiol,58*(4),1089-1094.

Slater, S.C., Voige, W.H., and Dennis, D.E. (1988). Cloning and expression in *Escherichia coli* of the *Alcaligenes eutrophus* H16 poly-beta-hydroxybutyrate biosynthetic pathway. *J. Bacteriol.,170*(10),4431-4436.

Slepecky, R.A., and Law, J.H. (1961). Synthesis and degradation of poly-β-hydroxybutyric acid in connection with sporulation of *Bacillus megaterium*. *J Bacteriol,82*(1)37-42.

Solaiman, D.K.Y., Ashby, R.D., and Foglia, T.A. (2001). Production of polyhydroxyalkanoates from intact triacylglycerols by genetically engineered *Pseudomonas*. *Appl. Microbiol. Biotechnol.,56*(5-6),664-669.

Song, J.Y., and Kim, B.S. (2005). Characteristics of poly(3-hydroxybutyrate-co-4-hydroxybutyrate) production by *Ralstonia eutropha* NCIMB 11599 and ATCC 17699. *Biotechnol. Bioprocess Eng,10*(6), 603-606.

Steinbüchel, A., and Füchtenbusch, B. (1998). Bacterial and other biological systems for polyester production. *Trends Biotechnol, 16*(10), 419-427.

Steinbüchel, A., and Schlegel, H.G. (1991). Physiology and molecular genetics of poly(beta-hydroxy-alkanoic acid) synthesis in *Alcaligenes eutrophus*. *Mol. Microbiol, 5*(3),535-542.

Steinbüchel, A., and Valentin, H.E. (1995). Diversity of bacterial poly-hydroxyalkanoic acids. *FEMS Microbiol. Lett, 128,* 219-228.

Steinbüchel, A., and Wiese, S. (1992). A *Pseudomonas* strain accumulating polyesters of 3-hydroxybutyric acid and medium-chain-length 3-hydroxyalkanoic acids. *Appl Microbiol Biotechnol.,37*(6),691-697.

Sudesh, K., Abe, H., and Doi, Y. (2000) Synthesis, structure and properties of polyhydroxyalkanoates: biological polyesters. *Prog. Polym. Sci, 25,* 1503-1555.

Sujatha, K., Mahalakshmi, A., and Shenbagarathai, R. (2007). Molecular characterization of *Pseudomonas* sp. LDC-5 involved in accumulation of poly 3-hydroxybutyrate and medium-chain-length poly 3-hydroxyalkanoates. *Arch. Microbiol.,188*(5),451-462.

Sun, Z., Ramsay, J.A., Guay, M., and Ramsay, B.A. (2006). Automated feeding strategies for high-cell-density fed-batch cultivation of *Pseudomonas putida* KT2440. *Appl. Microbiol. Biotechnol.,71*(4),423-431.

Sun, Z., Ramsay, J.A., Guay, M., and Ramsay, B.A. (2007). Fermentation process development for the production of medium-chain-length poly-3-hyroxyalkanoates. *Appl. Microbiol. Biotechnol.,75*(3),475-485.

Sutherland, I.W. (1982). Biosynthesis of microbial exopolysaccharides. *Adv. Microb. Physiol., 23,* 79-150.

Sutherland, I.W. (1998). Novel and established applications of microbial polysaccharides. *Trends Biotechnol.,16*(1), 41-6.

Szewczyk E, and Mikucki J. (1989). Poly-β-hydroxybutyric acid in staphylococci. *FEMS Microbiol. Lett.,52*(3),279-284.

Taguchi, S., Nakamura, H., Kichise, T., Tsuge, T., Yamato, I., and Doi, Y. (2003). Production of polyhydroxyalkanoate (PHA) from renewable carbon sources in recombinant *Ralstonia eutropha* using mutants of original PHA synthase. *Biochem. Eng. J.,16*(2),107-113.

Tajima, K., Igari, T., Nishimura, D., Nakamura, M., Satoh, Y., and Munekata, M. (2003). Isolation and characterization of *Bacillus* sp. INT005 accumulating polyhydroxyalkanoate (PHA) from gas field soil. *J. Biosci. Bioeng,95*(1),77-81.

Tamer, I. M., Moo-Young, M., and Chisti, Y. (1998). Disruption of *Alcaligenes latus* for recovery of poly(β-hydroxybutyric acid): Comparison of high-pressure homogenization, bead milling, and chemically induced lysis. *Ind. Eng Chem. Res.,37*(5),1807-1814.

Tavares, L.Z., Da Silva, E.S., and Da Cruz Pradella, J.G. (2004). Production of poly(3-hydroxybutyrate) in an airlift bioreactor by *Ralstonia eutropha*. *Biochem. Eng. J.,18*(1),21-31.

Tian, S.-J., Lai, W.-J., Zheng, Z., Wang, H.-X., and Chen, G.-Q. (2005). Effect of over-expression of phasin gene from *Aeromonas hydrophila* on biosynthesis of copolyesters of 3-hydroxybutyrate and 3-hydroxyhexanoate. *FEMS Microbiol Lett.,244*(1),19-25.

Tsuge, T., Fukui, T., Matsusaki, H., Taguchi, S., Kobayashi, G., Ishizaki, A., and Doi, Y. (2000). Molecular cloning of two (R)-specific enoyl-CoA hydratase genes from *Pseudomonas aeruginosa* and their use for polyhydroxyalkanoate synthesis. *FEMS Microbiol Lett.,184*(2),193-198.

Tsuge, T., Taguchi, K., Seiichi, T., and Doi, Y. (2003). Molecular characterization and properties of (R)-specific enoyl-CoA hydratases from *Pseudomonas aeruginosa*: metabolic tools for synthesis of polyhydroxyalkanoates via fatty acid beta-oxidation. *Int J. Biol. Macromol.,31*(4-5),195-205.

Tsuge, T., Watanabe, S., Sato, S., Hiraishi, T., Abe, H., Doi, Y., and Taguchi, S. (2007). Variation in copolymer composition and molecular weight of polyhydroxyalkanoate generated by saturation mutagenesis of *Aeromonas caviae* PHA synthase. *Macromolecular Biosci,7*(6),846-854.

Valappil, S.P., Boccaccini, A.R., Bucke, C., and Roy, I. (2007). Polyhydroxyalkanoates in Gram-positive bacteria: insights from the genera *Bacillus* and *Streptomyces*. *Antonie Van Leeuwenhoek, 91*(1),1-17.

Valappil, S.P., Misra, S.K., Boccaccini, A.R., and Roy, I. (2006). Biomedical applications of polyhydroxyalkanoates: an overview of animal testing and in vivo responses. *Expert Rev. Med Devices,3*(6),853-868.

Valentin HF, and Dennis D. (1996). Metabolic pathway for poly(3-hydroxybutyrate-co-3-hydroxyvalerate) formation in *Nocardia corallina*: inactivation of *mut*B by chromosomal integration of a kanamycin resistance gene. *Appl. Environ Microbiol.,62*(2),372-379.

Valentin, H.E., and Dennis, D. (1997). Production of poly(3-hydroxybutyrate-*co*-4-hydroxybutyrate) in recombinant *Escherichia coli* grown on glucose. *J. Biotechnol.,58*(1),33-38.

Vandamme, P., and Coenye, T. (2004). Taxonomy of the genus *Cupriavidus*: a tale of lost and found. *Int. J. Syst Evol. Microbiol.,54*(6),2285-2289.

Wältermann, M., and Steinbüchel, A. (2005). Neutral lipid bodies in prokaryotes: recent insights into structure, formation, and relationship to eukaryotic lipid depots. *J. Bacteriol., 187*(11), 3607-3619.

Williams, D. (2005). The proving of polyhydroxybutyrate and its potential in medical technology. *Med. Device Technol. ,16*(1),9-10.

Wong, A.L., Chua, H., and Yu, P.H. (2000). Microbial production of polyhydroxyalkanoates by bacteria isolated from oil wastes. *Appl. Biochem. Biotechnol.,86*,843-57.

Wong, H.H., and Lee, S.Y. (1998). Poly-(3-hydroxybutyrate) production from whey by high-density cultivation of recombinant *Escherichia coli. Appl. Microbiol. Biotechnol.,50*(1),30-33.

Xie, W.P., and Chen, G.-Q. (2008). Production and characterization of terpolyester poly(3-hydroxybutyrate-co-4-hydroxybutyrate-co-3-hydroxyhexanoate) by recombinant *Aeromonas hydrophila* 4AK4 harboring genes *phaPCJ. Biochem. Eng. J.,38*(3),384-389.

Yagi, K., Miyawaki, I., Kayashita, A., Kondo, M., Kitano, Y., Murakami, Y., Maeda, I., Umeda, F., Miura, Y., Kawase, M., and Mizoguchi, T. (1996). Biosynthesis of poly(3-hydroxyalkanoic acid) copolymer from CO2 in *Pseudomonas acidophila* through introduction of the DNA fragment responsible for chemolithoautotrophic growth of *Alcaligenes hydrogenophilus. Appl. Environ. Microbiol.,62*(3),1004-1007.

Yao, J., Zhang, G., Wu, Q., Chen, G.Q., and Zhang, R. (1999). Production of polyhydroxyalkanoates by *Pseudomonas nitroreducens. Antonie Van Leeuwenhoek,75*(4),345-349.

Yasotha, K., Aroua, M.K., Ramachandran, K.B., and Tan, I.K.P. (2006). Recovery of medium-chain-length polyhydroxyalkanoates (PHAs) through enzymatic digestion treatments and ultrafiltration. *Biochem Eng J,30*(3),260-268.

Yu, H., Shi, Y., Zhang, Y., Yang, S., and Shen, Z. (2002). Effect of *Vitreoscilla* hemoglobin biosynthesis in *Escherichia coli* on production of poly(β-hydroxybutyrate) and fermentative parameters. *FEMS Microbiol Lett.,214*(2),223-227.

Yu, S.T., Lin, C.C., and Too, J.R. (2005). PHBV production by *Ralstonia eutropha* in a continuous stirred tank reactor. *Process Biochem.,40*(8),2729-2734.

Zhang, B., Carlson, R., and Srienc, F. (2006). Engineering the monomer composition of polyhydroxyalkanoates synthesized in *Saccharomyces cerevisiae. Appl. Environ. Microbiol., 72*(1), 536-543.

Zinn, M., Witholt, B., and Egli, T. (2001). Occurrence, synthesis and medical application of bacterial polyhydroxyalkanoate. *Adv. Drug Deliv. Rev.,53*(1),5-21.

In: Biotechnology: Research, Technology and Applications   ISBN 978-1-60456-901-8
Editor: Felix W. Richter   © 2008 Nova Science Publishers, Inc.

Chapter 3

Chemoenzymatic Synthesis of Glycoconjugates and Biological Applications

F. Javier Muñoz, José V. Sinisterra and María J. Hernáiz[*]

Grupo de Biotransformaciones. Departamento de Química Orgánica y Farmacéutica.
Universidad Complutense (UCM). Pz/ Ramón y Cajal s/n. 28040 Madrid, Spain
Servicio de Biotransformaciones Industriales, Parque Científico de Madrid
Tres Cantos, 28760, Madrid, Spain.
Servicio de Interacciones Biomoleculares. Facultad de Farmacia
Parque Científico de Madrid. Pz/ Ramón y Cajal s/n. 28040 Madrid, Spain

Abstract

Oligosaccharides play important roles in a number of biological events. To elucidate the biological functions of oligosaccharides, sufficient quantities of structurally defined oligosaccharides are required and one of the main limitation for this, is the availability of these molecules by tradicional purification methods. Hence, chemical and enzymatic synthesis of oligosaccharides are becoming increasingly important in glycobiology and glycotechnology. In addition, oligosaccharides often occur as glycoconjugates attached to proteins or lipids on the cell membranes.

Glycoconjugates can be synthesized by both chemical and enzymatic methods. The former approach involves the careful design of protecting groups, catalysts, reaction conditions, donor leaving groups, and acceptors. The enzymatic approach has several advantages over its chemical counterpart. Enzymatic glycosidation often takes place stereo- and regioselectively under mild reaction conditions without elaborated procedures such as protection and deprotection of hydroxyl groups or activation of anomeric position. Enzymatic approaches can be divided into two major categories based on glycosidases and glycosyltransferases.

[*] Corresponding_Author: María J. Hernáiz, Departamento de Química Orgánica y Farmacéutica. Facultad de Farmacia. Universidad Complutense de Madrid. Pz/ Ramón y Cajal s/n.28040 Madrid. Spain. Phone: (+34) 913947208. Fax: (+34) 913941822. E-mail: mjhernai@farm.ucm.es.

Glycosyltransferases, which are responsible for the biosynthesis of oligosaccharides and glycoconjugates in nature, catalyze the efficient and specific transfer of a saccharide from a sugar nucleotide donor to an acceptor. But, both the enzyme and the sugar nucleotide are expensive and the glycosylation pathway is often subject to feedback inhibition from the nucleoside phosphate that is generated. Recently, these problems have been rapidly resolved by the increasing of availability of some recombinant glycosyltransferases and the use of a nucleotide sugar regeneration system and engineered bacteria. Hence, glycosyltransferases are expected to become a useful class of enzymes for the synthesis of oligosaccharides and glycoconjugates.

Glycosidases have been used to synthesize oligosaccharides via transglycosidation and condensation. Although the regioselectivity of glycosidases is rarely absolute and their yields are lower than those glycosyltransferases, glycosides are relatively expensive and are utilized because they are relatively available as glycosyl donor substrates. Therefore, glycosidases have been considered a powerful tool for the practical synthesis of oligosaccharides.

As we review here, the development of simple and effective chemoenzymatic methods for the synthesis of oligosaccharides and glycoconjugates, is essential for a better understanding of glycosylation in biology and for the development of new diagnostic and therapeutic strategies.

Introduction

The knowledge of the role of oligosaccharides and glycoconjugates (glycoproteins and glycolipids) as mediators of complex cellular events has been challenging biologists for decades. Many different theories have been proposed about the different biological roles of carbohydrates. These functions are being elucidated in more detail, but these discoveries have been slow compared to the development performed in nucleic acids or protein studies. The structural diversity and complexity of carbohydrates enormously exceeds both protein and nucleic acids. This ability allows them to encode information for specific molecular recognition that includes bacterial or viral infection, cancer metastasis or blood-clotting and to serve as determinants of protein post-translational modifications, and many other crucial intercellular recognition events [1-4].

In contrast to linear oligopeptides and oligonucleotides, oligosaccharides are often complex branched molecules and the glycan core is commonly attached to proteins and lipids. In nature, three major classes of glycans exist: N-linked glycans, O-linked glycans and glycosylphosphatidylinositol (GPI) anchors.

This biological relevance led the investigators to advance in the methodologies and alternatives for the synthesis of natural and modified oligosaccharides and glycoconjugates, because the isolation of carbohydrates from natural sources is extremely difficult [5].

Chemical synthesis of carbohydrates presents technical difficulties due to the need of protecting groups, activating reagents, catalysts and deprotecting chemistry. Therefore, improvements and innovations are required for the synthesis of complex glycoconjugates, and many advances have been described in this area[5-7]. New alternatives like the solid-phase synthesis of carbohydrates in a similar way to the common used for proteins[8, 9], or the automated carbohydrate chemical synthesis[10], shows the future pathway for the synthesis of oligosaccharides and glycoconjugates.

On the contrary the chemical approach, enzymes are powerful tools in the synthesis and modification of carbohydrates because most of the relevant reactions require high degrees of chemo-, regio- and stereo-selectivity. These carbohydrates need to be connected through glycosidic linkages to carbohydrates or other biomolecules, such as lipids, proteins or metabolites. The main problem to be afforded is the high functionalization of monosaccharides that present several chiral centres requires the asymmetric synthesis. This is difficult to afford by the traditional organic synthesis, but enzymes are highly regio-, chemio- and stereoselective. Two groups of enzymes have been used for this purpose, glycosidases and glycosyltransferases [11-15].

In this chapter, the applicability of glycosyl hydrolases (or most commonly called glycosidases) and glycosyltransferases is presented. Biocatalysis is a great synthetic tool to create complex carbohydrates and glycoconjugates. In addition, the importance of all these tools in order to carry out carbohydrate-protein interaction studies is also remarked and different strategies for the application of glycoconjugates in glycobiology.

Enzymatic Synthesis of Oligosaccharides

The availability of structurally well defined oligosaccharides and glycoconjugates is of main importance for studying their biological roles, joined to the difficulty to isolate these compounds from natural sources, different chemoenzymatic approaches for the synthesis of complex carbohydrates have been developed [4, 15-19]. Too many obstacles have been described for the synthesis of these compounds. First, the high functionalization of carbohydrates requires complex synthetic route, with many different steps. And secondly, biological activity of carbohydrates is directly correlated to their specific configuration, for this reason an exhaustive control of the stereo- and regioselectivity of the product has to take in consideration. In that sense, enzymatic methodology is shown to be better alternative for an easier and efficient synthesis of carbohydrates and glycoconjugates [6, 20, 21]. In consequence, biocatalysis has spread in carbohydrate chemistry, avoiding complex intermediate steps [18, 22, 23].

Enzymes involved in carbohydrate chemistry belong to two main families: glycosyl hydrolases or glycosidases (EC 3.2) and glycosyltransferases (EC 2.4) [24]. Apart from these two main groups, in the recent years, a new group of synthetic enzymes called glycosynthases has been developed through selective mutation of specific residues in the active site [25]. The advantages of these new enzymes are, for example, that the product is not hydrolysed once it has been synthesized, and both yield and selectivity of the reactions are increased. At laboratory scale, glycosynthase derived from β-glucosidase from *Agrobacterium* sp. has been used for the synthesis of xylose oligosaccharides [26]. Industrial scaled-up has not been attempted yet, although the characteristics of glycosynthases point to the potential application.

In this sense, the industrial application of carbohydrate synthesis has been performed, and extremophile organisms have been used for the scale-up of these processes [27-29], for instance, an α-L-fucosidase from a hyperthermophilic microorganism *Sulfolobus solfataricus* has been characterized [30], and also thermostable α-glucosidase from *Thermus thermophilus* has been applied to starch industrial processing [31]. Furthermore, new libraries of thermophilic recombinant enzymes have been developed owing to the search of β-galactosidase activities in thermo- or extremophilic microorganisms through genetic engineering [32]. And finally, the industrial interest in these procedures related to carbohydrates has attempt the application of whole cells in the saccharide synthesis [33].

Glycosidases

Glycosidases or glycosyl hydrolases are a group of enzymes that catalyze the glycosidic linkage cleavage during biological carbohydrate catabolism. These enzymes are ubiquitously distributed in nature and have been isolated from microorganisms, plants or animals [11, 22, 23]. In detail, microbial glycosidases (from bacteria, yeast or fungi) have received main attention due to their industrial application and accessibility. Most of glycosidases belong to the group of exo-glycosidases, enzymes that only cleave the terminal glycosidic residues, although there is the other group, endo-glycosidases, that hydrolyse glycosidic linkages in a oligosaccharide chain.

Most of glycosidases have been already purified and sequenced in the past, cofactor is not necessary and are very selective to the glycosidic residue and the nature of the glycosidic bond. The huge number of structures or sequences available for glycosidases makes the setting of classifications hard. Henrissat and Davies have published several attempts for a structural and sequence-based classification of glycosylhydrolases [34-36].

The mechanism of glycosidases has been widely studied [25, 37-39], and the most accepted theory describes the formation of an oxonium intermediate. Glycosidases active site presents two glutamic residues (Glu^1, Glu^2) acting like either an acid or a base respectively. Firstly, Glu^1 acts as an acid giving a proton to the anomeric oxygen, while the glycosyl residue binds to the Glu^2 like oxonium ion. Then, the leaving group ROH is substituted by a nucleophile (NuH), which is water in the natural milieu. In a second step, the nucleophile is deprotonated by Glu^1 and attacks by the same face. As both steps are via S_N2, the anomeric configuration is retained in transglycosylation reactions (scheme 1).

On the other hand, there are glycosyl hydrolases that invert the configuration (scheme 2). In this case, the reaction takes place in only one step, where the nucleophile directly replaces the leaving group ROH and leading consequently leads to a reverse configuration.

Glycosidases are widely spread in nature, more than 2500 glycosyl hydrolases have been described [11], and it is relatively easy indeed to isolate them from natural sources. Commercial availability of these enzymes from laboratories all over the world has increased mainly because of the extended application and also low costs of the enzymes. Perhaps the most important application of glycosidases is the carbohydrate synthesis.

As described above, the water is the natural nucleophile involved in the catalyzed reaction. But, it is possible to replace this natural substrate by other nucleophiles such as glycosides, alcohol, thiol, or amine residues (scheme 1). For that purpose, two approaches have been described for the glycoside synthesis:

1. Thermodynamic approach. In this case, the equilibrium state is favoured to the synthesis and not for hydrolysis. For that purpose, unmodified substrates are used, and it is also called "direct glycosylation" or "reverse hydrolysis".
2. Kinetic strategy. Also called "transglycosylation", it uses activated glycosides that are bound to a nucleophile. This procedure receives this name because originally, the saccharide was cleaved and the glycosidic leaving residue was substituted by another glycosidic acceptor. However, donor substrates are functionalized with good leaving groups, e.g., glycosyl fluorides, azide or *p*-nitrophenyl groups.

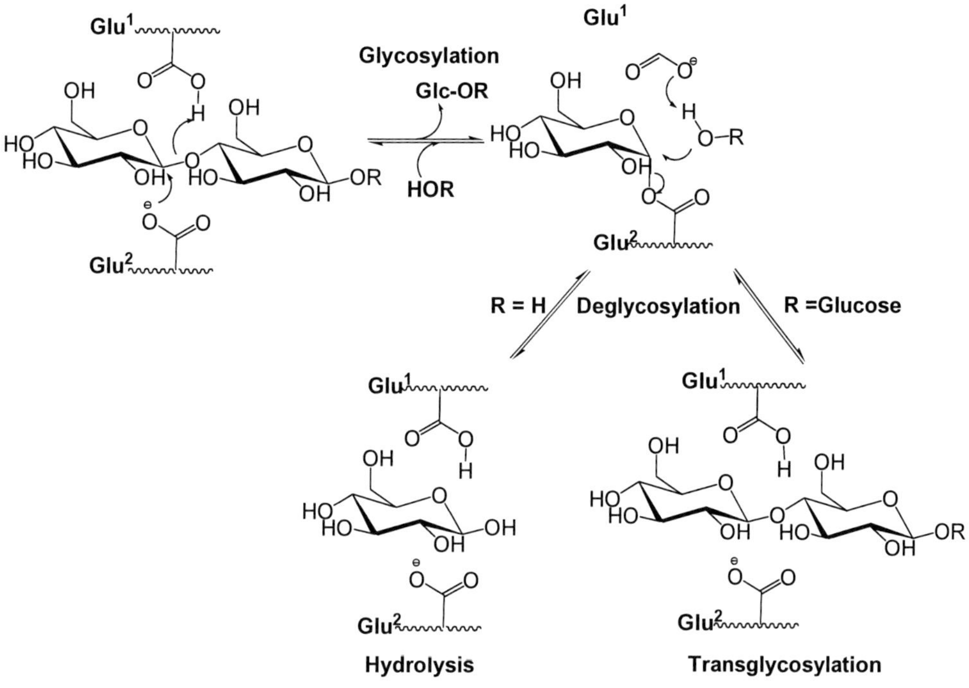

Scheme 1. General mechanism of catalysis by retaining glycosidases. The first step (glycosylation of the enzyme) results in the formation of a covalent glycosyl-enzyme intermediate that subsequently undergoes either hydrolysis or transglycosylation.

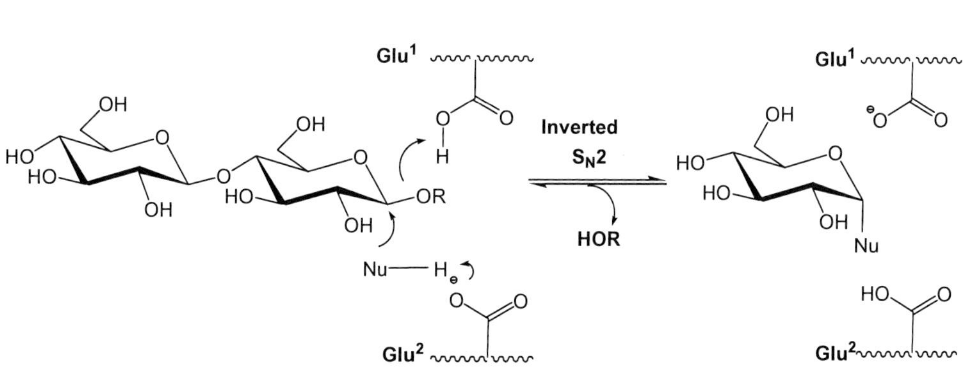

Scheme 2. General glycosidase mechanisms for an inverting β-glycosidase.

Major advantage of the biocatalyzed synthesis of carbohydrates using glycosidases is the lack of intermediate steps, required in the chemical synthesis, and the stereoselectivity control of the product by the proper enzyme, although in some cases it has been described the formation of minor products with undesired confirmations [15, 18, 33, 40]. In addition, glycosidases are also active in strong adverse conditions, which make them suitable for the organic synthesis and ease the manipulation of the enzymes [15, 41-43]. There are many examples in the literature for the application of glycosidases in organic synthesis; just few examples (scheme 3) would be the application in the synthesis of glycoconjugates coupled to aminoacids like serine [44, 45], the use of β-galactosidases from different origins for the synthesis of multivalent oligosaccharides (glycocluster) involved in molecular recognition roles [46], and not only glycosylation takes place with hydroxyl groups, but also galactosylation of thiol groups is possible [47].

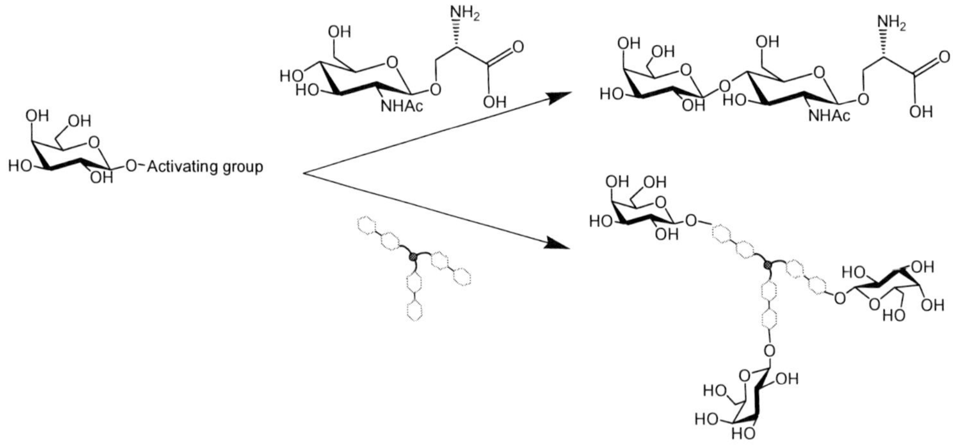

Scheme 3. Enzymatic synthesis of serine derivatives and glycoclusters mediated by glycosyltransferases.

In this context, examples for the catalysis of glycosidases in the presence of organic solvents or biphasic systems have been described [18, 33, 48]. Reverse hydrolysis is applied for the synthesis of β-mannosides and benzyl-β-D-glucopyranosides in the presence of allyl and benzyl alcohol forming aquo-organic systems [49]. Glycosidases are resistant to the presence of organic solvents in the reaction media, which is important for the reverse hydrolysis mechanism. This is the case of the glycosylation of glucose monosaccharide with 2-hydroxybenzyl and octyl derivatives catalyzed by the β-glucosidase from almond [50] (Scheme 4).

One step further, glycosidases are not only actives in conventional organic solvents, but also there have been approaches to the application of glycosidases in non-conventional media such as ionic liquids. For example, the addition of 25% of (1-methyl-3-methylimidazolium)·MeSO$_4$ to a β-galactosidase catalyzed reaction led to the elimination of the hydrolysis of the product synthesized and increased yield from 30% to 58%[51, 52] (Scheme 5).

One of the main challenging problems is the synthesis of glycolipids. The different polarity between carbohydrates and aliphatic lipid chains is hard to combine; in addition, it has been observed the need of an aqueous milieu around the active site of the enzyme. Recently, the use of liposome or reverse micelles to enclose the enzyme has been proposed. For instance, Rodríguez-Nogales et al. have studied the stability and activity of an encapsulated β-galactosidase from *Escherichia coli* in a liposome[53]; or Chen et al., who have described the ability of this enzyme to catalyze the enzymatic synthesis of galactooligosaccharides when it is enclosed into reverse micelles when the ammount of water in the reaction is controlled[54].

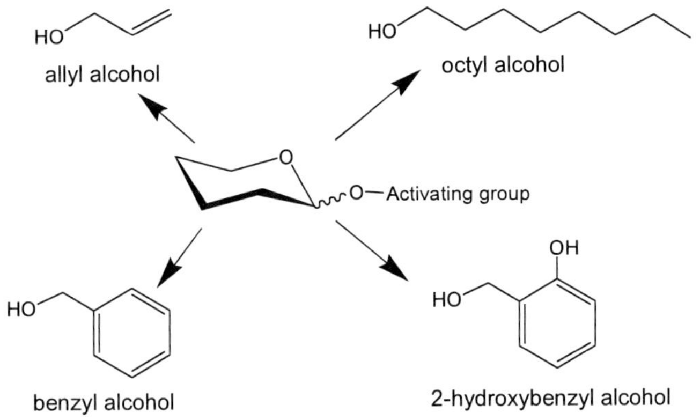

Scheme 4. Reverse hydrolysis has been applied for the synthesis of functionalized glycoconjugates. Different glycosylation reactions have been carried out by dissolving the glycosidases in the proper organic alcohol to afford glycosyl derivatives.

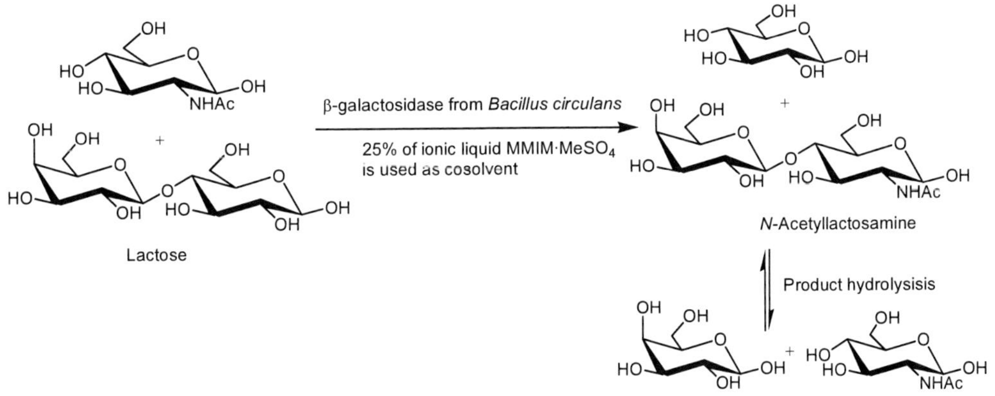

Scheme 5. The addition of 25% of ionic liquid to the synthesis of *N*-acetyllactosamine catalyzed by the β-galactosidase from *Bacillus circulans* increased the yield and decreased the hydrolysis of the product.

Recently, some researches are interested in the genetic modification of the active site of glycosidases to focus their main activity to synthetic and not to hydrolytic purposes. To put an example, mutagenesis and directed evolution are being attempted to obtain these modified glycosidases [11, 17].

Glycosyl Transferases

Glycosyl transferases are the biocatalysts responsible for the saccharide biosynthesis *in vivo*, classified into group E.C. 2.4 of enzymes. Glycosyltransferases are transmembrane enzymes located mainly in Golgi apparatus or plasmatic membrane [55]. There are many different glycosyltransferases families, of note, sialyltransferases, fucosyltransferases, *N*-acetylgalactosaminyltransferases and galactosyltransferases involved in many important biological reactions. In spite of many glycosyltransferases recognize the same donor or acceptor substrates, there are only few peptide sequences conserved among the families. An exception is constituted by sialyltransferases and fucosyltransferases. These families, present short consensus peptide motifs, and share structural and catalytic similarities in the conserved regions. Galactosyltransferases belong to a more complex group, where sequence similarities are found among families and other glycosyltransferases, and also an acidic motif occurs in almost all families of galactosyltransferases. These highly conserved sequences are closely involved in the catalytic activity and substrate binding [56]. In this, sense for better understanding of the mechanism of glycosyltransferases, crystal structures are being studied, and despite the structural differences showed, an specific donor and acceptor binding sites are retrieved [57].

The catalytic mechanism of glycosyltransferases requires nucleotide activated donors such as uridin-5'-diphospho-galactopyranoside (UDP-Gal) or cytidine-5'-monophospho-*N*-acetylneuraminic acid (CMP-NANA). There have been described two different pathways depending on the activating group of the anomeric position. The first one is called Leloir-pathway in which the glycoside donor is activated by a nucleotide. On the other hand, the non-Leloir pathway requires only a phosphate group as activating group (scheme 6) [59].

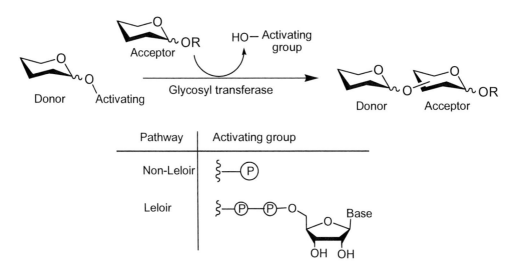

Scheme 6. Synthesis of oligosaccharides catalized by glycosyltransferases [58].

Glycosyltranferases are highly substrate-selective, extremely specific for the glycosidic linkage formed and very efficient. For all these reasons, glycosyltransferases are the preferred biocatalysts for the synthesis of oligosaccharides and glycoconjugates. We find considerable number of examples in the literature where glycosyltransferases have been chosen for the synthesis of oligosaccharides and glycoconjugates. They are very commonly used as biocatalysts for the synthesis of oligosaccharides [60-65], sialyl Lewis[x] derivatives [66], conjugation to peptides [67], or more complex structures like branched tetrasaccharides[59] or biantennary structures [68](Scheme 7).

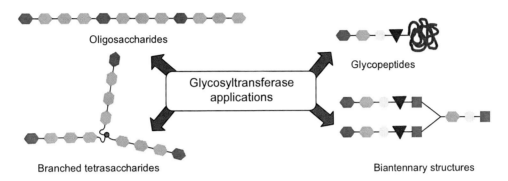

Scheme 7. Glycosyltransferases are widely used to catalyze the synthesis of a wide variety of glycoconjugates.

However, the high selectivity for the substrates could limit indeed the application of glycosyltransferases for the synthesis of carbohydrates. As an example, the β-(1→4)-galactosyltransferase from bovine milk binds specifically N-acetylglucosamine (GlcNAc) as acceptor and its derivatives, preferably β-derivatives. Glucosides are also accepted, although α-glucosides need the presence of lactalbumin to be recognized [69]. So clearly, the specificity precludes this enzyme to form derivatives with GalNAc or Gal as acceptors.

The sugar moiety is important for the binding of the acceptor to the recognition site, so that linker chains modifying the natural acceptor do not alter the glycosylation reaction. That is the case of Palcic et al., who have described a methodology to determine the glycosyltransferase activity. For that purpose, the authors proposed acceptor analogues functionalized with hydrophobic linker chains, octyl chains, to facilitate the purification of the product and further determination of enzymatic activity [70, 71].

The high substrate-selectivity is determined by a binding site restricted by structural properties of the protein. Small structural changes are able to modify the recognition of the substrates by the glycosyltransferase. As described by Wong and co-workers, the presence of sulphate groups in tyrosine residues of glycopeptides PSGL-1 decreases galactosyl and sialyltransferase activity in terms of the recognition of their respective donors [72].

But the main disadvantage of glycosyltransferase is the low commercial availability of both nucleotide donors and enzymes themselves. Although most of them are currently offered by commercial sources, they are still expensive and sold in very small amounts. Firstly, the availability of substrates is a problem to be solved. New methodologies are focused in the regeneration of donor substrate by a coupled enzymatic system, or even the substitution of traditional nucleotide glycosides by more accessible donors [56, 63, 73, 74]. On the other hand, for the last decade some researches have looked for new sources of glycosyltransferases with different regioselectivity and better kinetic or stability properties to improve the versatility and applicability of these enzymes for the synthesis of oligosaccharides and glycoconjugates. For example, β-(1$\rightarrow$3)-galactosyltransferase was isolated from the snail *Lymnaea stagnalis* [75], or the α-glucosyltransferase from *Talaromyces duponti* was isolated to apply it for the synthesis of alkyl-glucosides [76]. However, the isolation of glycosyltransferases is still complexes an enzimes obtaines are very unstable. Recently, two different strategies are being developed to improve the availability of these enzymes and therefore, synthetic procedures. Genetic engineering modification of the sequence of glycosyltransferases and the use of whole cells to carry out these transformations arc nowadays taking into consideration by some researchers [77, 78].

Solid Phase Synthesis as Tool for Glycomics

The complexity of oligosaccharides, glycans, and glycoconjugates require new techniques allowing a high-throughput synthesis, analysis or biological studies. Solid supports have played crucial roles in the development of genomics and proteomics. As proteins, peptides and nucleic acids, carbohydrates also need a well established methodology to develop what is recently called "glycomics". Methodologies for the immobilization of tools for protein and nucleic acids research are well-known, and this is a routine for a proteomic or genomic study. However, not many of these techniques are readily available in glycobiology or gycomics. Consolidated and emerging solid-phase approaches are presented in this chapter due to the need of a high-throughput strategy for the synthesis of oligosaccharides and glycoconjugates [9].

Similar to the chemical synthesis of oligosaccharides, enzymatic synthesis is led to an automation procedure through the use of solid phase. Solid-phase enzymatic synthesis of oligosaccharides presents an additional advantage over either enzymatic solution-synthesis or chemical solid-phase synthesis wich is the easy purification of products with stereo- and regiocontrol and without the need for intermediary protecting group manipulation. There are two different ways to carry out solid-supported synthesis: either the immobilization of acceptor saccharide, or the immobilization of the enzyme to a solid support (Scheme 8), although both enzymes and acceptors have been used in conjunction on differing supports.

Glycosidases and Glycosyltransferases Immobilized on Solid Supports

The use of bound enzymes is a well known strategy for the synthesis of oligosaccharides as it slows down the deactivation by proteases, natural denaturation and feedback inhibition. In addition, it makes the recovery of the products easier [13], or even it increases the stability of glycosidases in organic co-solvents [79]. Many examples of immobilization of enzymes are described. Hernáiz et al. immobilized a β-galactosidase from *Bacillus circulans* on Eupergit and described an improvement in the stability of the enzyme compared to the enzyme in solution for the same glycosylation reaction [80, 81]. Pessela et al. described the immobilization of a thermophilic glycosidase on Sepabeads [82, 83]. Many different solid supports are studied: Eupergit, Sepabeads, cellulose (batch or bed fluid), alginate or agarose, where glycosidases can be immobilized improving the stability and product recovery, although stereo- and regiochemical constraints are also observed and further development of this strategy is still necessary [80-85].

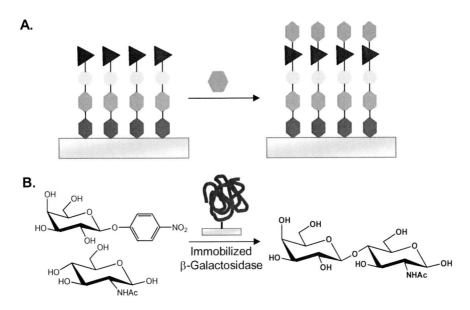

Scheme 8. A. Solid supported synthesis of oligosaccharides. The first strategy is the enzymatic glycosylation of immobilized carbohydrates. B. Glycosylation strategy catalyzed by solid-supported β-galactosidase.

Immobilization of Oligosaccharides and Glycoconjugates on Solid Supports

Microarray technologies are becoming vital tools in the 'omics' areas, such as genomics, proteomics and, most recently, glycomics. In answer to issues concerning biocompatibility and selectivity of reactions, enzymes are attractive catalysts for generating carbohydrate structures on solid surfaces [11, 86]. There are many different strategies for the immobilization of carbohydrates or glycoconjugates that can be afforded. Here we described different methodologies, from non-covalent immobilization of carbohydrates to well known procedures for covalent binding to surfaces through chemical linkages. In addition we cite applications of well accepted systems in glycobiology for the study of glycoconjugates roles like polymer supported, microtiter plates and glycoarrays, nanoparticles and biosensor chips.

Method for the Immobilization of Unmodified Carbohydrates

Perhaps, the first thought when immobilizing carbohydrates would be to try to avoid their functionalization, but at the end this turns tricky. There are two ways for the attachment of carbohydrates to a surface: non-covalent and covalent.

The non-covalent adsorption is the least technical demanding strategy to immobilize carbohydrates onto surfaces. It allows the adsorption of a wide variety of carbohydrates, including glycoproteins and glycolipids onto the surface without any previous requirement or reactive group. Natural glycolipids are readily adsorbed onto surfaces through the aliphatic chain that can be adsorbed on a lipid-coated surface (scheme 9A). This lipid chain interacting with the molecule provides good presentation properties for molecular recognition studies [87].

Covalent immobilization of unmodified carbohydrates takes place through the reactive groups that the carbohydrate presents in its structure. The main reactive function on carbohydrates is the aldehyde group at the reducing end. As this group is very reactive, amino (amine, hydrazide, etc.) functionalized surfaces are used to carry out the immobilization through reductive amination procedure (scheme 9B). This is a very well known reaction and it takes place in a very easy and efficient reaction. The main problem showed by this approach is the lack of the cyclic conformation of the carbohydrate at the reducing end with possible consequences in its functionality [88, 89]. On the other hand, the amino group of glycosylamines is another linking function allowing the immobilization (scheme 9C). Tolborg and Jensen described a methodology to readily anchor glycosylamines to polystyrene surfaces by using tris(alcoxy)-benzylamine as linker [90].

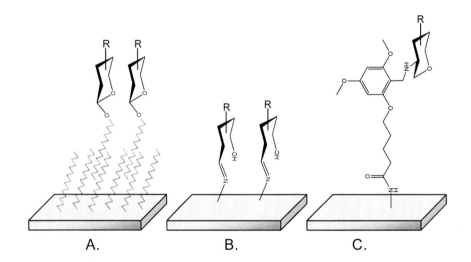

Scheme 9. A. Immobilization of glycolipids by adsorption on lipid-coated surfaces. B. Immobilization of carbohydrates through the aldehyde group of its reducing end. C. Immobilization of glycosylamines through the amino group.

Method for the Immobilization of Modified Carbohydrates

Modified carbohydrates present a wide variety of strategies to be immobilized onto surfaces. From non-covalent to covalent strategies, here we present the most used strategies for the immobilization of carbohydrates and their application to the enzymatic synthesis.

The first approach for a non-covalent methodology is the adsorption to surfaces. Wong and co-workers have described a methodology where amine glycoconjugates reacted with isocyanate functionalized hydrophobic chain without any further reagent forming a stable urea. The long hydrophobic chain allows the adsorption to microtiter plates [91].

The other non-covalent strategy is perhaps the most extended methodology for the immobilization of glycoconjugates. The avidity of biotin to avidin/streptavidin family of proteins is very strong, similar to covalent binding (K_D 10^{-14} M)[92, 93], and make this system very suitable for the immobilization of glycoconjugates (scheme 10). There are many examples for the immobilization through biotin/streptavidin adsorption [9, 65, 86, 94-97].

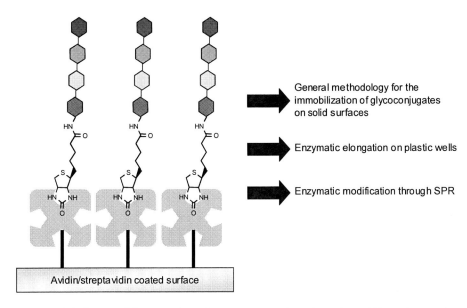

General methodology for the immobilization of glycoconjugates on solid surfaces

Enzymatic elongation on plastic wells

Enzymatic modification through SPR

Scheme 10. Avidin/streptavidin coated surfaces are general tools to immobilize glycoconjugates on surfaces. Here, two examples of the application in biocatalysis procedures of this strategy are shown.

Gabius and coworkers [98] have immobilized biotinylated glycoconjugates on plastic wells, and a sensitive enzyme-linked lectin assay (ELLA) was carried out to determine the activity of an α-1,3-galactosyltransferase (scheme 10). The enzymatic elongation takes place when the immobilized glycoconjugate (acceptor) is incubated in the presence of the enzyme and the donor substrate (UDPGal). Afterwards, a fluorescent tagged lectin is added to the plate to evaluate the efficiency of the reaction, which is directly related to the activity of the galactosyltransferase.

A different strategy is described by Hernáiz et al. These authors reported the immobilization of biotinylated Heparan Sulfate on a streptavidin-coated sensor chip for surface plasmon resonance (SPR) studies [99]. This immobilized glycoconjugate was further modified by a 3-O-sulfotransferase isoform-1 through the SPR to carry out interaction studies with antithrombin III. The advantage of this approach is that the enzymatic reaction was monitored on-line and the product was readily available for carbohydrate recognition studies (scheme 10).

Covalent attachment is often preferred over other types of immobilization because linkages are more stable and the efficiency of the reaction is easier to control. There are several requirements to consider: the spacer between the glycan and the surface must gives the optimal presentation; the choice of derivatization method depends on the system we want to study and finally the surface where the glyconjugates are immobilized on. Since solid-phase started, there are different immobilization methodologies.

The first approach is the immobilization by reductive amination, through the imine formation between the aldehyde of the reducing end and the amino group of the surface followed by selective reduction. Seo et al. have functionalized carbohydrates with an aromatic thiol by reductive amination and consequently the opening of the carbohydrate ring, and further direct immobilization on a gold surface. The aromatic rings is a common residue to reach the appropriate orientation of glycoconjugate chains[100].

There are different ways to functionalize glycoconjugates without the lack of their cyclic conformation. Many examples are found in literature for the immobilization of these types of compounds (scheme 11). Hydrazide functionalization of glycoconjugates is much extended in solid-phase carbohydrate chemistry. Matsumoto and co-workers described the immobilization of glycoconjugates on methyl vinyl ether-maleic anhydride through hydrazine chain[101]. An example for solid supported carbohydrate chemistry is the synthesis of 6-deoxysaccharides through sulfonation of carbohydrates on a Merrifield resin[102], or the building of β-linked oligosaccharides by coupling of glycal derived thioethyl glycosyl donors[103], or the mannosidation of a saccharide immobilized on a polymer support through alkoxybenzyl groups[104], covalent immobilization of S- and O-alkylated polystyrene/divinyl benzene copolymer[105], or finally, one of the most extended immobilization strategies is to functionalize carbohydrates with a maleimide linker. Shin and co-workers described the immobilization of maleimide-linker glycoconjugates on thiol functionalized glass slides for carbohydrate-protein interaction studies[106]. And finally, Wong and co-workers have developed the 1,3-dipolar cycloaddition between alkynes and azides, a well-known reaction for the formation of [1,2,3]-triazole ring for the immobilization of carbohydrates[107, 108].

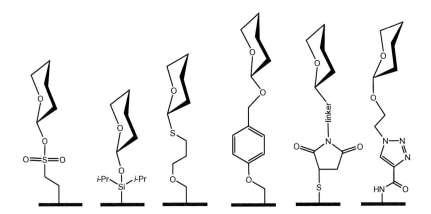

Scheme 11. Different strategies for the covalent immobilization of carbohydrates onto functionalized surfaces.

The most studied glycoconjugates are glycopeptides. These conjugates have multiple functional groups that allow a variety of immobilization options, and then immobilize onto surfaces for the synthesis of more complex glycopeptides or to study carbohydrate-protein interactions by different techniques as for instance SPR [109, 110].

The robust methodologies developed for the immobilization of functionalized carbohydrates are the key starting point for the solid-phase enzymatic synthesis of oligosaccharides and glycoconjugates. Different enzymes have been used for the solid supported synthesis of glycoconjugates (scheme 12).

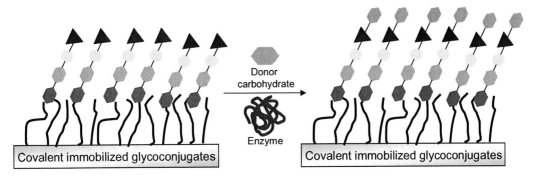

Scheme12. General methodology for the enzymatic glycosylation of covalent-immobilized glycoconjugates on solid surfaces. The donor substrate is added together with the enzyme to carry out the elongation of glycoconjugate chains. The efficiency of the reaction is monitored by different techniques like fluorescence detection [111] or SPR studies[112].

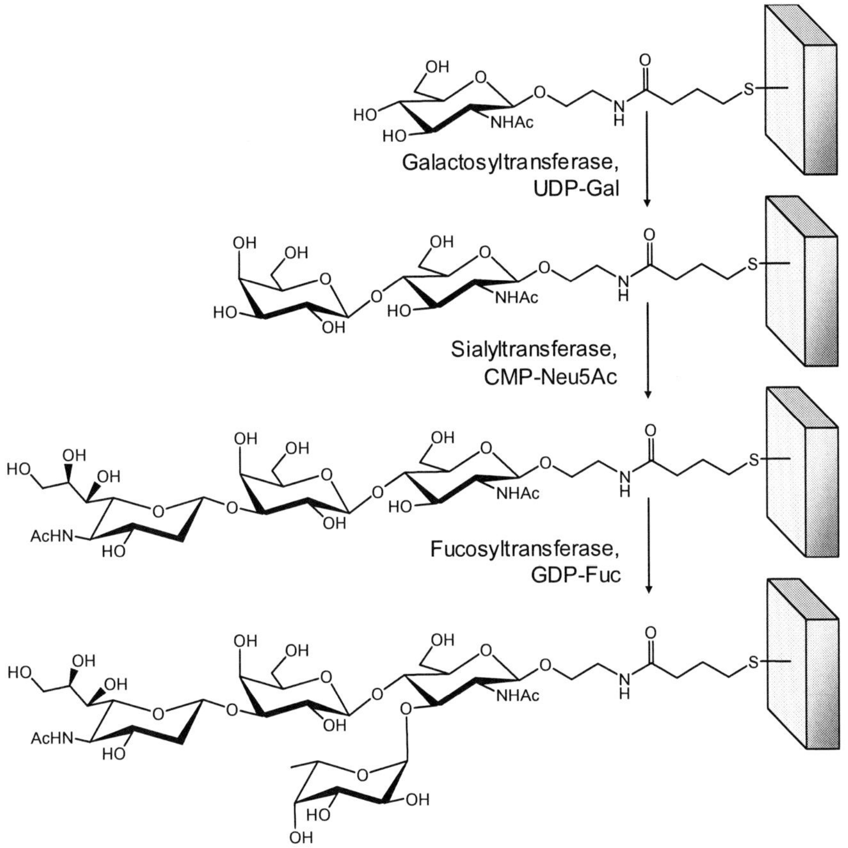

Scheme 13. Sequential solid supported synthesis of sialyl Lewis X on a sepharose gel [114].

Glycosynthases have catalyzed the glycosylation of immobilized glycoconjugates on PEG [113]. Synthesis of the oligosaccharide sialyl Lewis X has focused many attempts on solid-phase biocatalytic synthesis due to its biological roles. Wong and co-workers, using the 1,3-dipolar cycloaddition methodology to immobilize the acceptor, have synthesized sialyl Lewis X with α-1,3-fucosyltransferase and GDP-Fucose [108]. On the other hand, Blixt and

Norberg have described the immobilization of a monosaccharide on sepharose through thiol bridge formation, and further synthesis of sialyl Lewis X in three consecutive steps using three different glycosyltransferases: galactosyltransferase, sialyltransferase and finally fucosyltransferase [114].

Mrksich and coworkers[111] have described the immobilization of monosaccharides on gold-coated glass slides. This methodology was performed by the covalent functionalization of carbohydrates with alkanethiols. These alkanethiol chains were directly attached to the gold surface to form a self-assembled monolayer. After that, the carbohydrate was then coupled to the self-assembled monolayer in a very easy step, avoiding extra reagents and in high yields. In addition, one of the most important goals is the enzymatic synthesis of immobilized disaccharides in the presence of β-(1$\rightarrow$4)-galactosyltransferase.

Finally, an interesting approach is the one described by Plath et al. for assaying the sialyltransferase activity by SPR. In this work, the authors immobilized three different glycoconjugates on a biosensor SPR surface and several injections of sialyltransferase were monitored on-line. The relative response obtained by SPR is directly related to the activity of the sialyltransferase, in a similar way to the most commonly used radioactivity assay [112].

Glycoarrays as Tool for Biological Studies

The combination of enzymatic synthesis and immobilization techniques is a potent tool for the development of glycomics. The determination of biological roles of oligosaccharides and glycoconjugates requires more sophisticated technology, more sensitive, precise, without derivatization and demanding small amounts of analytes. Many efforts are directed to develop techniques capable to cover all these limitations. As consequence, glycoarrays have been prepared for high-throughput analysis of biological functionality of carbohydrates, the SPR provides kinetic and affinity information about molecular interaction procedures or nanoparticles enhancing the multivalency of glycoconjugates presented on their surfaces for the detection of weak interactions [115, 116].

Glycomic revolution really starts with the development and spreading of carbohydrate microarrays. The high specialization in carbohydrate chemistry provides a wide range of tools to functionalize glycoconjugates and to further immobilize them on surfaces [117]. With the appearance of glycoarrays, high-throughput identification, synthesis or analysis of glycoconjugates is possible in an easy and efficient way. In the last decade, the number of glycoarrays has increased exponentially, and many researchers have focused their work in the creation of glycoarrays as tool for a rapid understanding of biological roles of carbohydrates, as it happened firstly with genomics and proteomics. Many examples are found in literature for the preparation of glycoarrays, mainly for the molecular recognition studies between immobilized glycoconjugates and glycan binding proteins [118-125]. For example, Seeberger and co-workers have used this approach for the immobilization of several aminoglycosides onto arrays to study antibiotic resistance[126] to perform the study of HIV disease by the immobilization of gp120[127]. However, not only glycoarrays are important for the interaction, but also for the solid phase synthesis [128]. Alternatives to glycoarrays are also emerging, like for instance, fiber-optic microsphere arrays for the study of carbohydrate-protein interaction developed by Adams et al. [129]; the immobilization on a polyacrylamide chip for the fluorescent study of the interaction between biotinylated glycoconjugates and several antibodies carried out by Bovin and co-workers [130]; or finally, a high-throughput enzymatic assay on a multichannel microchip based on the determination of hydrolysis of fluorescein-β-galactopyranoside catalyzed by a β-galactosidase [131].

Microtiter plates are also well accepted solid supports for routine assays as ELISA. For that purpose, different strategies are also described for the immobilization of carbohydrates on microtiter plates [132, 133].

The very young SPR technique provides full information about kinetics and affinity parameters in biomolecular interactions. The sensitivity of this technique, together with the development of immobilization methodologies and the on-line monitorization of the injection, make SPR suitable for the study of carbohydrate-protein interactions [94, 109, 110], e. g. interaction of β-galactosidase with 1G40 phage through SPR[134].

The importance of multivalency in the presentation of glycoconjugates to the glycan binding protein prompted the appearance of 3D structures based on the same principles valid for the immobilization of glycoconjugates on layer solid supports. These glyconanoparticles are polyvalent 3D-models to study carbohydrate-protein or carbohydrate-carbohydrate interactions [135-137]. These systems have been applied for the immobilization of different carbohydrate patterns, for instance mannose encapsulated gold nanoparticles [138, 139], Lewis X [140], and Lewis Y [141] or the tumour related Thomsen-Friedenreich antigen on gold nanoparticles [142].

Conclusion

The structures of glycans present in glycoconjugates generally depend on the cell type and development stage. In addition, different copies of a single polypeptide backbone can be modified with scores of distinct glycans. These features of glycans make it difficult to isolate individual species for the study of their structure and function.

In this context, the development of efficient synthetic methods for glycoconjugates and oligosaccharides on a practical scale has received considerable attention over the past several years as their biological impact. As described above, several useful methods of obtaining oligosaccharides and glycoconjugates have been developed through enzymatic and chemoenzymatic approaches, making possible to obtain well-defined glycans in sufficient amounts.

Oligosaccharides and glycoconjugates can be prepared rapidly in a solid phase approach that should lead to large libraries of oligosaccharide arrays for biological studies. At the same time, new methods have been developed through mutagenesis and directed evolution to produce modified glycosidases and glycosyltransferases. Together these two approaches provide a platform for glycobiology research and for the production of glycoprotein and other novel glycomaterials with therapeutic applications.

References

[1] Dove, A., The bittersweet promise of glycobiology. *Nature Biotech*, 2001. 19: p. 913-917.

[2] Varki, A., Biological roles of oligosaccharides: all of the theories are correct. *Glycobiol,* 1993. 3(2): p. 97-130.

[3] Yarema, K.J., Bertozzi, C. R., Chemical approaches to glycobiology and emerging carbohydrate-based therapeutic agents. *Curr Opin Chem Biol*, 1998. 2: p. 49-61.

[4] Bertozzi, C.R., Kiessling, L. L., Chemical glycobiology. *Science,* 2001. 291: p. 2357-2369.

[5] Hölemann, A., Seeberger, P.H., Carbohydrate diversity: synthesis of glycoconjugates and complex carbohydrates. *Curr Opin Biotech*, 2004. 15: p. 615-622.

[6] Nicolaou, K.C., Mitchell, H. J., Adventures in carbohydrate chemistry: New synthetic technologies, chemical synthesis, molecular design, and chemical biology. *Angew Chem Int Ed*, 2001. 40: p. 1576-1624.

[7] Schmidt, R.R., New methods for the synthesis of glycosides and oligosaccharides: are there alternatives to de Koenigs-Knorr method? *Angew Chem Int Ed*, 1986. 25: p. 212-235.

[8] Bartolozzi, A., Seeberger, P.H., New approaches to the chemical synthesis of bioactive oligosaccharides. *Curr Opin Struct Biol*, 2001. 11: p. 587-592.

[9] Larsen, K., Thygesen, M.B., Guillaumie, F., Willats, W.G.T., Jensen, K.J., Solid-phase chemical tools for glycobiology. *Carbohydr Res,* 2006. 341: p. 1209-1234.

[10] Seeberger, P.H., Automated carbohydrate synthesis to drive chemical glycomics. *Chem Commun*, 2003. 10: p. 1115-1121.

[11] Daines, A.M., Maltman, B. A., Flitsch, S. L., Synthesis and modifications of carbohydrates, using biotransformations. *Curr Opin Chem Biol*, 2004. 8: p. 106-113.

[12] Watt, G.M., Lowden, P.A.S., Flitsch, S., Enzyme-catalyzed formation of glycosidic linkages. *Curr Opin Struct Biol*, 1997. 7: p. 652-660.

[13] Hanson, S., Best, M., Bryan, C., Wong, C-H., Chemoenzymatic synthesis of oligosaccharides and glycoproteins. TRENDS *Biochem Sci*, 2004. 29(12): p. 656-663.

[14] Flitsch, S.L., Chemical and enzymatic synthesis of glycopolymers. *Curr Opin Chem Biol*, 2000. 4: p. 619-625.

[15] Scigelova, M., Singh, S., Crout, D. H. G., Glycosidases: a Great Synthetic Tool. *J Mol Cat B: Enzymatic,* 1999. 6: p. 483-494.

[16] Riva, S., Biocatalytic modification of natural products. *Curr Opin Chem Biol*, 2001. 5: p. 106-111.

[17] Wymer, N., Toone, E. J., Enzyme-catalyzed synthesis of carbohydrates. *Curr Opin Chem Biol*, 2000. 4: p. 110-119.

[18] Crout, D.H.G., Vic, G., Glycosidases and glycosyl transferases in glycoside and oligosaccharide synthesis. *Curr Opin Chem Biol*, 1998. 2: p. 98-111.

[19] Rich, J.O., Michels, P. C:, Khmelnitsky, Y. L., Combinatorial biocatalysis. *Curr Opin Chem Biol*, 2002. 6: p. 161-167.

[20] Nilsson, K.G.I., Enzymatic synthesis of di- and tri-saccharide glycosides, using glycosidases and β-D-galactoside 3-a-sialyltransferase. *Carbohydr Res*, 1989. 188: p. 9-17.

[21] Davis, B.G., Recent developments in oligosaccharide synthesis. *J Chem Soc*, Perkin Trans. 1, 2000: p. 2137-2160.

[22] Koeller, K.M., Wong, C-H., Synthesis of complex carbohydrates and glycoconjugates: enzyme-based and programmable one-pot strategies. *Chem Rev*, 2000. 100: p. 4465-4493.

[23] Gijsen, H.J.M., Qiao, L., Fitz, W., Wong, C-H., Recent advances in the chemoenzymatic synthesis of carbohydrates and carbohydrate mimetics. *Chem Rev*, 1996. 96: p. 443-473.

[24] Trincone, A., Giordano, A., Glycosyl hydrolases and glycosyl transferases in the synthesis of oligosaccharides. *Curr Org Synth*, 2006. 10: p. 1163-1193.

[25] Moracci, M., Trincone, A., Rossi, M., Glycosynthases: new enzymes for oligosaccharide synthesis. *J Mol Cat B: Enzymatic*, 2001. 11: p. 155-163.

[26] Kim, Y.-W., Chen, H., Withers, S.G., Enzymatic transglycosylation of xylose using a glycosynthase. *Carbohydr Res*, 2005. 340: p. 2735-2741.

[27] Haki, G.D., Rakshit, S. K., Developments in industrially important thermostable enzymes: a review. *Bioresource Technol*, 2003. 89: p. 17-34.

[28] Van der Burg, B., Extremophiles as a source for novel enzymes. *Curr Opin Microbiol.*, 2003. 6: p. 213-218.

[29] Eichler, J., Biotechnological uses of archaeal extremozymes. *Biotech Adv*, 2001. 19: p. 261-278.

[30] Rosano, C., Zuccotti, S., Cobucci-Ponzano, B., Mazzone, M., Rossi, M., Moracci, M., Petoukhov, M. V., Svergun, D. I., Bolognesi, M., Structural characterization of the nonameric assembly of an Archaeal α-L-fucosidase by synchroton small angle X-ray scattering. *Biochem Biophys Res Comm*, 2004. 320: p. 176-182.

[31] Zdzieblo, A., Synowiecki, J., New source of the thermostable α-glucosidase suitable for single step starch proceeing. *Food Chem*, 2002. 79: p. 485-491.

[32] Li, J., Wang, P. G., Chemical and enzymatic dynthesis of glycoconjugates 2. High yielding regioselective synthesis of *N*-acetyllactosamine by use of recombinant thermophilic glycosidases library. *Tetrahedron Lett* 1997. 38(46): p. 7967-7970.

[33] Palcic, M.M., Biocatalytic synthesis of oligosaccharides. *Curr Opin Biotech*, 1999. 10: p. 616-624.

[34] Henrissat, B., Davies, G., Structural and sequence-based classification of glycoside hydrolases. *Curr Opin Struct Biol*, 1997. 7: p. 637-644.

[35] Bourne, Y., Henrissat, B., Glycoside hydrolases and glycosyltransferases: families and functional modules. *Curr Opin Struct Biol*, 2001. 11: p. 593-600.

[36] Davies, G.J., Gloster, T. M., Henrissat, B., Recent structural insights into expanding world of carbohydrates-active enzyme. *Curr Opin Struct Biol*, 2005. 15: p. 637-645.

[37] Blanchard, J.E., Withers, S. G., Rapid screening of the aglycone specificity of glycosidases: applications to enzymatic synthesis of oligosaccharides. *Chem Biol*, 2001. 8: p. 627-633.

[38] Vasella, A., Davies, G. J., Böhm, M., Glycosidase mechanisms. *Curr Opin Chem Biol*, 2002. 6: p. 619-629.

[39] Rye, C.S., Withers, S.G., Glycosidase mechanisms. *Curr Opin Chem Biol*, 2000. 4: p. 573-580.

[40] Harrison, J.A., Kartha, K. P. R., Turnbull, W. B., Scheuerl, S. L., Naismith, J. H., Schenkman, S., Field, R. A., Hydrolase and sialyltransferase activities of *Trypanosoma cruzi* trans-sialidase towards NeuAc-α-2,3-Gal-β-O-PNP. *Bioorg Med Chem Lett*, 2001. 11: p. 141-144.

[41] Houseman, B.T., Mrksich, M., Model systems for studying polyvalent carbohydrate binding interactions. *Top Curr Chem*, 2002. 218: p. 1-44.

[42] Loughlin, W.A., Biotransformations in organic synthesis. *Biores Techn*, 2000. 74: p. 49-62.

[43] Nilsson, K.G.I., Pan, H., Larsson-Lorek, U., Syntheses of modified carbohydrates with glycosidases: stereo- and regiospecific syntheses of lactosamine derivatives and related compounds. *J Carbohydr Chem*, 1997. 16(4-5): p. 459-477.

[44] Nilsson, K.G.I., Synthesis of Galβ1-4GlcNAcb- and Galβ1-3GalNAca-O-L-Serine derivatives employing glycosidases. *Biotech Lett*, 1996. 18(7): p. 791-794.

[45] Nilsson, K.G.I., Ljunger, G., Melin, P. M., Glycosidase-catalysed synthesis of glycosylated amino acids: synthesis of GalNAca-Ser and GlcNAcb-Ser derivatives. *Biotech Lett*, 1997. 19(9): p. 889-892.

[46] Santiago, R., Fernández-Mayoralas, A., García-Junceda, E., Enzymatic synthesis of disaccharides by β-galactosidase-catalyzed glycosylation of a glycocluster. *J Mol Cat B: Enzymatic*, 2000. 11: p. 71-79.

[47] Nakano, H., Shizuma, M., Kiso, T., Kitahata, Galactosylation of thiol group by β-galactosidase. *Biosci Biotechnol Biochem*, 2000. 64(4): p. 735-740.

[48] Giacomini, C., Irazoqui, G., Gonzalez, P., Batista-Viera, F., Brena, B. M., Enzymatic synthesis of galactosyl-xylose by *Aspergillus oryzae* β-galactosidase. *J Mol Cat B: Enzymatic*, 2002. 19-20: p. 159-165.

[49] Kildemark, N., Nilsson, K. G. I., Enzymatic synthesis of β-D-mannopyranosides and benzyl β-D-glucopyranoside by reverse hydrolysis in aqueous-organic systems. *Carbohydr Lett*, 1998. 3(3): p. 211-216.

[50] Vic, G., Thomas, D., Crout, D. H. G., Solvent effect on enzyme-catalyzed synthesis of β-D-glucosides using the reverse hydrolysis method: Application to the preparative-scale synthesis of 2-hydroxybenzyl and octyl β-D-glucopyranosides. *Enz Microb Tech*, 1997. 20: p. 597-603.

[51] Park, S., Kazlauskas, R. J., Biocatalysis in ionic liquids: advantages beyond green technology. *Curr Opin Biotech*, 2003. 14: p. 432-437.

[52] Kaftzik, N., Wasserscheid, P., Kragl, U., Use of ionic liquids to increase the yield and enzyme stability in the β-galactosidase catalysed synthesis of *N*-acetyllactosamine. *Org Proc Res Dev*, 2002. 6: p. 553-557.

[53] Rodríguez-Nogales, J.M., Delgadillo, A., Stability and catalytic kinetics of microencapsulated β-galactosidase in liposomes prepared by the dehydration-rehydration method. *J Mol Cat B: Enzymatic*, 2005. 33: p. 15-21.

[54] Chen, C.W., Ou-Yang, C-C., Yeh, C-W., Synthesis of galactooligosaccharides and transglycosylation modeling in reverse micelles. *Enz Microb Technol*, 2003. 33: p. 497-507.

[55] Shur, B.D., Glycosyltransferases as cell adhesion molecules. *Curr Opin Cell Biol*, 1993. 5: p. 854-863.

[56] Breton, C., Imberty, A., Structure/function studies of glycosyltransferases. *Curr Opin Struct Biol* 1999. 9: p. 563-571.

[57] Ramakrishnan, B., Balaji, P. V., Qasba, K., Crystal structure of β1,4-galactosyltransferase complex with UDP-Gal reveals an oligosaccharide acceptor binding site. *J Mol Biol*, 2002. 318: p. 491-502.

[58] Faber, K., *Biotransformations in organic chemistry*. 5th ed. 2004, Berlin Heidelberg New York: Springer-Verlag.

[59] Niggemann, J., Kamerling, J. P., Vliegenthart, F. G., β-1,4-Galactosyltransferase-catalyzed synthesis of the branched tetrasaccharide repeating unit of *Streptococcus pneumoniae* type 14. *Bioorg Med Chem*, 1998. 6: p. 1605-1612.

[60] Dudziak, G., Zeng, S., Berger, E. G., Gutierrez Gallego, R., Kamerling, J. P., Kragl, U., Wandrey, C., In situ generated *O*-glycan Core 1 structure as substrate for Gal(β1-3)GalNAc β-1,6-GlcNAc transferase. *Bioorg Med Chem Lett*, 1998. 8: p. 2595-2598.

[61] Hirschbein, B.L., Mazenod, F. P., Whitesides, G. M., Synthesis of phosphoenolpyruvate and its use in adenosine triphosphate cofactor regeneration. *J Org Chem*, 1982. 47: p. 3765-3766.

[62] Hoh, C., Dudziak, G., Liese, A., Optimization of the enzymatic synthesis of *O*-glycan core 2 structure by use of a genetic algorithm. *Bioorg Med Chem Lett*, 2002. 12: p. 1031-1034.

[63] Wong, C.-h., Haynie, S. L., Whitesides, G. M., Enzyme-catalyzed synthesis of *N*-acetyllactosamine with in situ regeneration of uridine 5'-diphosphate glucose and uridine 5'-diphosphate galactose. *J Org Chem*, 1982. 47: p. 5416-5418.

[64] Michalik, D., Vliegenthart, J. F. G., Kamerling, J. P., Chemoenzymatic synthesis of oligosaccharide fragments of the capsular polysaccharide of *Streptococcus pnemoniae* type 14. *J Chem Soc, Perkin Translation 1*, 2002: p. 1973-1981.

[65] Zeng, X., Sun, Y., Ye, H., Liu, J., Xiang, X., Zhou, B., Uzawa, H., Effective chemoenzymatic synthesis of *p*-aminophenyl glycosides of sialyl *N*-acetyllactosaminide and analysis of their interactions with lectins. *Carbohydr Res*, 2007. 342: p. 1244-1248.

[66] Elling, L., Zervosen, A., Gutiérrez Gallego, R., Nieder, V., Malissard, M., Berger, E. G., Vliegenthart, J. F. G., Kamerling, J. P., UDP-*N*-Acetyl-α-D-glucosamine as acceptor substrate of β-1,4-galactosyltransferase. Enzymatic synthesis of UDP-*N*-acetyllactosamine. *Glycoconj J*, 1999. 16: p. 327-336.

[67] Gutiérrez Gallego, R., Dudziak, G., Kragl, U., Wandrey, C., Kamerling, J. P., Vliegenthart, J. F. G., Enzymatic synthesis of the core-2 sialyl Lewis X *O*-glycan on the tumor-associated MUC1a' peptide. *Biochimie*, 2003. 85: p. 275-286.

[68] Unverzagt, C., André, S., Seifert, J., Kojima, S., Fink, C., Srikrishna, G., Freeze, H., Kayser, K., Gabius, H.-J., Structure-activity profiles of complex biantennary glycans with core fucosylation and with/without additional 2,3/2,6 sialylation: synthesis of

neoglycoproteins and their properties in lectin assays, cell binding, and organ uptake. *J Med Chem*, 2002. 45: p. 478-491.

[69] Wong, C.-H., Ichikawa, Y., Krach, T., Gautheron-Le Narvor, C., Look, G. C., Probing the acceptor specificity of β-1,4-galactosyltransferase for the development of enzymatic synthesis of novel oligosaccharides. *J Am Chem Soc*, 1991. 113(21): p. 8137-8145.

[70] Palcic, M.M., de Heerze, L. D., Pierce, M., Hindsgaul, O., The use of hydrophobic synthetic glycosides as acceptors in glycosyltransferase assays. *Glycoconj J*, 1988. 5: p. 49-63.

[71] Kamath, V.P., Seto, N. O., Compston, C. A., Hindsgaul, O., Palcic, M. M., Synthesis of the acceptor analog α-Fuc(1→2)αGal-*O*(CH$_2$)$_7$ CH$_3$: a probe for the kinetic mechanism of recombinant human blood group B glycosyltransferase. *Glycoconj J*, 1999. 16(10): p. 599-606.

[72] Koeller, K.M., Smith, M. E. B., Wong, C-H., Chemoenzymatic synthesis of PSGL-1 glycopeptides: sulfation on tyrosine affects glycosyltransferase-catalyzed synthesis of the *O*-glycan. *Bioorg Med Chem*, 2000. 8: p. 1017-1025.

[73] Srivastava, G., Hindsgaul, O., Palcic, M. M., Chemical synthesis and kinetic characterization of UDP-2-deoxy-D-lyxo-hexose ("UDP-2-deoxy-D-galactose"), a donor-substrate for β-(1-4)-D-galactosyltransferase. *Carbohydr Res*, 1993. 245: p. 137-144.

[74] Palcic, M.M., Hindsgaul, O., Flexibility in the donor substrate specificity of β-1,4-galactosyltransferase: application in the synthesis of complex carbohydrates. *Glycobiol*, 1991. 1(2): p. 205-209.

[75] Mulder, H., Schachter, H., de Jong-Brink, M., van der Ven, J. G., Kamerling, J. P., Vliegenthart, J. F. G., Identification of a novel UDP-Gal:GalNAcb1-4GlcNAc-Rb1-3-galactosyltransferase in the connective tissue of the snail *Lymnaea stagnalis*. *Eur J Biochem*, 1991. 201: p. 459-465.

[76] Bousquet, M.-P., Willemot, R-M., Monsan, P., Boures, E., Production, purification, and characterization of thermostable α-transglucosidase from *Talaromyces duponti*: application to α-alkylglucoside synthesis. *Enz Microb Technol*, 1998. 23: p. 83-90.

[77] Baisch, G., Öhrlein, R., Streiff, M., Kolbinger, F., On the preparative use of recombinant β-(1-3)galactosyl-transferase. *Bioorg Med Chem Lett*, 1998. 8: p. 751-754.

[78] Herrmann, G.F., Elling, L., Krezdorn, C. H., Kleene, R., Berger, E. G., Wandrey, C., Use of transformed whole yeast cells expressing β-1,4-galactosyltransferase for the synthesis of *N*-acetyllactosamine. *Bioorg Med Chem Lett*, 1995. 5(7): p. 673-676.

[79] Giacomini, C., Irazoqui, G., Batista-Viera, F., Brena, B. M., Influence of the immobilization chemistry on the properties of immobilized β-galactosidases. *J Mol Cat B: Enzymatic*, 2001. 11: p. 597-606.

[80] Hernáiz, M.J., Crout, D. H. G., A highly selective synthesis of *N*-acetyllactosamine catalyzed by immobilised β-galactosidase from *Bacillus circulans*. *J Mol Cat B: Enzymatic*, 2000. 10: p. 403-408.

[81] Hernáiz, M.J., Crout, D. H. G., Immobilization/stabilization on Eupergit C of the β-galactosidase from *B. circulans* and an α-galactosidase from *Aspergillus oryzae*. *Enz Microb Technol*, 2000. 27: p. 26-32.

[82] Pessela, B.C.C., Mateo, C., Fuentes, M., Vian, A., García, J. L., Carrascosa, A. V., Guisán, J. M., Fernández-Lafuente, R., The immobilization of a thermophilic β-galactosidase on Sepabeads supports decreases product inhibition . Complete hydrolysis of lactose in dairy products. *Enz Microb Technol*, 2003. 33: p. 199-205.

[83] Pessela, B.C.C., Fernández-Lafuente, R., Fuentes, M., Vián, A., García, J. L., Carrascosa, A. V., Mateo, C., Guisán, J. M., Reversible immobilization of a thermophilic β-galactosidase via ionic adsorption on PEI-coated Sepabeads. *Enz Microb Technol*, 2003. 32: p. 369-374.

[84] Roy, I., Gupta, M. N., Lactose hydrolysis by Lactozym ™ immobilized on cellulose beads in batch and fluidized bed modes. *Proc Biochem*, 2003. 00: p. 1-8.

[85] Tanriseven, A., Dogan, S., A novel method for the immobilization of β-galactosidase. *Proc Biochem*, 2002. 38: p. 27-30.

[86] Love, K.R., Seeberger, P.H., Carbohydrate arrays as tools for glycomics. *Angew Chem Int Ed*, 2002. 41(19): p. 3583-3586.

[87] Lopez, P.H. and R.L. Schnaar, Determination of glycolipid-protein interaction specificity. *Methods Enzymol*, 2006. 417: p. 205-20.

[88] Lee, M., Shin, I., Facile preparation of carbohydrate microarrays by site-specific, covalent immobilization of unmodified carbohydrates on hydrazide-coated glass slides. *Org Lett*, 2005. 7(19): p. 4269-4272.

[89] Zhi, Z., Powell, A.K., Turnbull, J.E., Fabrication of carbohydrate microarrays on gold surfaces: direct attachment of nonderivatized oligosaccharides to hydrazide-modified self-assembled monolayers. *Anal Chem*, 2006. 78(14): p. 4786-4793.

[90] Tolborg, J.F., Jensen, K.J., Solid-phase oligosaccharide synthesis with tris(alkoxy)benzyl amine (BAL) safety-catch anchoring. *Chem Commun*, 2000: p. 147-148.

[91] Fazio, F., Bryan, M.C., Lee,H.-K., Chang, A., Wong, C.-H., Assembly of sugars on polystyrene plates: a new facile microarray fabrication technique. *Tetrahedron Lett*, 2004. 45: p. 2689-2692.

[92] Gitlin, G., Bayer, E.A., Wilchek, M., Studies on the biotin-binding site of avidin. Lysine residues involved in the active site. *Biochem J*, 1987. 242: p. 923-926.

[93] Gitlin, G., Bayer, E.A., Wilchek, M., Studies on the biotin-binding site of avidin. Tryptophan residues involved in the active site. *Biochem J*, 1988. 250: p. 291-294.

[94] Muñoz, F.J., Rumbero, A., Sinisterra, J.V., Santos, J.I., André, S., Gabius, H.-J., Jiménez-Barbero, J., Hernáiz, M.J., Versatile strategy for the synthesis of biotin-labelled glycans, their immobilization to establish a bioactive surface and interaction studies with a lectin on a biochip. *Glycoconj J*, 2008. DOI: 10.1007/s10719-008-9115-y.

[95] Tao, L., Geng, J., Chen, G., Xu, Y., Ladmiral, V., Mantovani, G., Haddleton, D.M., Bioconjugation of biotinylated PAMAM dendrons to avidin. *Chem Commun*, 2007: p. 3441-3443.

[96] Herranz, F., Santa María, M. D., Claramunt, R. M., Molecular recognition: improved binding of biotin derivatives with synthetic receptors. *J Org Chem*, 2006. 71: p. 2944-2951.

[97] Grün, C.H., van Vliet, S. J., Schiphorst, W. E. C. M., Bank, C. M. C., Meyer, S., van Die, I., van Kooyk, Y., One-step biotinylation procedure for carbohydrates to study carbohydrate-protein interactions. *Anal Biochem*, 2006. 354: p. 54-63.

[98] Khraltsova, L.S., Sablina, M.A., Melikhova, T.D., Joziasse, D.H., Kaltner, H., Gabius, H.-J., Bovin, N.V., An enzyme-linked lectin assay for α-1,3-galactosyl. *Anal Biochem*, 2000. 280: p. 250-257.

[99] Hernáiz, M.J., Liu, J., Rosenberg, R.D., Linhardt, R.J., Enzymatic modification of heparan sulfate on a biochip promotes its interaction with antithrombin III. *Biochem Biophys Res Comm*, 2000. 276(1): p. 292-297.

[100] Seo, J.H., Adachi, K., Lee, B.K., Kang, D.G., Kim, Y.K., Kim, K.R., Lee, H.Y., Kawai, T., Cha, H.J., Facile and rapid direct gold surface immobilization with controlled orientation for carbohydrates. *Bioconj Chem*, 2007. 18: p. 2197-2201.

[101] Satoh, A., Kojima, K., Koyama, T., Ogawa, H., Matsumoto, I., Immobilization of saccharides and peptides on 96-well microtiter plates coated with methyl vinyl ether-maleic anhydride copolymer. *Anal Biochem*, 1998. 260: p. 96-102.

[102] Hunt, J.A., Roush, W.R., Solid-phase synthesis of 6-deoxyoligosaccharides. *J Am Chem Soc*, 1996. 118(41): p. 9998-9999.

[103] Zheng, C., Seeberger, P.H., Danishefsky, S.J., Solid support oligosaccharide synthesis: construction of β-linked oligosaccharides by coupling of glycal derived thioethyl glycosyl donors. *J Org Chem*, 1998. 63(4): p. 1126-1130.

[104] Ito, Y., Ogawa, T., Intramolecular aglycon delivery on polymer support: gatekeeper monitored glycosylation. *J Am Chem Soc*, 1997. 119(24): p. 5562-5566.

[105] Rademann, J., Schmidt, R.R., Repeptitive solid phase glycosylation on an alkyl thiol polymer leading to sugar oligomers containing 1,2-trans- and 1,2-cis-glycosidic linkages. *J Org Chem*, 1997. 62(11): p. 3650-3653.

[106] Park, S., Lee, M., Pyo, S.-J., Shin, I., Carbohydrate chips for studying high-throughput carbohydrate-protein interactions. *J Am Chem Soc*, 2004. 126(15): p. 4812-4819.

[107] Bryan, M.C., Fazio, F., Lee, H.-K., Huang, C.-Y., Chang, A., Best, M.D., Calarese, D.A., Blixt, O., Paulson, J.C., Burton, D., Wilson, I.A., Wong, C.-H., Covalent display of oligosaccharide arrays in microtiter plates. *J Am Chem Soc*, 2004. 126(28): p. 8640-8641.

[108] Fazio, F., Bryan, M.C., Blixt, O., Paulson, J.C., Wong, C.-H., Synthesis of sugar arrays in microtiter plate. *J Am Chem Soc*, 2002. 124(48): p. 14397-14402.

[109] Vila-Perelló, M., Gutiérrez Gallego, R., Andreu, D., A simple approach to well-defined sugar-coated surfaces for interaction studies. *ChemBioChem*, 2005. 6: p. 1831-1838.

[110] Roberge, J.Y., Beebe, X., Danishefsky, S.J., Convergent synthesis of *N*-linked glycopeptides on a solid support. *J Am Chem Soc*, 1998. 120(16): p. 3915-3927.

[111] Houseman, B.T., Mrksich, M., Carbohydrate arrays for the evaluation of protein binding and enzymatic modification. *Chem Biol*, 2002. 9: p. 443-454.

[112] Probert, M.A., Milton, M. J., Harris, R., Schenkman, S., Brown, J. M., Homans, S. W., Field, R. A., Chemoenzymatic synthesis of GM$_3$, Lewis x and sialyl Lewis x oligosaccharides in ^{13}C-enriched form. *Tetrahedron Lett*, 1997. 38(33): p. 5861-5864.

[113] Tolborg, J.F., Petersen, L., Jensen, K.J., Mayer, C., Jakeman, D.L., Warren, R.A.J., Withers, S.G., Solid-phase oligosaccharide and glycopeptide synthesis using glycosynthases. *J Org Chem*, 2002. 67(12): p. 4143-4149.

[114] Blixt, O., Norberg, T., Solid-phase enzymatic synthesis of a sialyl Lewis X tetrasaccharide on a sepharose matrix. *J Org Chem*, 1998. 63: p. 2705-2710.

[115] Ratner, D.M., Adams, E. W., Disney, M. D., Seeberger, P. H., Tools for glycomics: mapping interactions of carbohydrates in biological systems. *ChemBioChem,* 2004. 5: p. 1375-1383.

[116] Ratner, D.M., Seeberger, P. H., Carbohydrate microarrays as tools in HIV glycobiology. *Curr Pharm Design*, 2007. 13: p. 173-183.

[117] Disney, M.D., Seeberger, P. H., Carbohydrate arrays as tools for the glycomics revolution. *Drug Discov Today: TARGETS*, 2004. 3(4): p. 151-158.

[118] Blixt, O., Head, S., Mondala, T., Scanlan, C., Huflejt, M.E., Alvarez, R., Bryan, M.C., Fazio, F., Calarese, D., Stevens, J., Razi, N., Stevens, D.J., Skehel, J.J., van Diel, I., Burton, D.R., Wilson, I.A., Cummings, R., Bovin, N., Wong, C.-H., Printed covalent glycan array for ligand profiling of diverse glycan binding proteins. *PNAS*, 2004. 101(49): p. 17033-17038.

[119] Kiessling, L.L., Cairo, C.W., Hitting the sweet spot. *Nature Biotech*, 2002. 20: p. 234-235.

[120] Fukui, S., Feizi, T., Galustian, C., Lawson, A.M., Chai, W., Oligosaccharide microarrays for high-throughput detection and specificity assignments of carbohydrate-protein interactions. *Nature Biotech*, 2002. 20: p. 1011-1017.

[121] Feizi, T., Chai, W., Oligosaccharide microarrays to decipher the glyco code. *Nature Rev. Mol Cell Biol*, 2004. 5: p. 582-588.

[122] Rifai, A., Wong, S. S., Preparation of phosphorylcholine-conjugated antigens. *J Immunol Meth*, 1986. 94: p. 25-30.

[123] Park, S., Shin, I., Fabrication of carbohydrate chips for studying protein-carbohydrate interactions. *Angew Chem Int Ed*, 2002. 41(17): p. 3180-3182.

[124] de Boer, A.R., Hokke, C.H., Deelder, A.M., Wuhrer, M., General microarray technique for immobilization and screening of natural glycans. *Anal Chem*, 2007. 10.1021/ac071187g.

[125] Ortiz Mellet, C., García Fernández, J.M., Carbohydrate microarrays. *ChemBioChem*, 2002. 3: p. 819-822.

[126] Disney, M.D., Magnet, S., Blanchard, J.S., Seeberger, P. H., Aminoglycoside microarrays to study antibiotic resistance. *Angew Chem Int Ed*, 2004. 43: p. 1591-1594.

[127] Adams, E.W., Ratner, D.M., Bokesch, H.R., McMahon, J.B., O'Keefe, B.R., Seeberger, P.H., Oligosaccharide and glycoprotein microarrays as tools in HIV glycobiology: glycan-dependent gp120/protein interactions. *Chem Biol*, 2004. 11: p. 875-881.

[128] Feizi, T., Fazio, F., Chai, W., Wong, C-H., Carbohydrate microarrays: a new set of technologies at the frontiers of glycomics. *Curr Opin Struct Biol*, 2003. 13: p. 637-645.

[129] Adams, E.W., Ueberfeld, J., Ratner, D.M., O'Keefe, B.R., Walt, D.R., Seeberger, P.H., Encoded fiber-optic microsphere arrays for probing protein-carbohydrate interactions. *Angew Chem Int Ed*, 2003. 42: p. 5317-5320.

[130] Galanina, O.E., Mecklenburg, M., Nifantiev, N.E., Pazynina, G.V., Bovin, N.V., GlycoChip: multiarray for the study of carbohydrate-binding proteins. *Lab Chip*, 2003. 3: p. 260-265.

[131] Xu, H., Ewing, A.G., High-throughput enzyme assay on a multichannel microchip using optically gated sample introduction. *Electrophoresis,* 2005. 26(24): p. 4711-4717.

[132] Satoh, A., Fukui, E., Yoshino, S., Shinoda, M., Kojima, K., Matsumoto, I., Comparison of methods of immobilization to enzyme-linked immunosorbent assay plates for the detection of sugar chains. *Anal Biochem*, 1999. 275: p. 231-235.

[133] Bryan, M.C., Plettenburg, O., Sears, P., Rabuka, D., Wacowich-Sgarbi, S., Wong, C.-H., Saccharide display on microtiter plates. *Chem Biol*, 2002. 9: p. 713-720.

[134] Nanduri, V., Balasubramanian, S., Sista, S., Vodyanoy, V.J., Simonian, A.L., Highly sensitive phage-based biosensor for the detection of β-galactosidase. *Anal Chim Acta*, 2007. 589: p. 166-172.

[135] Reynolds, A.J., Haines, A.H., Russell, D.A., Gold glyconanoparticles for mimics and measurement of metal ion-mediated carbohydrate-carbohydrate interactions. *Langmuir,* 2006. 22(3): p. 1156-1163.

[136] Fuente, J.M., Barrientos, A.G., Rojas, T.C., Rojo, J., Cañada, J., Fernández, A., Penadés, S., Gold glyconanoparticles as water-soluble polyvalent models to study carbohydrate interactions. *Angew Chem Int Ed*, 2001. 40(12): p. 2258-2261.

[137] Otsuka, H., Akiyama, Y., Nagasaki, Y., Kataoka, K., Quantitative and reversible lectin-induced association of gold nanoparticles modified with α-lactosyl-w-mercapto-poly(ethylene glycol). *J Am Chem Soc*, 2001. 123(34): p. 8226-8230.

[138] Lin, C.-C., Yeh, Y.-C., Yang, C.-Y., Chen, C.-L., Chen, G.-F., Chen, C.-C., Wu, Y.-C., Selective binding of mannose-encapsulated gold nanoparticles to Type 1 pili in Escherichia coli. *J Am Chem Soc*, 2002. 124(14): p. 3508-3509.

[139] Hone, D.C., Haines, A.H., Russell, D.A., Rapid, qantitative colorimetric detection of a lectin using mannose-stabilized gold nanoparticles. *Langmuir,* 2003. 19(17): p. 7141-7144.

[140] Fuente, J.M., Eaton, P., Barrientos, A.G., Menéndez, M., Penadés, S., Thermodynamic evidence for Ca^{2+}-mediated self-aggregation of Lewis X gold glyconanoparticles. A model for cell adhesion via carbohydrate-carbohydrate interaction. *J Am Chem Soc*, 2005. 127(17): p. 6192-6197.

[141] Paz, J.-L., Ojeda, R., Barrientos, A. G., Penadés, S., Martín-Lomas, M., Synthesis of a Le^y neoglycoconjugate and Le^y-functionalized gold glyconanoparticles. *Tetrahedron: Asymmetry*, 2005. 16: p. 149-158.

[142] Svarovsky, S.A., Szekely, Z., Barchi Jr, J.J., Synthesis of gold nanoparticles bearing the Thomsen-Friedenreich disaccharide: a new multivalent presentation of an important tumor antigen. *Tetrahedron: Asymmetry*, 2005. 16: p. 587-598.

In: Biotechnology: Research, Technology and Applications   ISBN 978-1-60456-901-8
Editor: Felix W. Richter   © 2008 Nova Science Publishers, Inc.

Chapter 4

Hydrolase-Based Synthesis of Enantiopure α-Hydroxy Ketones: From Racemic Resolutions to Chemo-Enzymatic Dynamic Kinetic Resolutions

*Pilar Hoyos[a], José Vicente Sinisterra[a,b], Pablo Domínguez de María[c] and Andrés R. Alcántara[a,b]**

[a]Grupo de Biotransformaciones, Departamento de Química Orgánica y Farmacéutica, Facultad de Farmacia, Universidad Complutense de Madrid, Plaza de Ramón y Cajal, s/n. 28040 Madrid, Spain
[b]Unidad de Biotransformaciones Industriales. Parque Científico de Madrid PTM, C/ Santiago Grisolía, 2. 28760 Tres Cantos, Madrid, Spain
[c]AkzoNobel BV. Process and Product Technology Department (RTC-CPT). Velperweg 76, P.O. Box 9300. 6800 SB, Arnhem. The Netherlands

Abstract

Biotransformations – that is, the use of biocatalysts in organic synthesis – constitutes a relevant area within White Biotechnology, especially significant in the industrial manufacture of enantiopure drugs or building blocks. In this area, hydrolases are powerful biocatalysts for the production of enantiopure α-hydroxy ketones. These chemicals are valuable structures for the asymmetric synthesis of biologically active compounds, such as antidepressants (i.e., bupropion), inhibitors of amyloid-β protein production – for Alzheimer´s disease treatment, farnesyl-transferase inhibitors Kurasoin A and B, and antitumor antibiotics (Olivomycin A and Chromomycin A_3).

Originally, the hydrolase-based kinetic resolution (KR) of α-hydroxy ketones *via* hydrolysis and/or acylation led to positive results using different lipases, i.e., *Candida antarctica* lipase B (CAL-B), *Candida rugosa* lipase (CRL), *Burkholderia cepacia* lipase

---

* Corresponding author: Dr. Andrés Alcántara. Tel.: +34 91 394 1820; fax: +34 91 394 1822. E-mail address: andresr@farm.ucm.es

(BCL), or *Pseudomonas stutzeri* lipase (PSL). Later on, to overcome intrinsic drawbacks derived from the application of KR′s – where reaction yields are obviously limited to 50 % –, chemo-enzymatic dynamic kinetic resolutions (DKR) have been developed as well. In that case the enzymatic resolution is combined with a(n) (*in situ*) racemization of the starting material. Thus, enantiopure α-hydroxy ketones can be obtained in high yields when a lipase-based resolution is properly combined with a ruthenium catalyst racemization *via* hydrogen transfer. For this DKR approach, lipases from *Pseudomonas stutzeri* (PSL) and *Candida antarctica* lipase B (CAL-B) have been reported as promising biocatalysts.

Taking into account the emerging importance of this topic, the present chapter aims to provide a comprehensive view on the field of hydrolases and α-hydroxy ketones, by focussing on the type of substrates, enzymes, and optimum operational conditions used.

1. Introduction

Within the emerging field of White Biotechnology, Biotransformations can be defined as the use of enzymes and / or whole cells as biocatalysts for the production of different (enantiopure) chemicals [1]. Since these bio-catalyzed reactions are usually conducted under very mild reaction conditions – and with high atom efficiency –, such (bio-)technological reactions can facilitate sustainable processing and reduce environmental impacts of chemical processes [1, 2]. Moreover, due to the high chemo- and enantio-selectivities that enzymes commonly display, the biotransformations field has gained an enormous interest for the production of natural products, pharmaceuticals, as well as chiral building blocks. As a consequence, a broad number of biocatalytic-based processes have been implemented at industrial scale, thus showing that environment and economic drivers can be perfectly combined in the same chemical process [1, 3]. As a matter of fact, the useful catalytic promiscuity that many enzymes display – both in terms of substrate acceptance and type of reaction catalyzed –, leads to the practical use of these biocatalysts in processes that are indeed quite far from the natural *actual* catalytic role of those proteins [4]. In addition, the impressive development of molecular biology techniques has widely facilitated the way to find and characterize new enzymes (i.e., *via* metagenomic assessments) [5], and is enabling the development of so-called tailor-made biocatalysts (i.e., *via* directed evolution and analogous strategies) [6]. As a consequence, biotransformations nowadays represent a highly multi-disciplinary field, in which expertises from different research areas are required, to assure the successful set-up of a biocatalytic-based process.

Within the herein defined field of biotransformations, hydrolases (especially lipases and esterases) represent an outstanding example of powerful biocatalysts. These enzymes are very versatile for a plethora of different types of reactions and substrates, usually leading to highly chemo- and / or enantio-selective processes [7]. Furthermore, their ability to perform reactions without the need for additional cofactors, their availability at commercial scale, their capacity to catalyze reactions both in aqueous and non-conventional reaction media, as well as their widely diverse catalytic promiscuity, are important assets for their practical application. Although the natural role of such hydrolases consists of the cleavage of (fatty) ester bonds, many other different promiscuous applications have been reported as well [1, 3,

4, 7]. Even more, apart from those *classic* synthetic / hydrolytic (promiscuous) applications of hydrolases [1, 3, 4, 7], recently other totally unexpected hydrolase-based performances were published, like catalysis of C-C bond formation, or Michael addition reactions [8]. Likewise, the smart combination of these hydrolases with molecular biology strategies – i.e., metagenomics and directed evolution –, is currently opening up new important research areas. As an example of such combinations, very promising results have been published in the hydrolase-based asymmetric resolution of tertiary alcohols [9], structures that are useful chemical synthons for a vast number of natural products, and for which chemical synthesis is particularly challenging [10].

Lipases and esterases have been successfully applied in the synthesis of enantiopure α-hydroxy ketones as well. These compounds are highly valuable molecules – as chemical precursors and building blocks –, for different synthetic procedures to afford natural products and / or drugs. For instance, these structures can be employed in the subsequent synthetic routes to furnish antidepressants (i.e., bupropion), selective inhibitors of amyloid-β protein production – for Alzheimer´s disease treatment –, farnesyl-transferase inhibitors Kurasoin A and B, and antitumor antibiotics (Olivomycin A and Chromomycin A$_3$), among many other examples [11]. Those α-hydroxy ketones occur in nature as well, for example, as chemical signals in bacteria and beetles [12], and thus their production might have other interests in natural products field. Therefore, addressing the relevance of these topics, both in terms of synthetic substrates and natural existence, considerable efforts have been made to develop competitive synthetic procedures, aiming to produce of some of these chemicals in an affordable and efficient manner [13].

The present chapter aims to provide a comprehensive overview on the hydrolase-based strategies for the enantioselective production of α-hydroxy ketones. To this end, both kinetic resolutions (KR´s) and chemo-enzymatic dynamic kinetic resolutions (DKR´s) will be discussed in detail in the following sections.

2. Hydrolase-Based Kinetic Resolutions (KR) for the Production of Chiral α-Hydroxy-Ketones

The well-known and widely described hydrolase-catalyzed kinetic resolution (KR) is based on the capability of enzymes to accept, as substrates, preferentially one of the enantiomers of a racemate. This enantioselectivity is traditionally quantified in terms of enantiomeric ratio, *E*, proposed several decades ago by Sih and co-workers [14]. Therefore, by means of very enantioselective enzymes, highly specific racemic kinetic resolutions can be implemented, in some cases even at commercial scale [1, 3]. Within the group of hydrolases, such KR's strategies have found enormous application especially in the resolution of secondary alcohols. Thus a large number of structures and enzymes have been studied [1, 3, 4, 7]. As depicted in Scheme 1 for these secondary alcohols, when hydrolases are selected as the biocatalysts, synthetic or hydrolytic approaches can be applied for such purpose, depending on the reaction media employed (aqueous or non-aqueous). If a hydrolytic process is envisaged (aqueous media), esters will be the substrates. Conversely, if a synthetic approach is set up (non-conventional media), free alcohols will be selectively acylated by the biocatalyst.

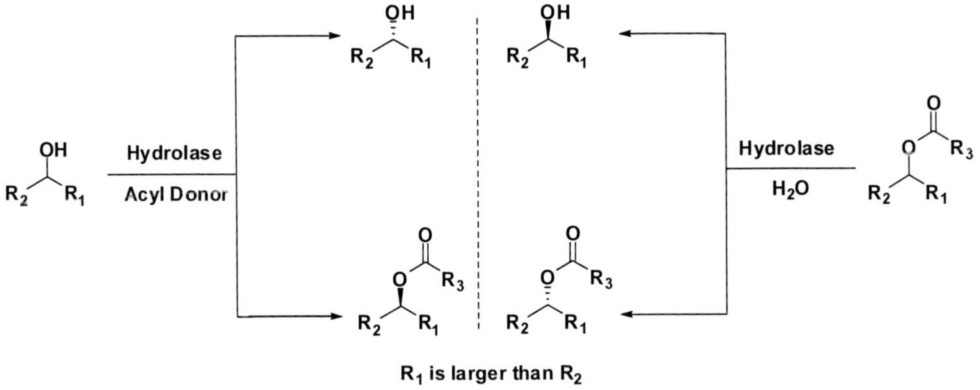

R₁ is larger than R₂

Scheme 1. Approaches for the hydrolase-based kinetic resolution of secondary alcohols. Left: Synthesis in non-conventional media. Right: hydrolysis in aqueous media. For further reading on the topic, see [1, 3, 4, 7].

The stereochemistry of the substrate structure recognized by most of lipases is that one defined by the so-called Kazlauskas rule [15], which relates the enantiopreference with the relative size of substituents around the stereocenter. This empirical rule has been rationalized by different studies on the nature of the substrate binding area [7c].

Following the enzymatic strategy depicted in Scheme 1, until today a considerable number of α-hydroxy ketones have been resolved by performing hydrolase-based kinetic resolutions, by means of different (commercial) hydrolases, solvents, and reaction conditions. For instance, some cyclic α-hydroxy ketones have been studied as potential substrates for biocatalysis. In this area, Tanyeli *et al.* have reported the hydrolytic resolution of several quaternary α-acetoxy α-substituted cyclic ketones by means of *Candida rugosa* lipase (CRL) (formerly *Candida cylindracea*) (Scheme 2) [16]. In this case, the addition of DMSO as cosolvent (up to 10 %) was necessary to assure proper enzymatic selectivities, as well as substrate solubility in the aqueous media. This addition of cosolvents in this type of enzymatic reactions (aqueous media) is relatively common and well stablished, as in many cases leads to improvements in terms of enantioselectivity, as well as in substrate solubility of organic structures in aqueous systems [1d].

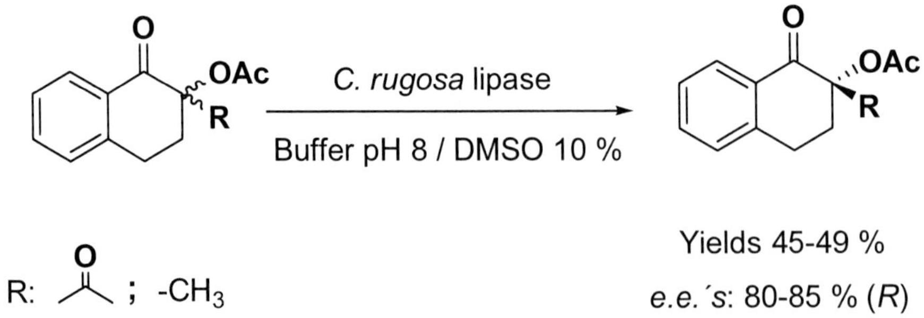

Yields 45-49 %

e.e.´s: 80-85 % (*R*)

Scheme 2. Enzymatic kinetic resolution of α-acetoxy α-substituted cyclic ketones by means of *Candida rugosa* lipase (CRL). Some other chemical structures as examples can be found in literature [16].

Importantly, that work represents also a further example of an enzyme-based resolution to afford certain type of chiral tertiary alcohols, which are quite useful as synthons for further asymmetric synthetic procedures [9]. Yet, in this especific case (Scheme 2) [16], though enantioselectivities were acceptably high, long reaction times (> 70 h) were needed to assure reasonable conversions for a kinetic resolution approach (> 45 %) (Scheme 2). Presumably, the high steric hindrance of those bulky substrates is the reason for the slow reaction rate. Surely further work in terms of directed evolution could provide more efficient biocatalysts for that reaction as well [6a, 9].

In the same area, when those cyclic α-hydroxy ketones are less sterically hindered, better results can be obtained. Thus, sucessful hydrolase-based kinetic resolution of 2-hydroxy-1-indanone and analogous compounds have been published, conducting hydrolysis in aqueous media or, conversely, applying irreversible transesterifications in organic solvents [17]. To this end, a screening of 84 commercial hydrolases was performed, showing that some lipases from *Pseudomonas sp.* were the best enzymes to resolve these substrates. Noteworthy, one of these hydrolases, the lipase from *Pseudomonas fluorescens* (PFL), was successfully employed as biocatalyst in the asymmetric resolution of the chromophoric aromatic part of Chromomycin A$_3$, structure widely used as antitumor antibiotic (Scheme 3) [18]. To produce that synthon properly, the irreversible transesterification by means of vinyl acetate as acyl donor was applied at 25 °C, thus leading to full conversion within 5 h. High enantiomeric excess of the desired enantiomer (> 95 % (S)) was achieved as the acylated isomer. Thus, after separation from the reaction mixture containing also the non-converted alcohol, a subsequent PFL-based hydrolysis of that previously acylated structure in buffer yielded the desired pure chiral enantiomer [18].

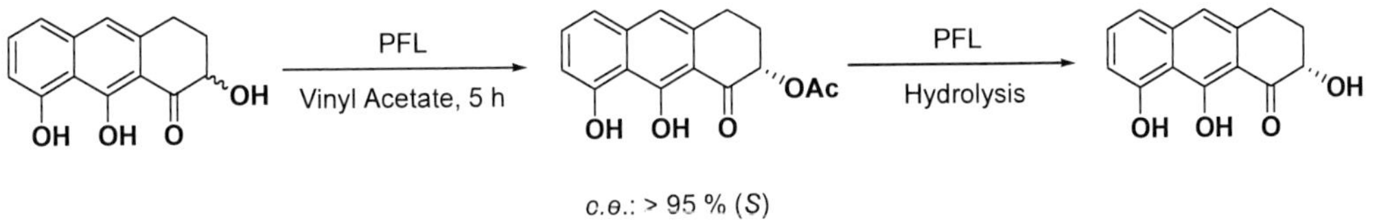

Scheme 3. PFL-mediated asymmetric resolution of the chromophoric part of the antibiotic Chromomycin A3 [18].

Likewise, some other hydrolases have been applied as useful biocatalysts for the asymmetric production of various α′-acetoxy-α,β-unsaturated cyclic ketones, which are, once again, very useful synthons for many natural products [19]. For this purpose, PS lipase (*Burkholderia cepacia* lipase, also refered as BCL), and pig liver esterase (PLE) are the most suitable enzymes. Some examples are shown in Figure 1.

Apart from the cyclic compounds described above, hydrolases have also found use in the kinetic resolution of some lineal acyloins [20, 21]. In this particular area, it is worth to mention some research that has been devoted to the production of chiral building blocks for the synthesis of several epothilones [21]. These compounds are natural polyketides that exhibit, *in vitro*, cytotoxic effects resembling those of taxol. Thus, the practical production of those structures (and analogous) may be of high interest in the pharmaceutical field. For this synthesis, lipase B from *Candida antarctica* (CALB) and lipase from *Burkholderia cepacia* (BCL or PS) were the most effective biocatalysts.

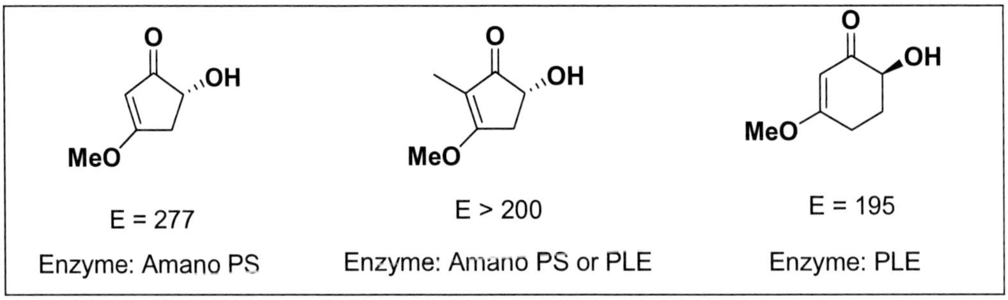

Figure 1. Examples of the hydrolase-based kinetic resolution of various α′-acetoxy-α,β-unsaturated cyclic ketones, by means of pig liver esterase (PLE) and *Burkholderia cepacia* lipase (PS or BCL). More examples and further information can be found in literature [19].

Other commercial hydrolases tested, like, for instance, lipase A from *Candida antarctica* (CAL-A), porcine pancreatic lipase (PPL), lipase from *Rhizopus oryzae* (ROL), or lipase from *Candida rugosa* (CRL), were able to accept those acyloins as substrates, but displayed low enantioselectivities. CALB-, and BCL-based reactions, however, were conducted in aqueous media at room temperature, thus yielding high enantioselectivities in competitive reaction times (ca. 30 min). Some relevant results of this work are briefly summarized in Table 1.

As can be observed in Table 1, the reported enzymatic kinetic resolutions represent a useful practical method to afford relevant chiral structures, with high selectivity and in short reaction times. In this specific area, also esterase from pig liver (PLE) has been reported as an active biocatalyst towards those substrates [21a]. Notably, huge differences in terms of enantioselectivities were found when different PLE′s – either recombinantly over-expressed in heterologous hosts, or from commercial suppliers –, were applied as biocatalysts towards these acyloins (*E* values ranging from 8 to 50) [21a]. That aspect obviously suggests largely different enantioselectivities of the different PLE isoenzymes for such structures. Since presently some recombinant PLE′s are available, probably many aspects will be clarified the next years [22, 23]. In addition, the fact that these recently reported rec-PLE′s are actually produced microbiologically – thus no longer coming from animal origin –, opens the possibility of using PLE as biocatalyst in pharmaceutical fields as well. Since the animal origin of PLE has clearly hampered the use of this powerful biocatalyst at that industrial (pharma) scale, many new practical PLE applications are expected to be implemented in the forthcoming years [23].

Within another different area, hydrolases have also been sucessfully applied in the resolution of aryl α-hydroxy ketones. As early as in the 1980s, Ohta *et al.* reported on the highly enantioselective hydrolysis of α-acetoxyacylophenones by means of cells of *Pichia miso* as the biocatalyst, presumably due to the presence of hydrolases [24]. Interestingly, the same research group described the same whole cell biocatalyst being able to hydrolyze selectively esters of tertiary alcohols [9]. Although hydrolases are not normally employed as whole cells, but rather as free enzymes, as in nature these enzymes are usually extracellular proteins, also Demir and co-workers have employed whole cells of *Rhizopus oryzae* to effectively hydrolyze different α-hydroxy aryl alkyl ketones with high enantioselectivities, though after long reaction times (50-160 h) [25]. In Figure 2 some examples of molecules produced *via* this (bio-)technology are depicted. As can be observed, a wide range of substrates can be accepted for that hydrolase. Apart from those whole cells approaches, some free commercial hydrolases have also been applied in the asymmetric resolutions of some of those α-hydroxy aryl alkyl ketones, with moderate-to-high enantioselections [26].

Table 1. Selected results of the hydrolase-based kinetic resolution of different acyloins. More information in literature [21]. CAL-B stands for lipase B from *Candida antarctica* and BCL for *Burkholderia cepacia* lipase

Substrate	Applied Enzyme	Enantioselectivity (*E*)
	CAL-B	165
	CAL-B	152
	CAL-B or BCL	> 200
	BCL	> 200
	CAL-B or BCL	> 200
	BCL	177

Within this field of chiral α-hydroxy ketones, benzoins (1,2-diaryl-2-hydroxyethanones) display a crucial role, as many uses of them as building blocks can be envisaged. As can be observed in Figure 2, kinetic resolution of benzoin esters by means of whole cells of *R. oryzae* led to moderate enantioselectivities (Figure 2, *e.e.* 70 %) [25]. Other approaches with free hydrolases, such as *C. rugosa* lipase in supercritical carbon dioxide, led to moderate enantioselectivities as well [27].

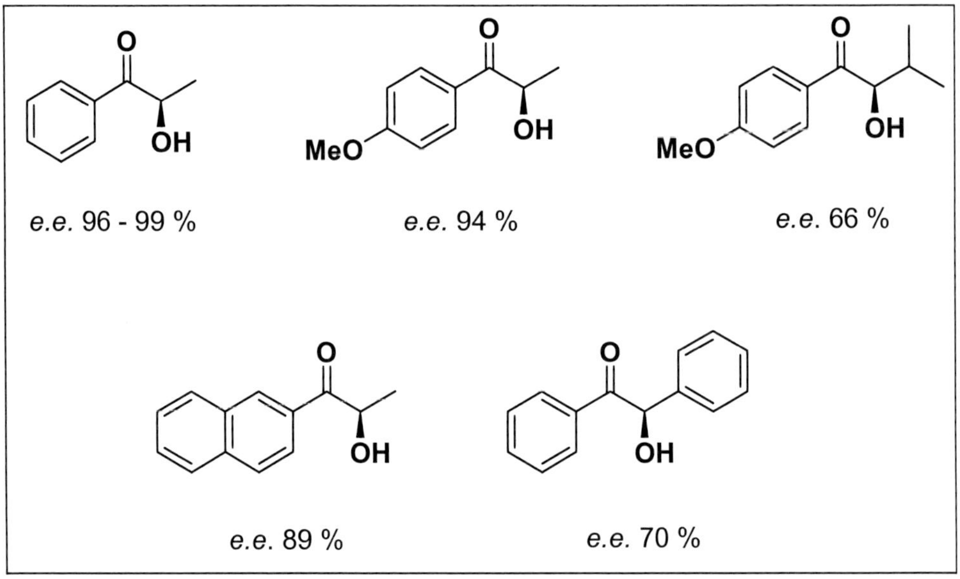

e.e. 96 - 99 %   *e.e. 94 %*   *e.e. 66 %*

e.e. 89 %   *e.e. 70 %*

Figure 2. Some α-hydroxy aryl alkyl ketones resolved by using whole cells of *Rhizopus oryzae* as the biocatalyst [25].

Remarkably, excellent results have been reported to afford chiral benzoins by using the lipase from *Pseudomonas stutzeri* (PSL, commercialized as Lipase TL® by *Meito Sangyo*) as the biocatalyst [28]. From a preliminary screening of commercially available hydrolases, only PSL was found to be effective. Thus, high enantioselections and conversions were achieved by implementing a PSL-based transesterification process in THF as solvent, and with vinyl acetate as the acyl donor (Scheme 4). As a drawback of this preliminary work, however, high enzymatic loadings were usually needed to perform the reaction efficiently.

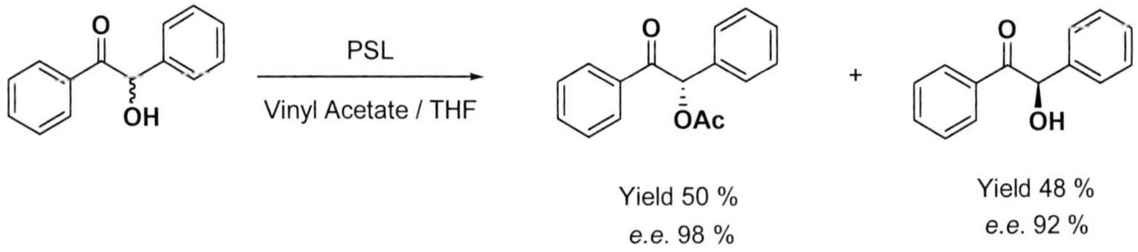

PSL

Vinyl Acetate / THF

Yield 50 %
e.e. 98 %

Yield 48 %
e.e. 92 %

Scheme 4. PSL-based transesterification of racemic benzoin, to perform an efficient kinetic resolution [28].

Since the herein described reaction (Scheme 4) may be obviously relevant from a practical viewpoint, the biotransformation was further optimized in terms of reaction conditions, to afford more competitive parameters than that "proof-of-principle", previously reported [28]. Moreover, the substrate scope of the reaction was extended to a wide range of different benzoin-like products [29], thus showing that the development of a versatile PSL-based platform for the synthesis of different chiral benzoins was perfectly feasible. Notably, maximum conversions and enantiomeric excesses could be obtained in short reaction times (4-24 h), employing vinyl butyrate as (the best) acyl donor, and carrying out the reaction at 37-50 °C. Interestingly, enantioselectivities remained high in the majority of the structures evaluated as substrates. Some of the most relevant results are shown in Table 2.

Table 2. Some relevant results of the kinetic resolution of different benzoins catalyzed by lipase from *Pseudomonas stutzeri* [29]

Substrate	T (°C)	Reaction time	E
	50	4 h	> 200
	37	24 h	125
	50	6 h	> 200
	50	6 h	> 200
	50	6 h	> 200

As can be seen in Table 2, the enzyme acylated preferentially the *S*-enantiomer, instead of the *R*-one, which would have been expected according Kazlauskas rule [15]. This special recognition had been previously pointed out by Martín-Matute and Bäckvall for lipase B from *C. antarctica* and recently the influence of a δ-keto group on the enantiorecognition of secondary alcohols by CALB has been examined by molecular dynamics by this group [30].In addition to the benzoin-like structures depicted in Table 2, some further experiments using PSL were conducted, aiming to study its capability to catalyze the kinetic resolution of other aromatic α-hydroxy ketones, even more sterically hindered [31]. Results achieved in these experiments confirmed the excellent enantioselectivities and reaction parameters (ca. 5 h, 50 °C) of this lipase towards other bulkier crossed benzoin-type structures, which clearly enhances even more the potentiality that PSL can have as biocatalyst for asymmetric syntheses in the coming future. Reactions and some results are shown in Scheme 5.

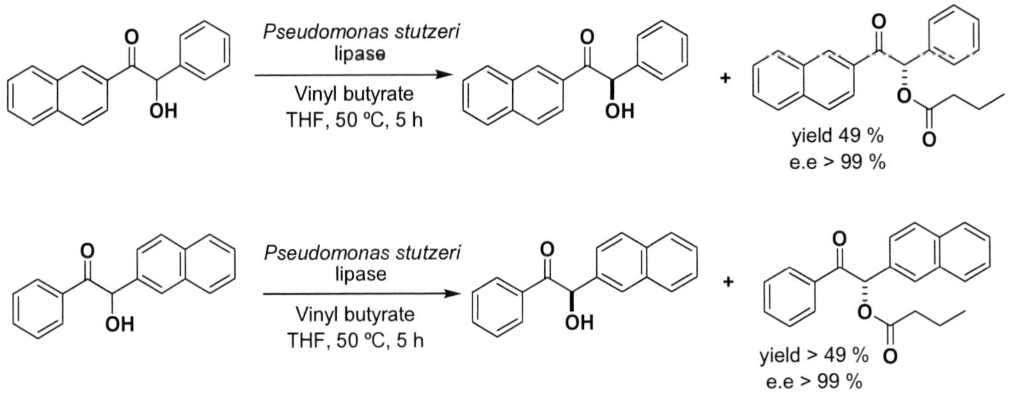

Scheme 5. Asymmetric crossed benzoins obtained through PSL-catalyzed kinetic resolution [31].

3. Overcoming Yield Issues:
Hydrolase-Based Dynamic Kinetic Resolutions
(DKR) to Afford (Theoretical) 100 % Conversion

The previous section of this chapter has clearly shown the tremendous potential that the practical use hydrolases as biocatalysts can have in the production of a wide number of chiral α-hydroxy ketones. Thus, very simple and powerful synthetic procedures can be implemented by the smart combination of the right hydrolase and the proper reaction media, applied to a specific substrate. In addition, when necessary, straightforward well-known approaches like directed evolution can even improve the enzymatic performance up to a desired *on-spec* application.

Yet, with these kinetic resolution approaches only an intrisic maximum (theoretical) yield of 50 % can be achieved. This is clearly a drawback from a practical viewpoint, as it implies that half of the substrate is not used. Such kind of problems may hamper in many cases a further implementation of a certain biocatalytic process at industrial level, as economic considerations are not properly fulfilled. Therefore, as general concept, to overcome this situation biocatalysis is currently being integrated by combining (*in situ*) chemical methods within the enzymatic step. For instance, very recently the first combination of an oxidoreductase platform with a chemical Wittig reaction was published, thus improving the already powerful oxido-reduction bio-approach [32]. As analogous *conceptual* approach, in the field of hydrolases, the importance of the adequate combination of transition metals with enzymes was already pinpointed years ago as well [33]. Obviously those new chemoenzymatic platforms confer biocatalysis a superior role, as much more competitive processes can be set up.

Within this conceptual strategy, and regarding the topic of this chapter, to circumvent problems derived from yields in KR strategies, the so-called Dynamic Kinetic Resolution (DKR) approach has been put forward [34]. Basically, such DKR encompasses a(n) (*in situ*) racemization of the remnant substrate, while the enzymatic reaction is being carried out. Conclusively, theoretical yields of up to 100 % can be achieved (Scheme 6).

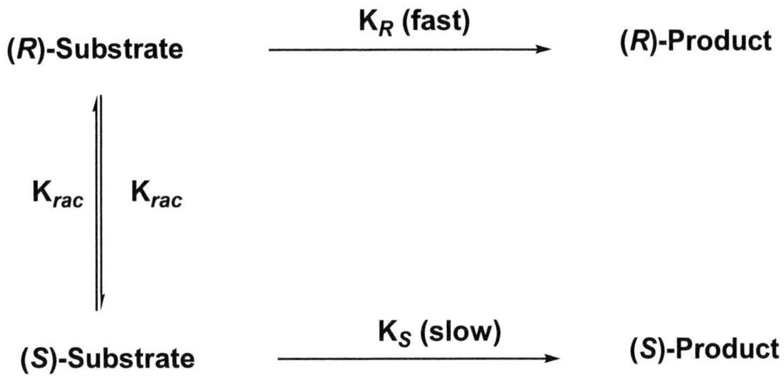

Scheme 6. General reaction pathway for a DKR [29].

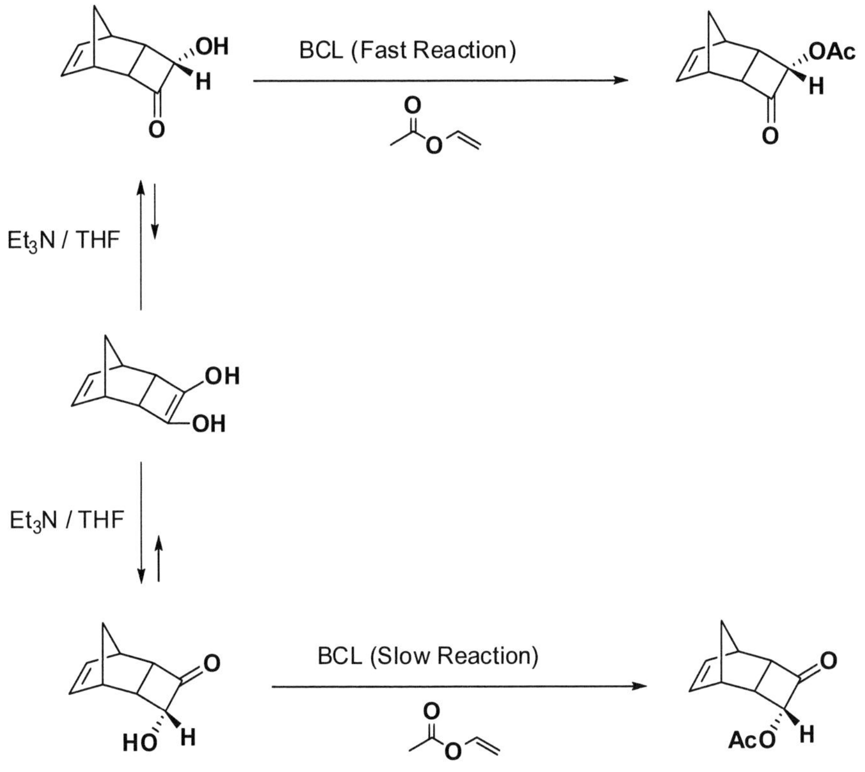

Scheme 7. First example of base-mediated DKR of some α-hydroxy-ketones, reported by Taniguchi et al. [35].

In the field of hydrolases and chiral synthesis of α-hydroxy ketones, the first DKR′s were reported a decade ago, by making use of intrinsic chemical properties of the specific substrates therein assayed [35]. In this regard, Taniguchi *et al.* reported on the *in situ* base-mediated racemization of some tricyclic acyloins (substrates), whereas lipase from *Burkholderia cepacia* (BCL) was able to acylate enantio-selectively one of the enantiomers in organic solvent (THF). The process is depicted in Scheme 7 [35].

Later on, considerable research has been conducted in developing strategies able to implement DKR′s in many types of reactions and substrates, that could confer a general utility to those technologies. However, such a general implementation of DKR′s is unfortunately not straightforward, as the intrinsic properties of the enzymes and substrates employed play a pivotal role. Therefore, each application deserves its own development, which usually differ widely from one case to others. To illustrate this challenging issue, when asymmetrical α-hydroxy ketones are used as substrates, the common combination of metal complexes, such as ruthenium catalysts, and enzymes – to set up an ideal *one pot* DKR –, is not feasible, as the fast formation of an intermediate diketone proceeds, thus leading to rearrangement of the starting material, as shown in Scheme 8. Consequently, if such highly attractive DKR′s are considered, other different methodologies need to be approached.

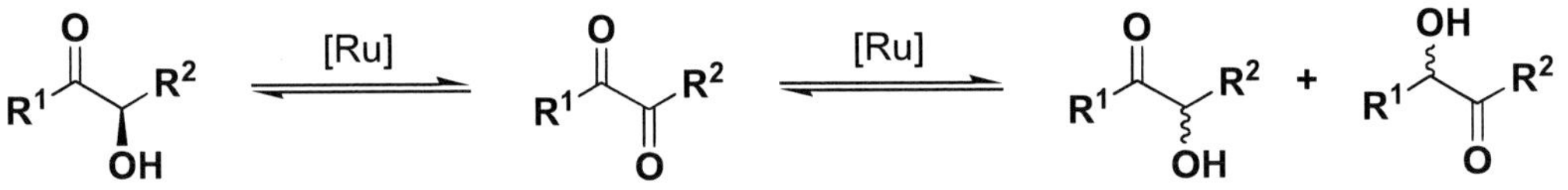

Scheme 8. Racemization of acyloins, catalyzed by a ruthenium complex.

To overcome these issues, several approximations have been put forth in the recent years. For instance, an efficient two-compartment DKR was disclosed [36]. To provide competitive reaction conditions, lipase B from *Candida antarctica* (CAL-B) was used as the biocatalyst, and vinyl butyrate was chosen as the acyl donor. For racemization of the remnant substrate, Amberlyst 15 was employed in a second compartment. Overall, conversions of up to > 90 % and enantiomeric excesses of up to > 91 % were achieved, starting from a racemate (Scheme 9).

Conversion > 90 %

e.e. > 91 %

Scheme 9. CAL-B-based two compartment DKR to afford chiral α-hydroxy ketones, recently published [36].

Another methodology aiming to set up DKRs in this area was recently developed by Bäckvall and co-workers. Such approach was based on a first one-pot hydrolase-based dynamic kinetic resolution of an allylic alcohol, as those structures tolerate Ru catalysts without showing the aforementioned chemical problems. Later on, a subsequent chemical oxidation of the enantiopure acetylated-allyl-structure formed, yields the desired chiral α-hydroxy-ketone [37]. Within this area, in the first DKR step, among the different commercially available lipases tested, lipase B from *Candida antarctica* (CALB) was found to be the most effective. Therefore, the CALB-based (kinetic) resolution was coupled with a ruthenium catalyst substrate racemization, thus affording the enantiomerically pure acylated secondary (allyl) alcohols in very high yields, wihtin competitive reaction times (ca. 24 h).

In Table 3 some of these outstanding results are depicted, covering the overall process reported: the first hydrolase-based one pot DKR – to generate the chiral center in the molecule with high enantiopurity –, and the further chemical oxidation to afford the corresponding protected acyloin. As can be observed, impressive yields (> 95 % in many cases) and enantiomeric excesses (> 99 %) were obtained.

Table 3. Chemo-enzymatic dynamic kinetic resolution of allylic alcohols and further oxidation to the corresponding acyl-protected acyloin [37]

Substrate	DKR product	Oxidative cleavage product	Yield (%)	e.e.$_p$ (%)
			98	> 99
			> 98	> 99
			96	> 99

Table 3. (Continued)

Substrate	DKR product	Oxidative cleavage product	Yield (%)	e.e$_p$ (%)
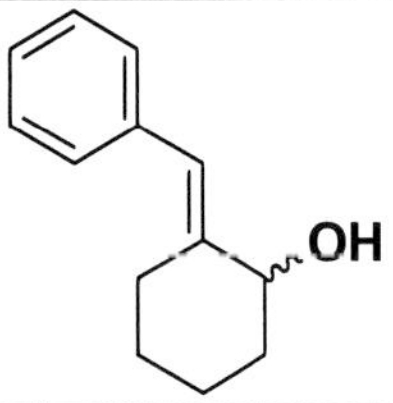	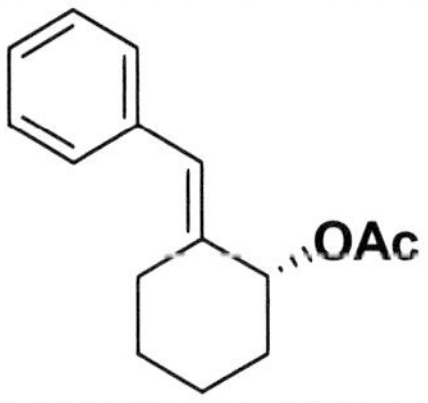	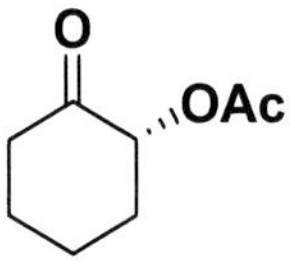	86	> 99

Furthermore, very relevant results have been achieved by using totally different synthetic strategies, though also aiming to produce different chiral benzoins in higher yields than the maximum 50 % accesible by a classic kinetic resolution. In this respect, Demir *et al.* reported that the enzymatic machinery of fungus *Rhizopus oryzae* catalyzes the inversion of the chirality of benzoin [38], thus affording both enantiomers depending on the pH of the medium, as depicted in Scheme 10.

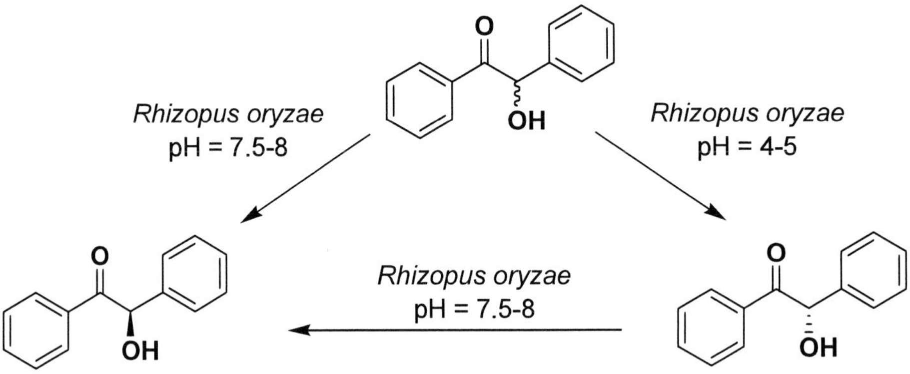

Scheme 10. Benzoin deracemization based on pH values, catalyzed by whole cells of *Rhizopus oryzae* [38].

With this above-depicted interesting synthetic approach, both enantiomers of benzoin could be obtained in high yields and enantiopurities, by simply switching the reaction acidic conditions to the desired pH value. Yet, reaction times required to complete the conversion were rather long (i.e., 76 % yield for (*R*)-benzoin after 21 days, or 71 % yield for (*S*)-benzoin after 15 days). This aspect clearly diminishes the potential practical use of this system. However, this enzymatic approach might be surely further improved by means of different molecular biology techniques – i.e., cloning and overexpressing the enzyme(s) involved in such processes –, thus providing a novel strategy for many chiral molecules. Once again, the biotransformations field would comprise a fascinating combination of different disciplines, with the aim to complement organic (asymmetric) syntheses by applying very mild reaction conditions.

Encompassing a different approach to afford the desired chiral benzoins in a proper practical manner, very high conversions in short reaction times have been recently described in the DKR of racemic benzoins [29]. In this case, the strategy combines a *Pseudomonas stutzeri* lipase-catalyzed kinetic resolution of benzoin, together with the racemization of the remnant alcohol mediated by a commercially available ruthenium catalyst (the so-called Shvo′s catalyst). Yet, the implementation of the herein described process was not straightforward, as the high temperature required for the activation of the ruthenium catalyst promoted the deactivation of the lipase. Once again, each DKR must be optimized specifically for each reaction system. In this case, in a first approach better results were obtained if the process was carried out in three steps: first a quick KR at 50 °C, using low lipase loadings and vinyl butyrate as acyl donor. Such vinyl butyrate was reported as the best acyl donor [39], but unfortunately is not useful for a DKR approach. Secondly, when ca. 30 % conversion was achieved, the remnant acyl donor was evaporated, and then Shvo′s catalyst, a new loading of fresh lipase, as well as trifluoroethyl butyrate, were added, thus starting the actual DKR process at 50 °C. Finally, a new loading of fresh enzyme was added, to compensate the enzyme deactivation at 50 °C. By implementing this methodology, moderate-to-high conversions values (50-92 %) were obtained when different diaryl α-hydroxy-ketones were tested, involving moderate-to-long reaction times (48-72 h). Remarkably, in all cases an excellent optical purity of the products was maintained. The overall three-step one-pot approach is depicted in Figure 3. This previously mentioned drawback of lipase deactivation at 50 °C was recently overcome by the enzyme entrapment in silicon spheres [40].

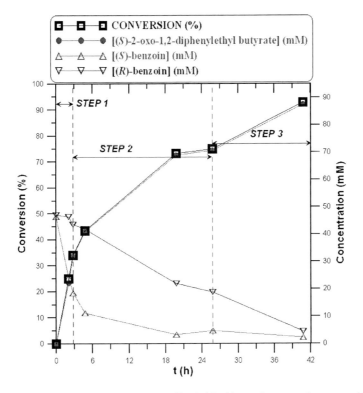

Figure 3. Three-step one-pot PSL-based DKR to afford chiral benzoin, recently reported [29].

The immobilization of the lipase from *Pseudomonas stutzeri* allowed a significant enhancement of enzyme stability, making possible the efficient development of a *one pot* DKR process of benzoin, coupling the entrapped lipase and Shvo´s catalyst, in 20 h. Furthermore, this immobilized biocatalyst could be reused several times, and thus different cycles of benzoin-DKR were carried out. Thus, the productivity of the process was improved in terms of reaction times, and in terms of enzyme loading. Therefore, those final DKR reaction conditions are rather close to those of a practical biocatalytic process, in which productivity, catalytic efficiency, and re-usability aspects are obviously key-factors. Notably, that lipase from *Pseudomonas stuzteri* seems to be a promising biocatalyst for such DKR´s, and thus very recently analogous DKR processes by means of that enzyme, with other different (sterically hindered) secondary alcohols as substrates, have been reported [41].

To complement the above-described chemical racemization of remnant substrates, during the last years another different biocatalytic methodology has been developed for the same purpose of racemization of α-hydroxy-ketones. In this particular case, microbial cells are used as the biocatalysts, what could be a useful and promising tool for a sustainable racemization of the remnant alcohol of an enzymatic resolution process [42]. Overall, that approach would include two enzymatic concomitant steps to perform the DKR, thus conducting the whole process under very mild reaction conditions. According to work recently disclosed, the racemization of different acyloins and benzoins was carried out using whole lyophilized cells of several bacteria, fungi and yeast strains. In Scheme 11 the process and some relevant examples of this novel approach are depicted. As can be observed, a wide substrate range of α-hydroxy-ketones can be efficiently racemized *via* this microbial alternative.

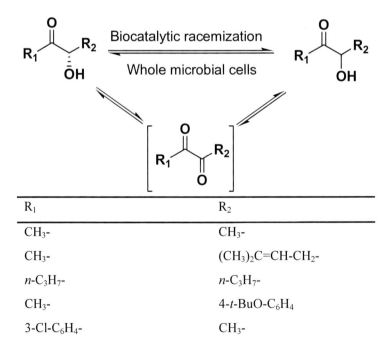

R₁	R₂
CH₃-	CH₃-
CH₃-	(CH₃)₂C=CH-CH₂-
n-C₃H₇-	*n*-C₃H₇-
CH₃-	4-*t*-BuO-C₆H₄
3-Cl-C₆H₄-	CH₃-

Scheme 11 (Continued).

3,5-di-F-C6H3-	CH3-
Ph-	CF_3
2-furyl-	2-furyl-
2-pyridyl-	2-pyridyl-
2-MeO-C_6H_4-	2-MeO-C_6H_4-
4-MeO-C_6H_4-	4-MeO-C_6H_4-
-$(CH_2)_4$-	

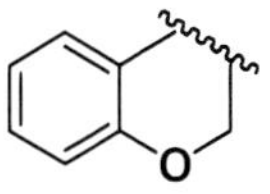

Scheme 11. Racemization of different aliphatic and aromatic α-hydroxy-ketones catalyzed by whole cells of bacteria, fungi or yeast strains, recently disclosed [42a].

Conclusion

The present chapter has especially focused on the use of hydrolases as powerful biocatalysts to afford chiral α-hydroxy-ketones, which are actually very important building blocks for many subsequent syntheses in different chemical and pharmacological fields. In particular, in this chapter the high versatility of hydrolases in the preparation of homochiral compounds has been illustrated by the recognition of these special, secondary alcohols, possessing a carbonyl group in α-position, both for either cyclic or lineal, aliphatic or aromatic structures. Also, as it has been widely discussed, the performance of those hydrolases proceeds commonly under very mild reaction conditions, as well as with high atom efficiency. Currently those statements represent a very important asset, as more sustainable reaction conditions can be set *via* these biocatalytic-based approaches. Moreover, fewer wastes are normally generated by implementing these bioprocesses, thus contributing to reach the increasingly important postulates of the Green Chemistry.

Within the field of chiral α-hydroxy-ketones production, we have described how in a first approach some commercial hydrolases display the capability of producing many different α-hydroxy-ketones in an enantiomerically pure form, *via* kinetic resolution approaches. Subsequently, the intrinsic limitation of such kinetic resolutions has been overcome by implementing efficient dynamic kinetic resolutions, leading to, not only high enantiopurities, but also to very high yields, thus making this approach very attractive for industrial applications. Since, presently, enzymes can also be artificially tailored for using them in specific applications, this is obviously an emerging field from which many more applications and nice innovative concepts are expected to be disclosed in the forthcoming years. In this sense, the combination of chemical knowledge with the new biocatalytic-based approaches is clearly a dynamic sub-area, from which many novel processes can be envisaged.

We hope that this contribution will prompt more research groups to work on the development of new, efficient, and competitive, DKR concepts.

References

[1]   Recent relevant books on Biotransformations: a) Liese, A; Seebach, K; Wandrey, C. *Industrial Biotransformations*, Wiley-VCH, Weinheim. 2006.b) Buchholz, K; Kasche, V; Bornscheuer, UT. *Biocatalysts and Enzyme Technologies*, Wiley-VCH, Weinheim. 2005.c) Bommarius, AS; Riebel-Bommarius, B. *Biocatalysis*, Wiley-VCH, Weinheim, 2004.d) Faber, K. *Biotransformations in Organic Chemistry*, Springer Verlag. 2004.e). Drauz, KH; Waldmann, H. *Enzyme Catalysis in Organic Synthesis*, Wiley-VCH, Weinheim. 2003.

[2]   See, for instance: Sheldon, RA. The E Factor: fifteen years on. *Green Chem.* 2007, *9*, 1273-1283, and references compiled therein.

[3]   Recent reviews and leading articles on the industrial use of Biotransformations:a) Ran, N; Zhao, L; Chen, S; Tao, J. Recent applications of biocatalysis in developing green chemistry for chemical synthesis at the industrial scale. *Green Chem.* 2008, (*DOI: 10.1039/b716045c*).b) Tao, J; Zhao, L; Ran, N. Recent advances in developing chemoenzymatic processes for active pharmaceutical ingredients. *Org. Process. Res. Dev.* 2007, *11*, 259-267.c) Hilterhaus, L; Liese, A. Building Blocks. *Adv. Biochem. Engin/Biotechnol.* 2007, *105*, 133-173.d) Panke, S; Wubbolts, M. Advances in biocatalytic synthesis of pharmaceutical intermediates. *Curr. Opin. Chem. Biol.* 2005, *2*, 188-194.e) Schoemaker, HE; Mink, D; Wubbolts, MG. Dispelling the Myths--Biocatalysis in industrial synthesis. *Science* 2003, *299*, 1694-1697.f) Schmid, A; Hollmann, F; Byung Park, J; Bühler, B. The use of enzymes in the chemical industry in Europe. *Curr. Opin. Biotechnol.* 2002, *13*, 359-366.g) Straathof, AJJ; Panke, S; Schmid, A. The production of fine chemicals by biotransformations. *Curr. Opin. Biotechnol.* 2002, *13*, 548-556.h) Liese, A; Villela Filho, M. Production of fine chemicals using biocatalysis. *Curr. Opin. Biotechnol.* 1999, *10*, 595-603.

[4]   Recent reviews and relevant articles dealing with enzymatic promiscuity applied to Biotransformations:a) Nath, A; Atkins, WM. A quantitative index of substrate promiscuity. *Biochemistry.* 2008, *47*, 157-166.b) Hult, K; Berglund, P. Enzyme promiscuity: mechanism and applications. *Trends Biotechnol.* 2007, *25*, 231-238.c) Carboni-Oerlemans, C; Domínguez de María, P; Tuin, B; Bargeman, G; van der Meer, A; van Gemert, RW. Hydrolase-catalysed synthesis of peroxycarboxylic acids: Biocatalytic promiscuity for practical applications. *J. Biotechnol.* 2006, *126*, 140-151. d) Kazlauskas, RJ. Enhancing catalytic promiscuity for biocatalysis. *Curr. Opin. Chem. Biol.* 2005, *9*, 195-201. e) Bornscheuer, UT; Kazlauskas, RJ. Catalytic promiscuity in biocatalysis: using old enzymes to form new bonds and follow new pathways. *Angew. Chem. Int. Ed.* 2004, *43*, 6032-6040. f) Copley, SD. Enzymes with extra talents: moonlighting functions and catalytic promiscuity. *Curr. Opin. Chem. Biol.* 2003, *7*, 265-272. g) O'Brien, PJ; Herschlag, D. Catalytic promiscuity and the evolution of new enzymatic activities. *Chem. Biol.* 1999, *6*, 91-105.

[5]   Recent reviews on Metagenomics and Biotransformations: a) Gabor, E; Liebeton, K; Niehaus, F; Eck, J; Lorentz, P. Updating the metagenomics toolbox. *Biotech. J.* 2007, *2*, 201-206. b) Lorentz, P; Eck, J. Metagenomics and industrial applications. *Nature Rev. Microbiol.* 2005, *3*, 510-516. c) Ferrer, M; Martínez-Abarca, F; Golyshin, PN. Mining genomes and 'metagenomes' for novel catalysts *Curr. Opin. Biotechnol.* 2005, *16*, 588-593.

[6]   Some recent relevant examples for the development of tailor-made biocatalysts: a) Bartsch, S; Kourist, R; Bornscheuer, UT. Complete inversion of enantioselectivity towards acetylated tertiary alcohols by a double mutant of a *Bacillus subtilis* esterase. *Angew. Chem. Int. Ed.* 2008, *47*, 1508-1511. b) Kourist, R; Bartsch, S; Bornscheuer, UT. Highly enantioselective synthesis of arylaliphatic tertiary alcohols using mutants of an esterase from *Bacillus subtilis*. *Adv. Synth. Catal.* 2007, *349*, 1393-1398. c) Reetz, MT; Carballeira, JD. Iterative saturation mutagenesis (ISM) for rapid directed evolution of functional enzymes. *Nature Prot.* 2007, *2*, 891-903. d) Reetz, MT. Evolution in the test-tube as a means to create selective biocatalysts. *Chimia.* 2007, *61*, 100-103. e) Reetz, MT; Carballeira, JD; Peyralans, J; Höbenreich, H; Maichele, A; Vogel, A. Expanding the substrate scope of enzymes: combining mutations obtained by CASTing. *Chem. Eur. J.* 2006, *12*, 6031-6038. f) Reetz, MT; Bocola, M; Carballeira, JD; Zha, D; Vogel, A. Expanding the range of substrate acceptance of enzymes: Combinatorial active-site saturation test. *Angew. Chem. Int. Ed.* 2005, *44*, 4192-4196. g) Reetz, MT. Asymmetric catalysis special feature Part II: Controlling the enantioselectivity of enzymes by directed evolution: Practical and theoretical ramifications. *Proc. Natl. Acad. Sci.* 2004, *101*, 5716-5722.

[7]   Reviews and books dealing with the practical use of hydrolases: a) Gotor-Fernández, V; Brieva, R; Gotor, V. Lipases: Useful biocatalysts for the preparation of pharmaceuticals. *J. Mol. Cat. B: Enzym.* 2006, *40*, 111-120. b) Hasan, F; Shah, AA; Hameed, A. Industrial applications of microbial lipases. *Enz. Microb. Technol.* 2006, *39*, 235-251. c) Bornscheuer, UT; Kazlauskas, RJ. *Hydrolases in Organic Synthesis*, Wiley-VCH, Weinheim. 2006. d) Thum, O. Enzymatic production of care specialties based on fatty acid esters.*Tenside. Surf. Det.* 2004, *41*, 287-293. e) Schmid, RD; Verger, R. Lipases: Interfacial enzymes with attractive applications. *Angew. Chem. Int. Ed.* 1998, *37*, 1608-1633.

[8]   See, for instance: a) Carlqvist, P; Svedendahl, M; Branneby, C; Hult, K; Brinck, T; Berglund, P. Exploring the active-site of a rationally redesigned lipase for catalysis of Michael-type additions. *ChemBioChem.* 2005, *6*, 331-336. b) Torre, O; Alfonso, I; Gotor, V. Lipase catalysed Michael addition of secondary amines to acrylonitrile. *Chem. Comm.* 2004, 1724-1725. c) Branneby, C; Carlqvist, P; Magnusson, A; Hult, K; Brinck, T; Berglund, P. Carbon-carbon bonds by hydrolytic enzymes. *J. Am. Chem. Soc.* 2003, *125*, 874-875.

[9]   The hydrolase-based production of chiral tertiary alcohols is currently a matter of active research, by making use of metagenomics and directed evolution strategies. For further reading in the field, see (and references disclosed therein): Kourist, R; Domínguez de María, P; Bornscheuer, UT. Enzymatic synthesis of optically active

tertiary alcohols: Expanding the biocatalysis toolbox. *ChemBioChem*. 2008, *9*, 491-498.

[10]   Reviews: a) Cozzi, PG; Hillgraf, R; Zimmerman, N. Enantioselective catalytic formation of quaternary stereogenic centers. *Eur. J. Org. Chem*. 2007, *36*, 5969-5994. b) García, C; Martín, VS. Asymmetric addition to ketones: enantioselective formation of tertiary alcohols. *Curr. Org. Chem*. 2006, *10*, 1849-1889. c) Christoffers, J; Baro, A. Stereoselective construction of quaternary stereocenters. *Adv. Synth. Catal*. 2005, *347*, 1473-1482. d) Ramón, DJ; Yus, M. Enantioselective synthesis of oxygen-, nitrogen- and halogen-substituted quaternary carbon centers. *Curr. Org. Chem*. 2004, *8*, 149-183.

[11]   Some selected examples on the use of α-hydroxy ketones as building blocks: a) Tanaka, T; Kawase, M; Tani, S. α-Hydroxyketones as inhibitors of urease. *Bioorg. Med. Chem.* 2004, *12*, 501-505. b) Wallace, OB; Smith, DW; Deshpande, MS; Polson, C; Felsenstein, KM. Inhibitors of Aβ production: solid-phase synthesis and SAR of α-hydroxycarbonyl derivatives. *Bioorg. Med. Chem. Lett*. 2003, *13*, 1203-1206. c) Fang, QF; Han, Z; Grover, P; Kessler, D; Senanayake, CH; Wald, SA. Rapid access to enantiopure bupropion and its major metabolite by stereospecific nucleophilic substitution on an α-ketotriflate. *Tetrahedron: Asymmetry* 2000, *11*, 3659-3663.

[12]   a) Higgins, DA; Pomianek, ME; Kraml, CM; Taylor, RK; Semmelhack, MF; Bassler, BL. The major Vibrio cholerae autoinducer and its role in virulence factor production *Nature*. 2007, *450*, 883-886. b) Parsek, MR. Microbiology - Bilingual bacteria. *Nature*. 2007, *450*, 805-807. c) Francke, W; Dettner, K. Chemical signalling in beetles. *Top. Curr. Chem*. 2005, *240*, 85-166. d) Shi, X; Leal, WS; Meinwald, J. Assignment of absolute stereochemistry to an insect pheromone by chiral amplification. *Bioorg. Med. Chem*. 1996, *4*, 297-303.

[13]   Relevant articles and reviews: a) Adam, W; Lazarus, M; Saha-Möller, CR; Schreier, P. Biocatalytic synthesis of optically active alpha-oxyfunctionalized carbonyl compounds *Acc. Chem. Res*. 1999, *32*, 837-845. b) Davis, FA; Chen, BC. Asymmetric hydroxylation of enolates with n-sulfonyloxaziridines.*Chem. Rev*. 1992, *92*, 919-934.

[14]   a) Sih, CJ; Wu, SH. Resolution of enantiomers via biocatalysis. *Top. Stereochem.* 1989, *19*, 63-125. b) Chen, CS; Fujimoto, Y; Girdaukas, G; Sih, CJ. Quantitative analyses of biochemical kinetic resolutions of enantiomers. *J. Am. Chem. Soc*. 1982, 104, 7294-7299.

[15]   Kazlauskas, RJ; Weissfloch, ANE; Rappaport, AT; Cuccia, LA. A rule to predict which enantiomer of a secondary alcohol reacts faster in reactions catalyzed by cholesterol esterase, lipase from *Pseudomonas cepacia*, and lipase from *Candida-rugosa*. *J. Org. Chem*. 1991, *56*, 2656-2665.

[16]   a) Tanyeli, C; Akhmedov, I; Iyigun, C. The first enzymatic resolution of quaternary α-acetoxy α-substituted cyclic ketones. *Tetrahedron: Asymmetry*. 2006, *17*, 1125-1128. b) Tanyeli, C; Ozdemirhan, F; Iyigun, C. The first enzymatic resolution of quaternary α'-acetoxy α,β-unsaturated cyclohexenones and cyclopentenones. *Tetrahedron: Asymmetry*. 2005, *16*, 4050-4055.

[17]   a) Kajiro, H; Mitamura, S; Mori, A; Hiyama, T. Enantioselective synthesis of 2-hydroxy-1-indanone, a key precursor of enantiomerically pure 1-amino-2-indanol. *Tetrahedron: Asymmetry* 1998, *9*, 907-910. b) Adam, W; Díaz, MT; Fell, R; Saha-

Möller, C. Kinetic resolution of racemic α-hydroxy ketones by lipase-catalyzed irreversible transesterification. *Tetrahedron: Asymmetry* 1996, *7*, 2207-2210.

[18] Silva, DJ; Kahne, D; Kraml, CM. Chromomycin A3 as a blueprint for designed metal complexes. *J. Am. Chem. Soc.* 1994, *116*, 2641-2642.

[19] a) Demir, AS; Findik, H; Köse, E. A new and efficient chemoenzymatic route to both enantiomers of α'-acetoxy-α-methyl and γ-hydroxy-α-methyl cyclic enones. *Tetrahedron: Asymmetry* 2004, *15*, 777-781. b) Demir, AS; Sesenoglu, O. A new and efficient chemoenzymatic access to both enantiomers of 4-hydroxycyclopent-2-en-1-one. *Tetrahedron: Asymmetry* 2002, *13*, 667-670. c) Demir, AS; Sesenoglu, O. A new and efficient chemoenzymatic route to both enantiomers of 4-hydroxycyclohex-2-en-1-one. *Org. Lett.* 2002, *4*, 2021-2023. d) Tanyeli, C; Demir, AS; Dikici, E. New chiral synthon from the PLE catalyzed enantiomeric separation of 6-acetoxy-3-methylcyclohex-2-en-1-one. *Tetrahedron: Asymmetry* 1996, *7*, 2399-2402.

[20] Adam, W; Díaz, MT; Saha-Möller, C. Lipase-catalyzed kinetic resolution of α,β-unsaturated α'-acetoxy ketones. *Tetrahedron: Asymmetry* 1998, *9*, 791-796.

[21] a) Scheid, G; Ruitjer, E; Konarzycka-Bessler, M; Bornscheuer, UT; Wessjohann, LA. Synthesis and resolution of a key building block for epothilones: a comparison of asymmetric synthesis, chemical and enzymatic resolution. *Tetrahedron: Asymmetry* 2004, *15*, 2861-2869. b) Scheid, G; Kuit, W; Ruitjer, E; Orru, RV; Henke, E; Bornscheuer, UT; Wessjohann, LA. A new route to protected acyloins and their enzymatic resolution with lipases. *Eur. J. Org. Chem.* 2004, 1063-1074. c) Wessjohann, LA; Scheid, G; Bornscheuer, UT; Henke, E; Kuit, W; Orru, R. WO2002032844. 2002. For other approaches on hydrolases and epothilones, see: d) Shioji, K; Kawaoka, H; Miura, A; Okuma, K. Synthesis of C1-C6 fragment for epothilone a via lipase-catalyzed optical resolution. *Synth. Comm.* 2001, *31*, 3569-3575. e) Bornscheuer, UT; Altenbuchner, J; Meyer, HN. Directed evolution of an esterase for the stereoselective resolution of a key intermediate in the synthesis of epothilones. *Biotechnol. Bioeng.* 1998, *58*, 554-559.

[22] a) Hermann, M; Kietzmann, MU; Ivancic, M; Zenzmaier, C; Luiten, RG; Skrack, W; Wubbolts, M; Winkler, M; Birner-Grünberger, R; Pichler, H; Schwab, H. Alternative pig liver esterase (APLE) – Cloning, identification and functional expression in *Pichia pastoris* of a versatile new biocatalyst. *J. Biotechnol.* 2008, *133*, 301-310. b) Hummel, A; Brüsehaber, E; Böttcher, D; Doderer, K; Trauthwein, H; Bornscheuer, UT. nzymes of Pig-Liver esterase reveal striking differences in enantioselectivities. *Angew. Chem. Int. Ed.* 2007, *46*, 8492-8494. c) Böttcher, D; Brüsehaber, E; Doderer, K; Bornscheuer, UT. Functional expression of the γ-isoenzyme of pig liver carboxyl esterase in *Escherichia coli*. *Appl. Microb. Biotechnol.* 2007, *73*, 1282-1289.

[23] For a recent review on PLE as biocatalyst, see: Domínguez de María, P; García-Burgos, CA; Bargeman, G; van Gemert, RW. Pig Liver Esterase (PLE) as biocatalyst in organic synthesis: From nature to cloning and to practical applications. *Synthesis.* 2007, *10*, 1439-1452; See also reference [7c].

[24] Ohta, H; Ikemoto, M; Il., H; Okamoto, Y; Tsuchihaschi, G. Preparation of optically-active alpha-acetoxyacylophenones via enzyme mediated hydrolysis. *Chem. Lett.* 1986, 1169-1172.

[25]   a) Demir, AS; Hamamci, H; Sesenoglu, O; Aydogan, F; Capanoglu, D; Neslihanoglu, R. Simple chemoenzymatic access to enantiopure pharmacologically interesting (R)-2-hydroxypropiophenones. *Tetrahedron: Asymmetry* 2001, *12*, 1953-1956. b) Demir, AS; Hamamci, H; Tanyeli, C; Akhmedov, IM; Doganel, F. Synthesis and Rhizopus oryzae mediated enantioselective hydrolysis of α-acetoxy aryl alkyl ketones. *Tetrahedron: Asymmetry* 1998, *9*, 1673-1677.

[26]   Gala, D; DiBenedetto, DJ; Clark, JE; Murphy, BL; Schumacher, DP; Steinman, M. Preparations of antifungal Sch 42427/SM 9164: Preparative chromatographic resolution, and total asymmetric synthesis via enzymatic preparation of chiral α-hydroxy arylketones. *Tetrahedron Letters*. 1996, *37*, 611-614.

[27]   Celebi, N; Yildiz, N; Demir, A. Calimli, A. Enzymatic synthesis of benzoin in supercritical carbon dioxide. *J. Supercrit. Fluids*. 2007, *41*, 386-390.

[28]   a) Aoyagi, Y; Iijima, A; Williams, RM. Asymmetric synthesis of [2,3-13C2,15N]-4-Benzyloxy-5,6-diphenyl-2,3,5,6-tetrahydro-4H-oxazine-2-one via Lipase TL-mediated kinetic resolution of benzoin: General procedure for the synthesis of [2,3-13C2,15N]-L-Alanine. *J. Org. Chem.* 2001, *66*, 8010-8014. b) Aoyagi, Y; Agata, N; Shibata, N; Horiguchi, M; Williams, RM. Lipase TL®-mediated kinetic resolution of benzoin: facile synthesis of (1R,2S)-erythro-2-amino-1,2-diphenylethanol. *Tetrahedron Lett.* 2000, *41*, 10159-10162.

[29]   Hoyos, P; Fernández, M; Sinisterra, JV; Alcántara, AR. Dynamic kinetic resolution of benzoins by lipase-metal combo catalysis. *J. Org. Chem.* 2006, *71*, 7632-7637.

[30]   a) Martín-Matute, B; Bäckvall, JE. Ruthenium- and enzyme-catalyzed dynamic kinetic asymmetric transformation of 1,4-diols: Synthesis of γ-Hydroxy ketones. *J. Org. Chem.* 2004, *69*, 9191-9195. b) Nyhlén, J; Martín-Matute, B; Sandström, A; Bocola, M; Bäckvall, JE. Influence of δ-functional groups on the enantiorecognition of secondary alcohols by *Candida antarctica* lipase B. *ChemBioChem* 2008, *9*, 1968-1974.

[31]   Hoyos, P. PhD Thesis, Complutense University of Madrid, 2008.

[32]   Krausser, M; Hummel, W; Gröger, H. Enantioselective one-pot two-step synthesis of hydrophobic allylic alcohols in aqueous medium through the combination of a Wittig reaction and an enzymatic ketone reduction. *Eur. J. Org. Chem.* 2007, 5175-5179.

[33]   Stürmer, R. Enzymes and transition metal complexes in tandem — a new concept for dynamic kinetic resolution. *Angew. Chem. Int. Ed. Engl.* 1997, *36*, 1173-1174.

[34]   Reviews on DKR's: a) Pellissier, H. Recent developments in dynamic kinetic resolution. *Tetrahedron* 2008, *64*, 1563-1601. b) Martín-Matute, B; Bäckvall, JE. Dynamic kinetic resolution catalyzed by enzymes and metals. *Curr. Op. Chem. Biol.* 2007, *11*, 226-232. c) Pàmies, O; Bäckvall, JE. Combination of enzymes and metal catalysts. A powerful approach in asymmetric catalysis. *Chem. Rev.* 2003, *103*, 3247-3262.

[35]   a) Taniguchi, T; Ogasawara, K. Lipase–triethylamine-mediated dynamic transesterification of a tricyclic acyloin having a latent meso-structure: a new route to optically pure oxodicyclopentadiene. *Chem. Comm.* 1997, 1399-1400. b) Taniguchi, T; Kanada, RM; Ogasawara, K. Lipase-mediated kinetic resolution of tricyclic acyloins, endo-3-hydroxytricyclo[4.2.1.02,5]non-7-en-4-one   and   endo-3-hydroxytricyclo [4.2.2.02,5]dec-7-en-4-one. *Tetrahedron: Asymmetry* 1997, *8*, 2773-2780. c)

Matsamoto, K; Suzuki, N; Ohta, H. Preparation of optically active α-hydroxy ketone derivatives by Enzyme-Mediated Hydrolysis of Enol Esters. *Tetrahedron Lett.* 1990, *31*, 7159-7162.

[36] Ödman, P; Wessjohann, LA; Bornscheuer, UT. Chemoenzymatic dynamic kinetic resolution of acyloins. *J. Org. Chem.* 2005, *70*, 9551-9555.

[37] Bògar, K; Hoyos, P; Alcántara, AR; Bäckvall, JE. Chemoenzymatic dynamic kinetic resolution of allylic Alcohols: A highly enantioselective route to acyloin acetates. *Org. Lett.* 2007, *9*, 3401-3404.

[38] Demir, AS; Hamamci, H; Sesenoglu, O; Neslihanoglu, R; Asikoglu, B; Capanoglu, D. Fungal deracemization of benzoin. *Tetrahedron Lett.* 2002, *43*, 6447-6449.

[39] Pámies, O; Bäckvall, JE. Combined metal catalysis and biocatalysis for an efficient deracemization process. *Curr. Opin. Biotechnol.* 2003, *14*, 407-413.

[40] Hoyos, P; Buthe, A; Ansorge-Schumacher, MB; Sinisterra, JV; Alcántara, AR. Highly efficient one pot dynamic kinetic resolution of benzoins with entrapped *Pseudomonas stutzeri* lipase. *J. Mol. Cat. B: Enzym.* 2008, *52-53*, 133-139.

[41] Kim, MJ; Choi, YK; Kim, S; Kim, D; Han, K; Ko, SB; Park, J. Highly enantioselective dynamic kinetic resolution of 1,2-diarylethanols by a lipase-ruthenium couple. *Org. Lett.* 2008, *10*, 1295-1298.

[42] a) Nestl, BM; Bodlenner, A; Stuermer, R; Hauer, B; Kroutil, W; Faber, K. Biocatalytic racemization of synthetically important functionalized α-hydroxyketones using microbial cells. *Tetrahedron: Asymmetry* 2007, *18*, 1465-1474. b) Nestl, BM; Voss, CV; Bodlenner, A; Ellmer-Schaumberger, U; Kroutil, W; Faber, K. Biocatalytic racemization of sec -alcohols and α-hydroxyketones using lyophilized microbial cells. *Appl. Microbiol. Biotechnol.* 2007, *76*, 1001-1008.

In: Biotechnology: Research, Technology and Applications   ISBN 978-1-60456-901-8
Editor: Felix W. Richter   © 2008 Nova Science Publishers, Inc.

Chapter 5

Stability and Kinetic Studies on Recombinant Human Pyroglutamyl Peptidase I and a Single-Site Variant, Y147F

Karima Mtawae, Brendan O'Connor and Ciarán Ó'Fágáin[*]

School of Biotechnology and National Centre for Sensors Research,
Dublin City University, Dublin 9, Ireland

Abstract

Human brain pyroglutamyl peptidase (PAPI; EC 3.4.19.3) is an omega exopeptidase which cleaves pyroglutamic acid from the N-terminus of bioactive peptides and proteins. It plays an important role in the processing and degradation of regulatory peptides such as thyrotropin releasing hormone (TRII) and luteinizing hormone releasing hormone (LHRH). To gain further insights into its performance *in vivo* and suggest possible applications, such as peptide processing or sequencing, this study focuses on the *in vitro* stability properties and Michaelis-Menten kinetics of the recombinant wild type enzyme and a single-site mutant, Tyr147→Phe (Y147F).

At 60°C in 50mM potassium phosphate buffer, pH 8.0, recombinant PAPI underwent a first-order decay constant with a k value of 0.046 ± 0.002 min^{-1} and a half-life of 15 min. PAPI was unstable to most of the water-miscible solvents tested (dimethyl sulphoxide, methanol, acetone, tetrahydrofuran, acetonitrile, dimethyl formamide and ethanol) even at low v/v concentrations. Methanol and dimethyl sulphoxide were the least injurious to PAPI activity: 56% and 50% residual PAPI activity remained at 10% v/v methanol and DMSO, respectively. Chemical modification with dimethyl suberimidate gave only 20% recovery of initial activity and did not stabilize the enzyme. Polyol and other stabilizing additives were investigated: activity and stability increased with xylitol but not with trehalose, glycerol or ammonium sulphate. PAPI displayed Michaelis-Menten kinetics with the fluorescent substrate pyroglutamyl 7-

[*] Author for correspondence. Tel + 353 1 700 5288; Fax + 353 1 700 5412; Email ciaran.fagan@dcu.ie

aminomethylcoumarin at pH 8.0: values for K_m and k_{cat} were 0.132 ± 0.024 mM and $2.68 \pm 0.11 \times 10^{-5}$ s^{-1} respectively.

Mutant Y147F was notably more thermostable, despite differing from wild type only by the absence of a hydroxyl. At 70°C, the Y147F first-order k value was 0.0028 ± 0.001 min^{-1} (half-life 25 min) compared with 0.079 ± 0.003 min^{-1} (half life 9 min) for wild type (the higher temperature was required to achieve timely inactivation of Y147F). Values of K_m and k_{cat} for Y147F (0.115 ± 0.019 mM and $2.45 \pm 0.05 \times 10^{-5}$ s^{-1} respectively) closely resembled those of wild type.

It appears that the *in vitro* stability of wild type PAPI might limit its potential applicability in peptide processing or other fields. Additives and chemical modification seem to have limited scope for enhancing its stability but the generation of stabilized mutant variants, or the use of a thermophilic counterpart, should be explored further.

Abbreviations

ACN	acetonitrile;
ALT	Alanine aminotransferase;
AMC	7-Amino-4-methyl-coumarin;
BCA	Bicinchoninic acid;
BSA	Bovine serum albumin;
C_{50}	Solvent concentration where 50% of enzyme activity in aqueous buffer remains;
DMF	Dimethylformamide;
DMSO	Dimethylsulphoxide;
DMS	Dimethyl suberimidate;
DTT	Dithiothreitol;
EDAC	N-(3-Dimethylaminopropyl)-N'-ethyl carbodiimide;
EDTA	Ethylene diamine tetra-acetic acid;
GRH	Gonadotropin releasing hormone;
HRP	Horseradish peroxidase;
HPLC	High performance liquid chromatography;
IPTG	Isopropyl-beta-D-thiogalactopyranoside;
k	First-order decay constant;
LB	Luria-Bertani medium;
LHRH	Luteinizing hormone-releasing hormone;
MeOH	Methanol;
MEROPS	the peptidase database at http://merops.sanger.ac.uk/;
N	Native state of protein;
β-NA	β-Naphthylamine;
NEM	N-Ethylmaleimide;
PAPI	Pyroglutamyl peptidase I (E.C.3.4.19.3);
PAPII	Pyroglutamyl peptidase II (E.C.3.4.19.6);
pGDMK	pGlu diazomethyl ketone;
pGCK	pGlu chloromethyl ketone;

pGlu	Pyroglutamic acid;
pGlu-AMC	Pyroglutamyl-7-amino-4-methyl coumarin;
pNa	p-Nitroanilide;
RIA	Radioimmunoassay;
SDS	sodium dedecyl sulfate;
T_{50}	Half-inactivation temperature;
$t_{1/2}$	Half-life $(0.693/k)$;
THF	Tetrahydrofuran;
TRH	Thyrotropin releasing hormone;
Y147F	PAPI mutant Tyr147→Phe;
Z	Benzyloxycarbonyl.

Introduction

Pyroglutamyl peptides, which often comprise up to 40 amino acids, possess an N-terminal pyroglutamic acid (pGlu) residue that influences their biological properties. Many reports describe the enzymatic formation of pGlu from glutamic acid and glutaminyl peptides (Orlowski and Meister 1971). Cyclisation of the N-terminal glutamic acid to pGlu affords these peptides a longer half-life than others of similar size (De Gandarias *et al*, 2000). Many biologically active peptides (thyrotropin-releasing hormone (TRH), luteinizing hormone-releasing hormone (LHRH), neurotensin, etc.) and proteins have pGlu residues at their N-termini. Only the pyroglutamyl aminopeptidases can cleave these amino terminal pGlu. Three types of pyroglutamyl peptidase have been found in a wide variety of bacteria and in many plant, animal, and human tissues. These are pyroglutamyl peptidase I (E.C 3.4.19.3; PAP I), pyroglutamyl peptidase II (E.C 3.4.19.6; PAP II) and serum thyroliberinase (E.C 3.4.19.6) (Cummins and O'Connor, 1998; Robert-Baudouy and Thierry, 1998). In the brain, the cytoplasmic cysteine peptidase PAP I (EC 3.4.19.3, the subject of this paper) hydrolyses pGlu-X bonds (where X is any amino acid except proline) and is quite distinct from the membrane-bound, tetrameric, exo-acting metallo peptidase PAP type II (EC 3.4.19.6) that degrades thyrotropin releasing hormone in a highly specific manner (O'Connor and O'Cuinn, 1985).

Classification and Occurrence of PAPI

Pyroglutamyl peptidase I (PAPI, EC 3.4.19.3) is a cysteine omega-exopeptidase that specifically removes the amino-terminal pyroglutamyl residue from oligopeptides and proteins. It has been used in protein sequencing to unblock proteins and polypeptides with pGlu amino-termini prior to Edman degradation and has had several different names, including pyrrolidonyl peptidase, pyrrolidone carboxyl peptidase, 5-oxoprolyl-peptidase, pyrase and pyroglutamyl aminopeptidase (Cummins and O'Connor, 1998).

PAPI has been found in various plant, animal, and human tissues, but none has yet been found in the *Saccharomyces cerevisiae* genome nor in any fungus. It occurs as a soluble, intracellular cytosolic cysteine peptidase with broad specificity for pGlu-substrates in many mammalian tissues including human cerebral cortex, kidney and skeletal muscle, rat, bovine and guinea pig brain, and in various rat organs including liver (Cummins and O'Connor, 1998). In vertebrates, the liver and kidney show relatively high PAPI activities compared with other tissues. PAPI was localized in the renal proximal tubules, while an immunohistochemical localization study of PAPI demonstrated intracellular distribution in the pituitary. Known mammalian PAPI sequences include those from human (*Homo sapiens*; Dando et al., 2003) mouse (*Mus musculus*; Dando et al., 2003) and rat (*Rattus norvegicus*; Abe et al., 2003).

PAPI activity has also been noted in non-mammalian sources such as bird, fish and amphibian tissues. Among plants, it occurs in parsley, carrot, bean, oats, wheat, cauliflower, and potato. Tissues tested included leaves, seeds, sprouts and roots (Szewczuk and Kwiatkowska, 1970).

The purification and study of PAPI has been reported from several prokaryotic species, including species of *Bacillus* (Yoshimoto *et al.*, 1993; Awadé *et al.*, 1992a) *Pseudomonas fluorescens* (Gonzalès and Robert-Baudouy, 1994), species of *Pyrococcus* (Sokabe *et al.*, 2002; Tsunasawa *et al.*, 1998), *Streptococcus pyogenes* (Awadé *et al.*, 1992b; Cleuziat *et al.*, 1992), *Mycobacterium bovis* (Kim *et al.*, 2001), *Staphylococcus aureus* (Patti *et al.*, 1995), and *Thermococcus litoralis* (Singleton *et al.*, 2000). The sequences of bacterial and archaeal PAPI enzymes place them in peptidase family C15 of the MEROPS classification (Barrett and Rawlings, 2001).

Substrates, Catalysis and Inhibitors

PAPI hydrolytically removes L-pGlu from L-pGlu-L-X, where X is any amino acid (except proline), a peptide or an arylamide such as AMC. PAPI has a broad pyroglutamyl substrate specificity, with its activity being influenced by the amino acid directly after the pGlu residue. Synthetic compounds and dipeptides such as pGlu-AMC, pGlu-Ala, pGlu-pNa, pGlu-βNA and pGlu-Val are also substrates for the enzyme but pGlu-Pro bonds are normally not hydrolysed by mammalian PAPI (Cummins and O'Connor, 1996; Browne and O'Cuinn, 1983). Abe et al. (2004a,b) have shown that PAPI can tolerate some single atom substitutions on the pGlu ring. Due to the catalytic importance of its cysteine thiol group, PAPI has an absolute requirement for a thiol-reducing agent such as DTT and loses activity when treated with micromolar concentrations of a standard thiol inhibitor such as N-ethylmaleimide (NEM) or iodoacetate (Cummins and O'Connor, 1996; Tsunasawa et al, 1998; Singleton et al, 2000; Singleton and Littlechild, 2001; Dando et al, 2003). The additional involvement of a histidine and of acidic residues in catalysis was indicated by chemical modification studies with diethylpyrocarbonate and N-(3-Dimethylaminopropyl)-N'-ethyl carbodiimide (EDAC) respectively (Le Saux et al, 1996). Several specific inhibitors have been synthesized to elucidate the biological significance of the enzyme (Fujiwara et al, 1981; Friedman et al, 1985), including pGlu chloromethyl ketone (pGCK), Z-pGlu chloromethyl ketone (Z-pGCK), and Z-pGlu diazomethyl ketone (Z-pGDMK). Charli et al. (1987) reported that pGDMK did not enhance brain TRH levels in vivo or in vitro, but Faivre-Bauman et al. (1986) observed increased levels of TRH when primary hypothalamic cell cultures were treated with Z-Gly-Pro-CHN$_2$, another PAPI inhibitor. Fujiwara et al. (1981) noted a decrease in enzyme activity upon addition of pGCK in the absence of DTT. A synthetic aldehyde analog of pGlu, 5-oxoprolinal, was a potent competitive inhibitor of PAPI (Friedman et al, 1985). Several other inhibitors of mammalian PAPI exist. Yamada and Mori (1990) noted the inhibitory effects of 1,10-phenanthroline, excess DTT and EDTA while Cummins and O'Connor (1996) reported 28% inhibition of bovine brain PAPI by 1mM 1,10-phenanthroline. Two compounds isolated from the genus *Streptomyces*, benarthin and pyrizinostatin, acted as inhibitors of PAPI (Aoyagi et al, 1992a,b) as did an oligosaccharide gum from *Hakea gibbosa* (Alur et al, 2001). Mantle et al. (1991) noted inhibitory effects of amastatin, arphamenine, chymostatin, elastinal and leupeptin while benzamidine inhibited the bacterial PAPI enzyme (Awadé et al, 1994). Several workers have used the selective reversible inhibitor 2-pyrrolidone to stabilize type I pyroglutamyl peptidases during purification and storage (Mudge and Fellows, 1973).

Characterization

The molecular weight of PAPI monomer has mostly been reported as around 24 kDa by gel filtration and denaturing gel electrophoresis. Optimal activity lies in the pH range 6.0 to 9.5 and, where reported, the isoelectric point (pI) is around 5.0. Optimum temperatures for activity of mesophilic prokaryotic PAPI have mostly been reported as ranging from 30 - 45°C (37°C has been widely used as the standard reaction temperature for eukaryotic PAPI) but two thermophilic enzymes from *Thermococcus litoralis* PAPI and *Pyrococcus furiosus* PAPI exhibit optimum activity at 70°C and 90°C, respectively. (Singleton and Littlechild, 2001; Ogasahara *et al*, 2001). Purification of PAPI from several strains of *Bacillus* has been well documented. *Bacillus amyloliquefaciens* PAPI has been particularly well studied, including cloning, sequencing and expression of its gene in *Escherichia coli*. The enzyme comprises 215 amino acid residues, has a homodimer structure with a deduced subunit molecular mass of 23.3kDa and a pH optimum of 6.5. (Yoshimoto *et al*, 1993; Ito et al., 2001).

Physiological Role(s)

The physiological role of PAPI currently remains unclear. Its cytosolic location excludes a significant role in extracellular peptide degradation (Charli et al., 1987; Abe et al., 2004a,b). O'Cuinn et al. (1990) suggested that PAPI may represent a mechanism for returning pGlu terminating neuropeptides, released from damaged or ageing vesicles, back to the cellular amino acid pool. PAPI may participate in the intracellular catabolism of peptides to free amino acids which are then re-incorporated into biosynthetic pathways. Thus, PAPI may function in regulating the cellular pool of free pGlu. In mammals, it may be involved in the inactivation of biologically active peptides with an N-terminal pGlu, such as TRH, LHRH, bombesin, neurotensin, gastrin, fibrinopeptides and anorexigenic peptide (Alba et al, 1995). PAPI may also contribute to the inactivation of neuropeptides that are produced in excess. Faivre-Bauman *et al.* (1986) showed that addition of specific PAPI inhibitors to cultured TRH-synthesising hypothalamic cells significantly increases TRH content and release from cells under both basal and K$^+$-stimulated conditions. PAPI is involved in the hydrolysis of some xenobiotic compounds with an L-pGlu or L-pGlu-related structures (Abe *et al*, 2003; 2004b) and may also be involved in detoxification of pGlu-peptides, since high levels of such peptides would abnormally acidify the cell cytoplasm. The high level of Pyroglutamyl Peptidase in the human brain cortex agrees with the observation that inactivation of TRH is higher in the human brain cortex than in other regions (Griffiths, 1985).

Roles in Health and Disease

A correlation between PAPI activity and TRH levels in mammalian brain has been observed in many studies (De Gandarias *et al*, 1998; 2000). A decrease in PAPI activity coincides with increasing levels of TRH as brain development progresses, indicating that PAPI plays a part in the normal development of mammalian brain. The wide distribution of TRH throughout the central nervous system and the findings of various biochemical, pharmacological and behavioural studies strongly imply that TRH may act as a neuromodulator or neurotransmitter in the extrahypothalamic brain. Also, during earlier stages of development, high PAPI activity is linked to elevated levels of cyclo(His-Pro) (Prasad *et al*, 1983). In addition to its better-known neurohormonal role, PAPI may also have an important part to play in memory and learning, in metabolism and the possible control of conditions such as Alzheimer's disease (Irazusta *et al*, 2002).

The effect of light intensity on the activity of PAPI in the functionally connected rat retina and hypothalamus was studied by Ramírez *et al*. (1991) and Sánchez *et al*. (1996). PAPI levels fluctuate periodically, coincident with environmental light and dark conditions, suggesting a possible function of PAPI within the human "body clock" (Sanchez *et al*. 1996). Substrates of PAPI (such as TRH or LHRH) may play a functional role in the retina, apart from their well-known role in the hypothalamus. The concentration of TRH in the hippocampus of elderly controls and AD patients was recorded by radioimmunoassay (RIA). He and Barrow (1999) reported that PAPI may be involved in the propensity of amyloid precursors to form insoluble plaques, resulting in Alzheimer-type diseases. Free pGlu is known to have pharmacological properties and these have been demonstrated in certain disease states (Lauffart *et al*., 1989; Mantle *et al*., 1990; 1991).

Active Site

The active sites of cysteine proteases typically have a catalytic triad, an oxyanion hole and a specificity pocket. The catalytic triad of *B. amyloquefaciens* PAPI comprises Cys144, His168 and Glu81 (Odagaki *et al*., 1999), as confirmed by site-directed mutagenesis of the appropriate amino acids. Previously, the two cysteine residues of *B. amyloquefaciens* PAPI, Cys68 and Cys144, had been mutated to Ser. Mutant Cys144→Ser had no detectable PAPI activity (Yoshimoto *et al*, 1993), while Cys68→Ser had wild type activity, implicating Cys144 as the active site thiol. Also, titration with 5,5'-dithio-bis-(2-nitrobenzoate) showed that Cys68 is located internally. His168 is also completely conserved, and thus PAPI was thought to be a cysteine protease with a Cys-His catalytic diad or Cys-His-Asp/Glu catalytic triad.

Le Saux *et al*. (1996) investigated several candidate active site residues of *Pseudomonas fluorescens* PAPI. Substitution of residues Cys144 and His166 by Ala and Ser, respectively, resulted in inactive enzymes. Proteins with changes of Glu-81 to Gln and Asp-94 to Asn were not detectable in crude extract and were probably unstable in bacteria. The results suggest that Cys-144 and His-166 constitute the nucleophilic and imidazole residues of the enzyme active site, while residue Glu-81, Asp-89, or Asp-94 might constitute the third part of the active site (see below). Tsunasawa *et al.* (1998) substituted Cys142 of *P. fluorescens* PAPI with Ser, resulting in inactive enzyme and showed, by sequence analysis, that the catalytic triad Cys142-His166-Glu79 corresponds to Cys144-His168-Glu81 of *B. amyloquefaciens* PAPI. The location of the Cys144 at the N-terminus of an α-helix could be important. Such a position could affect catalysis, as the helix dipole can depolarise the amide bond and enhance its reactivity.

PAPI does not have a well defined oxyanion hole; however, it is possible that a tetrahedral oxyanion could be produced by the contribution of Cys144 and Arg91 and their respective side chains. A hydrophobic region close to Cys144 provides a highly specific binding site for the pGlu of the enzyme's substrate. This pocket appears to have some conformational flexibility, hence allowing for maximum interaction with the substrate. PAPI appears to have only one pocket of specificity (Odagaki *et al.*, 1999). The catalytic residues Glu81, Cys144, and His166 of the *B. amyloliquefaciens* enzyme are conserved in the human sequence (Dando et al, 2003).

Structure

Native mammalian PAPI appears to be monomeric, but molecular mass values of 50–91 kDa, consistent with a dimer, have been reported for the bacterial enzyme. The recombinant *B. amyloliquefaciens* enzyme probably functions as a dimer (Yoshimoto *et al.*, 1993) but has been crystallised as a tetramer (Odagaki *et al.*, 1999) while crystalline PAPI from *Thermococcus litoralis* is also tetrameric (Singleton *et al.*, 1999a,b).

The first documented X-ray crystal structures of bacterial PAPI were at 1.6 Å for *Bacillus amyloliquefaciens* (Odagaki *et al.*, 1999), at 1.73 Å for *Pyrococcus furiosus* PAPI (Tanaka *et al.*, 2001), at 2.0 Å for *Thermococcus litoralis* PAPI (Singleton *et al.*, 1999a,b) and at 2.2 Å for *Pyrococcus horikoshii* PAPI (Sokabe *et al.*, 2002). The monoclinic crystal form of each one has four crystallographically independent copies of PAPI in the asymmetric unit, and comprises a tetramer of four identical subunits designated A-D. Each monomer of the tetramer makes contact with two other subunit monomers. The A-D interface of *Bacillus amyloliquefaciens* PAPI involves hydrophobic interactions and several ionic salt bridges between each monomer. The A-C interface does not have any ionic salt bridges and a thin layer of water mediates hydrogen bonds between the subunits (Odagaki *et al.*, 1999).

The polypeptide folds in an α/β globular domain with a hydrophobic core comprising a twisted β- sheet surrounded by five α-helices. This structure allows the function of most of the conserved residues in the PAPI family to be identified, and it seems that the catalytic triad comprises Cys144, His168 and Glu 81 (see above) situated inside the doughnut-shaped tetramer.

The *Thermococcus litoralis* PAPI tetramer has a central cavity of 6000 $Å^3$ (Singleton *et al.*, 1999a,b). The A-B interface has hydrophobic interactions and salt bridges involving

Arg81, Asp88, Asp101 and Arg119. The A-C interface is formed by an extended loop and a disulphide bridge exists between Cys190 of each monomer. This hydrophobic core may contribute towards the thermostability of *Thermococcus litoralis* PAPI. Other residues along this interface are generally hydrophobic.

The A-B interface of the *Pyrococcus horikoshii* PAPI tetramer also involves hydrophobic interactions but there are, in addition, inter-subunit ionic bonds while the A-C interface has hydrogen bonds that are entirely mediated by a thin layer of water; however, the enzyme was a dimer in solution (Sokabe *et al.*, 2002).

PAPI Wild Type and Mutant Y147F as Candidates for Study

Recombinant human PAPI may have use in protein sequencing and/or in processing of peptides, and possibly in enzymatic peptide synthesis for the attachment of N-terminal pGlu residues. A thorough understanding of its stability and catalytic properties would be needed for any such applications. This paper reports a study of recombinant human PAPI and its single-site mutant Y147F, previously cloned, characterized and over expressed in *E. coli* (Vaas, 2005), with regard to their stabilities at elevated temperatures and kinetics with the chromogenic substrate pGlu-AMC. In addition, the stability of the wild type recombinant to organic solvents, and the effects on it of additives and of a crosslinking reagent, were also investigated.

Materials and Methods

Fisher Scientific supplied glacial acetic acid, acetone, EDTA, ethanol (100% v/v), glycerol, HCl (37% v/v), NaCl, NaOH, Tris-(hydroxymethyl) aminomethane (Tris) and the HPLC grade solvents acetonitrile (ACN), dimethylformamide (DMF) and tetrahydrofuran (THF). BCA (Bicinchoninic acid) protein assay kit was from Pierce Chemicals while pyroglutamyl-7-amino-4-methyl coumarin (pGlu-AMC) was sourced from Bachem. Sigma-Aldrich supplied His-Select nickel affinity gel, 7-amino-4-methylcoumarin (AMC), bovine serum albumin (BSA), Bradford reagent, dimethylsulphoxide (DMSO, HPLC grade), D,L-dithothreitol (DTT), ampicillin, IPTG, imidazole, dimethyl suberimidate (DMS), N-ethylmaleimide (NEM) and any other chemicals mentioned, which were reagent grade unless otherwise noted.

We used *E. coli* XL-10 Gold cells to express PAPI activity. Yeast extract and tryptone for the LB growth medium were from Oxoid UK. The cells contained the 5235 bp plasmid pRV5 encoding *Homo sapiens* PAPI (GenBank accession number AJ278828) as a 680 bp *rHsa-pap1*-(His)$_6$ fusion (via a Gly-Ser linker) under control of the P*tac* promoter (Vaas, 2005). This plasmid is very similar to pZK3, encoding bovine PAPI, described by Kilbane et al. (2007). The differences are located within the 680 bp segment, bounded by *Eco*R1 and *Hind*III sites, that codes for the fusion protein: an *Eco*RV restriction site at 406 bp replaces the *Eag*I site at 553 bp in pZK3 and the corresponding human amino acid residues (in brackets) are present at positions 81 (Thr), 205 (Tyr) and 208 (Lys). Commercial DNA sequencing (MWG Biotech, Germany) verified the exact sequence of pRV5 and of the single-site PAPI mutant, Y147F.

Production and Purification of Recombinant PAPI

LB growth broth (100ml, containing ampicillin to $100\mu g.ml^{-1}$) was aseptically inoculated with 3ml of overnight culture of *E.coli* XL-10 Gold harboring pRV5 and incubated at 37°C. Once the cells reached exponential phase (A_{600} 0.3-0.5), the inducer IPTG was added to 0.05 mM. Recombinant protein production continued for 5 h or overnight. Cells were then centrifuged ($3200 \times g$), the supernatant decanted, and the cell pellet labelled and stored at 4°C until required. The pellet was re-suspended in 10ml of buffer (50mM potassium phosphate, pH 8.0). The sample was then sonicated on ice (2.5 pulses.s^{-1}, 220 W, amplitude 40, 15 min), centrifuged ($3200 \times g$) to remove cell debris and the clear protein lysate supernatant was retained. Ni^{2+} resin (1ml) was added to the protein lysate, which was mixed with shaking for 1 h then poured into a column. Potassium phosphate buffer (50mM, pH 8.0, 10ml) was then run through the column. Next, wash buffer (50mM potassium phosphate, pH 8.0, containing 20mM imidazole, 20ml) was applied. Finally, elution buffer (50mM potassium phosphate buffer, pH 8.0, containing 200mM imidazole, 20ml) was run through the column to release PAPI (wild type or mutant Y147F, as appropriate). PAPI-containing fractions were combined, mixed with glycerol (40% v/v final concentration) and stored at 4°C until required. Purified PAPI typically had a specific activity of 0.34 $U.mg^{-1}$ (where 1 U = 1 nanomol $AMC.min^{-1}.ml^{-1}$) and gave a single 23-24 kDa band on denaturing gel electrophoresis (Laemmli, 1970). Specific activity of mutant Y147F was 0.83 $U.mg^{-1}$. Protein concentrations were measured by either the standard BCA protein assay (Smith et al., 1985) or the Bradford method (Bradford, 1976).

Quantitative Fluorimetric Measurement of PAPI Activity

The PAPI activity assay was according to Fujiwara and Tsuru (1978) as modified by Browne and Ó'Cuinn (1983). Each PAPI sample ($25\mu L$, in triplicate) was placed into separate wells of a 96-well fluorometer microplate and $100\mu L$ of the assay solution (50 mM potassium phosphate buffer, pH 8.0, containing final concentrations of 10 mM DTT, 2 mM EDTA and 0.5 mM pGlu-AMC) was added to each well. (Residual concentration of DMSO (required to prepare the 10 mM pGlu-AMC stock solution) in the assay mix did not exceed 5% v/v). The microplate was incubated at 37°C for 30 min; then, 1.5M acetic acid ($100\mu L$) was added to each well to arrest enzyme activity. Liberated AMC was measured using a Perkin Elmer LS-50 fluorescence spectrometer with excitation at 370nm and emission read at 440nm. Fluorescence readings were converted to nanomol of AMC released per minute using an AMC standard curve prepared under identical conditions. In addition to buffer, standard curves were determined in the presence of 10% (v/v) DMF, culture medium and of imidazole and also using crude and purified PAPI suspensions. Experimental values were determined using the most appropriate standard curve. The assay was shown to be linear with respect to time and amount of added PAPI ($r^2 = 0.99$ in both cases).

Kinetic Analysis

pGlu-AMC (10mM in DMSO) was diluted to 0.5mM with 50mM potassium phosphate buffer, pH 8.0. This solution was further diluted in buffer to give a range of pGlu-AMC concentrations and purified PAPI was assayed at each substrate concentration. The Michaelis-Menten constant (K_m) and maximum velocity (V_{max}) values were determined using the Enzfitter programme (Biosoft, Cambridge, UK). A similar procedure was followed for mutant Y147F.

Active site titration was based on Turk et al. (1993) but used N-Ethylmaleimide (NEM) instead of their E-64. A range of NEM concentrations (0-2.5µM, prepared by dilution of a 1mM stock solution with ultra-pure water) were used to titrate activated PAPI. At each point, 25µl of purified PAPI (diluted appropriately with 50mM potassium phosphate buffer, pH 8.0) was mixed with an equal volume of NEM solution, brought to a total volume of 100 µl with 50mM potassium phosphate buffer, pH 8.0, and incubated at 37°C for 15 min. Residual activity was then determined in triplicate. Fluorescence intensity was plotted versus NEM concentration and the operational molarity of PAPI determined. A similar procedure was followed for mutant Y147F.

Thermal Stability

Samples of purified PAPI (0.45 mg.ml^{-1} stock diluted appropriately in 50mM potassium phosphate buffer pH 8.0 to yield a suitable range of fluorescence units in the activity assay) were incubated for 10 min at temperatures between 30°-80°C, then placed on ice prior to re-warming and assay of the remaining PAPI activity in triplicate at 37°C. Buffer-only blanks were also prepared. A plot of % residual activity against temperature (°C) was constructed to give a thermal profile and the T_{50} (half-inactivation temperature) determined by inspection.

PAPI samples, prepared similarly, were placed in a waterbath held at 60°C. Aliquots were removed at intervals, cooled rapidly on ice and the remaining PAPI activity assayed as above following re-warming. Residual activity (%, calculated from the sample's initial activity) was plotted versus time. Data were fitted to a first-order exponential decay using Enzfitter (Biosoft, Cambridge, UK) to estimate the rate constant (k) and, hence, the half-life ($t_{1/2}$; $0.693/k$). PAPI and mutant Y174F were tested similarly at 70°C in buffer containing 0.12 mg.ml^{-1} BSA to ensure a uniform protein concentration.

Tolerance of Organic Solvents

PAPI samples, prepared as above, were incubated for 1 h at room temperature with various solvents (ACN, DMF, DMSO, THF, methanol, ethanol) ranging in concentration from 0-90% (v/v) in 50mM potassium phosphate pH 8.0 in a final volume of 1ml. Enzyme-free blanks were also set up for each solvent at each v/v concentration. The samples were then assayed in triplicate under normal conditions and a plot of % remaining activity versus solvent concentration (% v/v) was constructed.

Chemical Modification

The protocol was based on de Renobales and Welch (1980). Purified PAPI (0.45mg.ml^{-1}) was diluted with 50 mM potassium phosphate, pH 8.0 (3ml). A stock solution of dimethyl suberimidate (DMS; 2.5 mg.ml^{-1}) was similarly diluted and was mixed with with an equal volume of PAPI (final concentrations: DMS 25 µg.ml^{-1}; PAPI 5 µg.ml^{-1}). PAPI-DMS and a control were incubated at room temperature for 30 min. A thermal profile was then determined for each and plots of % residual activity versus temperatures were prepared.

Effects of Additives

Ammonium sulfate, trehalose and xylitol were each added to 0.5 M final concentration to separate samples of PAPI (5 µg.ml^{-1} final concentration). Thermal profiles were determined and plots of % residual activity versus temperature prepared in each case. The effects of 10% v/v and 50% v/v glycerol were determined similarly.

Results and Discussion

Thermal Stability

The temperature profile (Figure 1) showed that PAPI activity declined above 45°C. The half-inactivation temperature (T$_{50}$, value where observed activity was 50% of maximal) was estimated by inspection to be 60°C ± 1°C.

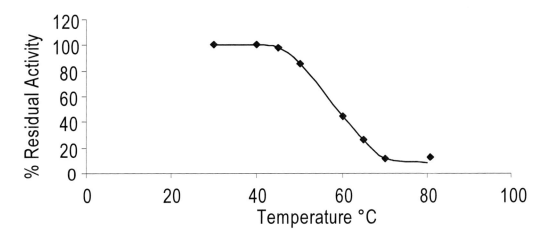

Figure 1. A temperature profile of PAPI. Plot of % residual activity versus temperature (°C), following 10 min incubations at each temperature. Activity is represented as a percentage of the 30°C value.

Data of % residual activity versus time at a uniform 60°C fitted well to a first order exponential decay to yield a k value of 0.046 ± 0.002 min^{-1} and, hence, a half-life ($t_{1/2}$) of 15 min. Human PAPI expressed in Sf9 insect cells was stable at temperatures up to 40°C for 4.5 h, but activity was almost completely lost within 30 min at 60°C, in line with the present results. Maximal activity was seen at 50°C in 10 min assays (Dando *et al.*, 2003), a somewhat higher temperature than our value of 45°C; the different values may arise from the use of different buffer compositions or protein concentrations. Many glycosylated polypeptides, both native and recombinant, are more heat-stable than their 'naked' counterparts expressed in *E. coli*. One potential N-glycosylation motif, Asn-Ala-Ser, occurs at positions 25-27 of human PAPI. Possibly, the insect cells used by Dando et al. (2003) perform a post-translational glycosylation of PAPI. Molecular mass values (Dando et al., 2003), however, suggest that glycosylation, if it indeed occurs, is not extensive.

Thermal stability of PAPI varies with the temperature preference of the source organism. The temperature-activity profile of PAPI from the hyperthermophilic archaeon *Thermococcus litoralis* has a maximum at 70°C, where it has a half-life of 1 h. Although considerably more thermostable than mesophilic enzymes, *T. litoralis* PAPI loses its activity rapidly at temperatures >80°C.

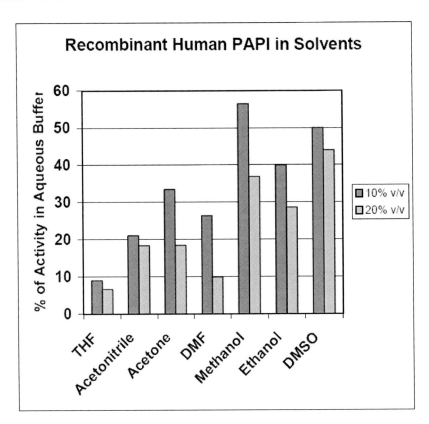

Figure 2. Effect of water-miscible solvents (10 and 20% v/v) on PAPI activity. Plot of % residual activity versus 10 or 20 (% v/v) solvents concentration. (■) Solvents concentration 10 % (v/v), (▣) solvents concentration 20 % (v/v).

This is likely to be due to destruction of the critical cysteine in the active site (Singleton *et al.*, 2000). PAPI from the hyperthermophilic archaeon *T. litoralis* showed enhanced thermal stability over that of mesophilic *B. amyloliquefaciens*, probably due to the presence of inter subunit disulphide bonds. *B. amyloliquefaciens* PAPI with an engineered inter-subunit disulphide bond showed increased thermal stability, without any decrease in enzymatic performance. However, pH stability was not altered (Kabashima *et al.*, 2001). The thermophilic *Pfu*PAPI exhibits optimum activity at 90°C (Ogasahara *et al.*, 2001).

Resistance to Solvents

PAPI was unstable in most of the water-miscible, hydrophilic solvents tested. The C_{50} values (the % v/v concentration leading to half-inactivation versus aqueous buffer) for the least injurious solvents DMSO and methanol were 10% v/v and 12% v/v, respectively (errors ± 0.5 % v/v).

THF, a strong denaturant, had a notably adverse effect even at 10% (v/v): activity declined sharply to about 1/10 of the aqueous level. Results with DMF were similar: at 10% (v/v) DMF, activity was 26% of that in aqueous buffer, while at 20% (v/v) DMF only 10% of the aqueous activity remained. PAPI activity declined to 21% of that in aqueous buffer at 10% (v/v) ACN. In 10% (v/v) acetone, residual activity was 33%, while above 50% (v/v) concentration, no significant PAPI activity remained. Ethanol at concentrations of 10% and 20% (v/v) reduced activity to 40% and 29%, respectively, of that in aqueous buffer. At ≥40% (v/v) ethanol, virtually no activity remained.

Chemical Modification with Dimethyl Suberimidate (DMS)

A temperature profile indicated that chemical modification with DMS rendered PAPI *less* thermostable than native at the same concentration. Only 20% of the initial PAPI activity remained following DMS treatment, a very low recovery. This poor outcome was unexpected: DMS is a very mild protein modifying reagent which should not lead to significant inactivation. DMS treatment has effectively stabilized other enzymes such as horseradish peroxidase (HRP, 4-fold increase in $t_{1/2}$ at 75°C; Ryan *et al.*, 1994) and alanine aminotransferase (16-fold increase in $t_{1/2}$ at 35°C; Moreno and Ó'Fágain, 1997). DMS usually reacts with Lys/ -NH$_2$ groups of proteins only and not with any other R-groups (Ji, 1983). There are 10 Lys residues in human PAPI, 3 of which lie within conserved sequences (R Larragy, unpublished); modification of a conserved Lys could compromise the enzyme's activity and/or stability.

Effects of Ammonium Sulfate and of Polyols on PAPI Activity and Stability

Contrary to expectation, ammonium sulfate had no stabilising or protective effect on PAPI at any of the temperatures tested. PAPI activity was significantly reduced at 0.5 M ammonium sulfate, with 35% less activity displayed than in buffer. Activity of a different brain peptidase, the proline-specific dipeptidyl peptidase IV (DPPIV; EC 3.4.14.5) from bovine brain, was also significantly reduced (80%) in presence of 1M ammonium sulfate compared with buffer alone (Buckley, 2001).

Glycerol also inhibited PAPI: only 50% of aqueous activity was manifested in 10% (v/v) glycerol and a mere 18% activity at 50% (v/v) glycerol. Except for a marginal effect >60°C at an inhibitory 50% (v/v), glycerol had no protective or stabilizing effect on PAPI at elevated temperatures.

It was hoped that PAPI would be stabilized against heat by the inclusion of xylitol or trehalose (Schein, 1990). Trehalose prevents proteins from denaturing at high temperature *in vitro*. It also suppresses the aggregation of denatured proteins, maintaining them in a partially-folded state from which they can be reactivated. The continued presence of trehalose, however, interferes with refolding (Singer and Lindquist, 1998) Trehalose (0.5M) did not stabilize PAPI.

Xylitol at 0.5M gave a significant increase in PAPI activity at 30°C and 40°C. With xylitol, PAPI retained >50% activity at 60°C while, in its absence, the T_{50} was 45°C. A protecting effect was also evident during thermoinactivations at 60°C where $t_{1/2}$ was 60% longer with xylitol (first-order k-values in presence and absence of xylitol were 0.05 ± 0.002 min^{-1} ($t_{1/2}$ 14 min) and 0.08 ± 0.003 min^{-1} ($t_{1/2}$ 9 min) respectively). Note that the protein concentrations in these experiments were less than in the thermal inactivations described above, hence the different values for the PAPI controls.

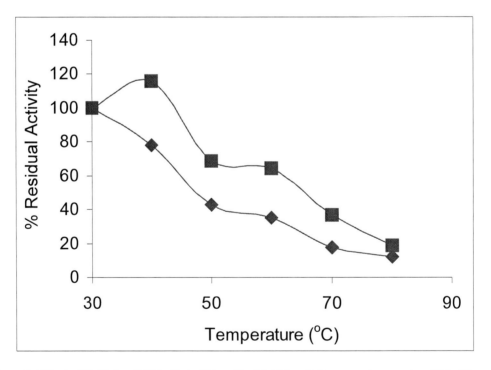

Figure 3. Effect of Xylitol on PAP1. Plot of % residual PAP1 activity versus temperature (°C). (♦) without xylitol; (■) with xylitol. Activity is represented as a percentage of the 30°C value in each case.

Thermal Stability and pGlu-AMC Kinetics of Mutant Y147F

Tyr 147 of PAPI is located in the substrate binding site, very close to the catalytic Cys 149 (Vaas, 2005). Comparison of 13 known eukaryotic PAPI amino acid sequences shows that both of these residues lie within a conserved sequence $DAGRY_{147}LC_{149}DFTYYTSLY$. At the equivalent position in prokaryotes, Phe occurs instead of Tyr in 25 of 37 sequences (R Larragy, unpublished). Both Phe and Tyr have phenyl R-groups but Tyr possesses a hydroxyl group on its benzene ring. Hydroxyls can participate in hydrogen bonding at moderate pH values or undergo ionization at strongly alkaline pH. We were curious, therefore, to investigate the effects of a Tyr147→Phe mutation.

Thermoinactivation of both wild type PAPI and Y147F (diluted appropriately in 50mM potassium phosphate buffer, pH 8.0, containing $0.12mg.ml^{-1}$ BSA to ensure a constant protein concentration) was studied at 70°C over 80 min. (The higher temperature was required to achieve timely inactivation of Y147F.) Under these conditions, $t_{1/2}$ for Y147F was 25 min, almost 3 times longer than that of wild type PAPI (9 min; see Table 1). Remarkably, removal of the –OH group from position 147 has a very stabilizing effect on PAPI, perhaps due to improved hydrophobic packing.

Michaelis-Menten Kinetics of PAPI and Y147F on pGlu-AMC

Both wild type PAPI and mutant Y147F utilized the convenient fluorimetric substrate pGlu-AMC and displayed Michaelis-Menten kinetics. Use of NEM, a thiol-directed reagent, allowed calculation of the active sites contents [E] and, hence, calculation of the respective k_{cat} values from the experimentally-determined V_{max}, since $k_{cat} = V_{max}/[E]$. Table 1 presents the calculated values. The amidase activity showed a k_{cat}/K_m value for PAPI of 0.202 ($s^{-1} M^{-1}$) while the corresponding value for Y147F was 0.212 ($s^{-1} M^{-1}$). These values are virtually identical, showing that loss of the –OH from position 147 does not affect cleavage of pGlu-AMC.

Table 1. Comparison of Native PAPI with Variant Y147F

	Native PAPI	Mutant Y147F
First-order k-value (min^{-1}), 70°C	0.079 ± 0.003	0.028 ± 0.001
Half-life ($t_{1/2}$), 70°C	9 min	25 min
Protein content (mg.ml^{-1})	0.12	0.12
K_m (mM)	0.13 ± 0.02	0.12 ± 0.02
V_{max} (μmol.min^{-1}.ml^{-1})	0.0013 ± 0.0001	0.0015 ± 0.0002
Active sites (μM; NEM method)	0.8	1.04
k_{cat} (s^{-1})	2.68×10^{-5}	2.45×10^{-5}
k_{cat}/K_m (s^{-1}.M^{-1})	0.202	0.212

Conclusion

Themal stability of recombinant human brain PAPI proved considerably higher than would be expected for this enzyme, since mammalian pyroglutamyl peptidases normally exist at the physiological body temperature of 37°C. PAPI was catalytically active up to quite elevated temperatures, with a $t_{1/2}$ of 15 min at 60°C. Nevertheless, the *in vitro* stability of wild type PAPI might limit its potential applicability in peptide processing or other fields, since PAPI was unstable to most of the water-miscible organic solvents tested. Methanol and DMSO were the least injurious to PAPI activity while THF was the most deleterious. Modification of PAPI with DMS, a very mild reagent, gave only 20% recovery of initial activity and did not stabilize the enzyme. PAPI activity and stability increased with xylitol but trehalose, glycerol and ammonium sulphate conferred no benefits. Xylitol, therefore, may be a better preservative for PAPI than glycerol, the additive currently used. It would seem that chemical modification and/or additives have limited scope to enhance PAPI stability but the use of a thermophilic counterpart, or the generation of stabilized mutant variants of recombinant PAPI, should be explored further. The latter may indeed be possible, since the single-site mutant Y147F, with a $t_{1/2}$ of 25 min at 70°C, is notably more thermostable than wild type PAPI. With pGlu-AMC as substrate, the kinetic parameters of the two enzymes were virtually identical.

Acknowledgements

Ms Karima Mtawae thanks the Libyan Peoples' Bureau for funding. Ms Ruth Larragy performed the sequence alignments mentioned in the Results section.

Bibliography

Abe, K; Saito, F; Yamada, M; Tokui, T. Pyroglutamyl Aminopeptidase I, as a Drug Metabolizing Enzyme, Recognises Xenobiotic Substrates Containing L-2-Oxothiazolidine-4-carboxylic Acid. *Biological and Pharmaceutical Bulletin* 2004a 27: 113-116.

Abe, K; Fukada, K; Tokui, T. Marginal Involvement of Pyroglutamyl Aminopeptidase I in Metabolism of Thyrotropin-Releasing Hormone in Rat Brain. *Biological and Pharmaceutical Bulletin* 2004b 27: 1197-1201.

Abe, K; Watanabe, N; Kosaka, T; Yamada, M; Tokui, T; Ikeda, T. Hydrolysis of Synthetic Substrate, L-Pyroglutamyl p-Nitroanilide is Catalyzed Solely by Pyroglutamyl Aminopeptidase I in Rat Liver Cytosol. *Biological and Pharmaceutical Bulletin* 2003 26: 1528-1533.

Alba, F; Arenas, C; Lopez MA. Comparison of soluble and membrane-bound pyroglutamyl-peptidase I activities in rat brain tissues in the presence of detergents. *Neuropeptides* 1995 29: 103-107.

Alur, HH; Desai, RP; Mitra, AK; Johnston, TP. Inhibition of a model protease - pyroglutamate aminopeptidase by a natural oligosaccharide gum from *Hakea gibbosa*. *International Journal of Pharmaceutics* 2001 212: 171-176.

Aoyagi, T; Hatsu, M; Kojima, F; Hayashi, C; Hamada, M; Takeuchi, T. Benarthin: A New Inhibitor Of Pyroglutamyl Peptidase I. Taxonomy, Fermentation, Isolation and Biological Activities. *The Journal of Antibiotics* 1992a 45: 1079-1083.

Aoyagi, T; Hatsu, M; Imada, C; Naganawa, H; Okami, Y; Takeuchi, T. Pyrizinostatin: A New Inhibitor Of Pyroglutamyl Peptidase. *The Journal of Antibiotics* 1992b 45: 1795-1796.

Awadé, A; Cleuziat, Ph; Gonzales, Th; Robert-Baudouy, J. Pyrrolidone Carboxyl Peptidase (Pcp): An Enzyme That Removes Pyroglutamic Acid (pGlu) From pGlu-Peptides and pGlu-Proteins. *Proteins: Structure, Function and Genetics* 1994 20: 34-51.

Awadé, A; Cleuziat, Ph; Gonzalès, Th; Robert-Baudouy, J. Characterisation of the pcp gene encoding the pyrrolidone carboxyl peptidase of *Bacillus subtilis*. *FEBS Letters* 1992a 305: 67-73.

Awadé, A; Gonzalès, Th; Cleuziat, Ph; Robert-Baudouy, J. One step purification and characterisation of the pyrrolidone carboxyl peptidase of *Streptococcus pyogenes* over-expressed in *Escherichia coli*. *FEBS Letters* 1992b 308: 70-74.

Barrett, AJ; Rawlings, ND. Evolutionary Lines of Cysteine Peptidases. *Biological Chemistry* 2001 382: 727-733.

Bradford, MM. Rapid and sensitive method for quantitation of microgram quantities of protein utilizing the principle of protein-dye binding. *Analytical Biochemistry* 1976 72: 248-254.

Browne, P; O'Cuinn. An evaluation of the role of a pyroglutamyl peptidase, a post-proline cleaving enzyme and a post-proline dipeptidyl amino peptidase, each purified from the soluble fraction of guinea-pig brain, in the degradation of thyroliberin *in vitro*. *European Journal of Biochemistry* 1983 137: 75-87.

Buckley, SJ. The purification and characterisation of prolyl oligopeptidase from human saliva and dipeptidyl peptidase IV from bovine serum. PhD thesis. Dublin, Ireland: Dublin City University; 2001.

Charli, JL; Mendez, M; Joseph-Bravo, P; Wilk, S. Specific Inhibitors Of Pyroglutamyl Peptidase I And Prolyl Endopeptidase Do Not Change The *In Vitro* Release Of TRH Or Its Content In Rodent Brain. *Neuropeptides* 1987 9: 373-378.

Cleuziat, P; Awadé, A; Robert-Baudouy, J. Molecular characterisation of pcp, the structural gene encoding the pyrrolidone carboxylyl peptidase from *Streptococcus pyogenes*. *Molecular Microbiology* 1992 6: 2051-2063

Cummins, PM; O'Connor, B. Pyroglutamyl peptidase: an overview of the three known enzymatic forms. *Biochimica et Biophysica Acta* 1998 1429: 1-17.

Cummins, PM; O'Connor, B. Bovine Brain Pyroglutamyl Aminopeptidase (Type-1): Purification and Characterisation of a Neuropeptide-Inactivating Peptidase. *The International Journal of Biochemistry and Cell Biology* 1996 28: 883-893.

Dando, PM; Fortunato, M; Strand, GB; Smith, TS; Barrett, AJ. Pyroglutamyl-peptidase I: cloning, sequencing, and characterisation of the recombinant human enzyme. *Protein Expression and Purification* 2003 28: 111-119.

DeGandarias, JM; Irazusta, J; Gil, J; Varona, A; Ortega, F; Casis, L. Subcellular ontogeny of brain pyroglutamyl peptidase I. *Peptides* 2000 21: 509-517.

DeGandarias, JM; Irazusta, J; Silio, M; Gil, J; Saitua, N; Casis, L. Soluble and membrane-bound pyroglutamyl-peptidase I activity in the developing cerebellum and brain cortex. *International Journal of Developmental Biology* 1998 42: 103-106.

DeRenobales, M; Welch, W. Chemical cross-linking stabilizes the enzymic activity and quaternary structure of formyltetrahydrofolate synthetase. *Journal of Biological Chemistry* 1980 255: 10460-10463.

Faivre-Bauman, A; Loudes, C; Barret, A; Tixier-Vidal, A; Bauer, K. Possible Role Of Neuropeptide Degrading Enzymes On Thyroliberin Secretion In Fetal Hypothalamic Cultures Grown In Serum Free Medium. *Neuropeptides* 1986 7: 125-138.

Friedman, TC; Kline, TB; Wilk, S. 5-Oxoprolinal: Transition-State Aldehyde Inhibitor of Pyroglutamyl-Peptide Hydrolase. *Biochemistry* 1985 24: 3907-3913.

Fujiwara, K; Tsuru, D. New chromogenic and fluorigenic substrates for pyrrolidonyl peptidase. *Journal of Biochemistry (Tokyo)* 1978 83: 1145-1149.

Fujiwara, K; Kitagawa, T; Tsuru, D. Inactivation of Pyroglutamyl Aminopeptidase by L-Pyroglutamyl Chloromethyl Ketone. *Biochimica et Biophysica Acta* 1981 655: 10-16.

Gonzalès, T; Robert-Baudouy, J. Characterisation of the pcp Gene of *Pseudomonas fluorescens* and of Its Product, Pyrrolidone Carboxyl Peptidase (Pcp). *Journal of Bacteriology* 1994 176: 2569-2576.

Griffiths, EC. Thyrotropin Releasing Hormone: Endocrine And Central Effects. *Psychoneuroendocrinology* 1985 10: 225-235.

He, W; Barrow, CJ. The Aβ 3-Pyroglutamyl and 11-Pyroglutamyl Peptides Found in Senile Plaque Have Greater β-Sheet Forming and Aggregation Propensities *in Vitro* than Full-Length Aβ. *Biochemistry* 1999 38: 10871-10877.

Irazusta, J; Larrinaga, G; González, J; Gil, J; Meana, J; Casis, L. Distribution of prolyl endopeptidase activities in rat and human brain. *Neurochemistry International* 2002 40: 337-345.

Ito, K; Inoue, T; Takahashi, T; Huang, HS; Esumi, T; Hatakeyama, S. The mechanism of substrate recognition of PAPI from *Bacillus amyloliquefaciens* as determined by X-ray crystallography and site-directed mutagenesis. *Journal of Biological Chemistry*. 2001 276: 18557-18562.

Ji, TH. Bifunctional reagents. In: Hirs, CHW, Timasheff, SN, editors. *Methods in Enzymology* 91:. NY: Academic Press; 1983; 580-609.

Kabashima, T., Li, Y., Kanada, N., and Yoshimoto, T. Enhancement of the thermal stability of pyroglutamyl peptidase I by introduction of an intersubunit disulfide bond. *Biochim Biophys Acta* 2001 1547: 214-220.

Kilbane, Z; Vaas, P-R; Ó'Cuív, P; O'Connor, B. Cloning and heterologous expression of bovine pyroglutamyl peptidase type-1 in *Escherichia coli*: purification, biochemical and kinetic characterisation. *Molecular and Cellular Biochemistry* 2007 297: 189-197.

Kim, JK; Kim, SJ; Lee, HG; Lim, JS; Kim, SJ; Cho, SH; Jeong, WH; Choe, IS; Chung, T; Paik, SG; Choe YK. Molecular Cloning and Characterisation of *Mycobacterium bovis* BCG pcp Gene Encoding Pyrrolidone Carboxyl Peptidase. *Molecules and Cells* 2001 12: 347-352.

Laemmli, UK. Cleavage of structural proteins during the assembly of the head of bacteriophage T4. *Nature* (London) 1970 227: 680-685.

Lauffart, B; McDermott, JR; Biggins, JA; Gibson, AM; Mantle, D. Purification and characterization of pyroglutamyl aminopeptidase from human cerebral cortex. *Biochemical Society Transactions* 1989 17: 207-208.

Le Saux, O; Gonzales, T; Robert-Baudouy, J. Mutational Analysis of the Active Site of *Pseudomonas fluorescens* Pyrrolidone Carboxyl Peptidase. *Journal of Bacteriology* 1996 178: 3308-3313.

Mantle, D; Lauffart, B; Gibson, A. Purification and characterisation of leucyl aminopeptidase and pyroglutamyl aminopeptidase from human skeletal muscle. *Clinica Chimica Acta* 1991 197: 35-46.

Mantle, D; Lauffart, B; McDermot, J; Gibson, A. Characterisation of aminopeptidases in human kidney soluble fraction. *Clinica Chimica Acta* 1990 187: 105-114.

Moreno, J; Ó'Fágáin, C. Activity and Stability of native and modified alanine aminotransferase in cosolvent systems and denaturants. *Journal of Molecular Catalysis B: Enzymatic* 1997 2: 271-279.

Mudge, AW; Fellows, RE. Bovine pituitary pyrrolidonecarboxylyl peptidase. *Endocrinologia* 1973 93: 1428–1434.

O'Connor B; O'Cuinn G. Purification of a kinetic studies on a narrow specificity synaptosomal membrane pyroglutamate aminopeptidase from guinea-pig brain. *European Journal of Biochemistry* 1985 150: 47-52.

O'Cuinn, G; O'Connor, B; Elmore, M. Degradation of Thyrotropin-Releasing Hormone and Luteinising Hormone-Releasing Hormone by Enzymes of Brain Tissues. *Journal of Neurochemistry* 1990 54: 1-13.

Odagaki, Y; Hayashi, A; Okada, K; Hirotsu, K; Kabashima, T; Ito, K; Yoshimoto, T; Tsuru, D; Sato, M; Clardy, J. The Crystal Structure Of Pyroglutamyl Peptidase I from *Bacillus amyloliquefaciens* reveals a new structure for a cysteine protease. *Structure* 1999 7: 399-411.

Ogasahara, K; Khcchinashvili, N; Nakamura, M; Yoshimoto, T; Yutani, K. Thermal stability of pyrrolidone carboxyl peptidase from the hyperthermophilic Archaeon, *Pyrococcus furiosus*. *European Journal of Biochemistry* 2001 268: 3233-3242.

Orlowski, M; Meister, A. Enzymology of Pyrrolidone Carboxylic Acid. In: Boyer, PD, editor. *The Enzymes, IV: Hydrolysis,*.NY: Academic Press; 1971; 123-151.

Patti, JM; Schneider, A; Garza, N; Boles, JO. Isolation and characterisation of pcp, a gene encoding a pyrrolidone carboxyl peptidase in *Staphylococcus aureus*. *Gene* 1995 166: 95-99.

Prasad, C; Mori, M; Pierson, W; Wilber, J; Ewards, R. Developmental changes in the distribution of rat brain pyroglutamate aminopeptidase, a possible determinant of endogenous cyclo(His-Pro) concentrations. *Neurochemical Research* 1983 8: 389-399.

Ramírez, M; Sanchez, B; Alba, F; Luna, J; Martínez, I. Diurnal variation and left-right distribution of pyroglutamyl peptidase I activity in the rat brain and retina. *Acta Endocrinologica* 1991 125: 570-573.

Robert-Baudouy, J., and Thierry, G. Bacterial pyroglutamyl peptidases. In: Barrett, AJ; Rawlings, ND; Woessner, F. editors. *Handbook of Proteolytic Enzymes.* San Diego: Academic Press; 1998; 791–795.

Ryan, O; Smyth, MR; Ó'Fágáin, C. Thermostabilized chemical derivatives of horseradish peroxidase. *Enzyme and Microbial Technology* 1994 16: 501-505.

Sánchez, B; Alba, F; Luna, JD; Martínez, I; Ramírez, M. Pyroglutamyl Peptidase I levels and their left–right distribution in the rat retina and hypothalamus are influenced by light-dark conditions. *Brain Research* 1996 731: 254-257.

Schein, CH. Solubility as a function of Protein Struction and Solvent Components. *Biotechnology* 1990 8: 308-317.

Singer, MA; Lindquist, S. Multiple effects of trehalose on protein folding in vitro and in vivo. *Molecular Cell* 1998 1: 639-648.

Singleton, MR; Littlechild, JA. Pyrrolidone Carboxylpeptidase from *Thermococcus litoralis.* In: Adams, MWW, Kelly, RM, editors. *Methods in Enzymology* 330: NY: Academic Press; 2001; 394-403.

Singleton, MR; Taylor, SJC; Parrat, JS; Littlechild, JA. Cloning, expression, and characterization of pyrrolidone carboxyl peptidase from the archaeon *Thermococcus litoralis.* *Extremophiles* 2000 4: 297-303.

Singleton, MR; Isupov, MN; Littlechild, JA. Crystallisation and preliminary X-ray diffraction studies of pyrrolidone carboxyl peptidase from the hyperthermophilic archaeon *Thermococcus litoralis.* *Acta Crystallographica* 1999a D55: 702-703.

Singleton, M; Isupov, M; Littlechild, J. X-ray structure of pyrrolidone carboxyl peptidase from the hyperthermophilic archaeon *Thermococcus litoralis.* Structure 1999b 7: 237-244.

Smith, PK; Krohn, RI; Hermanson, GT; Mallia, AK; Gartner, FH; Provenzano, MD; Fujimoto, EK; Goeke, NM; Olson, BJ; Klenk, DC. Measurement of protein using bicinchoninic acid. *Analytical Biochemistry* 1985 150: 76-85.

Sokabe, M; Kawamura, T; Sakai, N; Yao, M; Watanabe, N; Tanaka, I. X-ray crystal structure of pyrrolidone-carboxylate peptidase from hyperthermophilic *Pyrococcus horikoshii.* *Journal of Structural and Functional Genomics* 2002 2: 145-154.

Szewczuk, A; Kwiatkowska, J. Pyrrolidone Peptidase in Animal, Plant and Human Tissues. Occurrence and some Properties of the Enzyme. *European Journal of Biochemistry* 1970 15: 92-96.

Tanaka, H; Chinami, M; Mizushima, T; Ogasahara, K; Ota, M; Tsukihara, T; Yutani, K. X-Ray Crystalline Structures of Pyrrolidone Carboxyl Peptidase from a Hyperthermophile, *Pyrococcus furiosus*, and Its Cys-Free Mutant. *Journal of Biochemistry (Tokyo)* 2001 130: 107-118.

Tsunasawa, S; Nakura, S; Tanigawa, T; Kato, I. Pyrrolidone carboxyl peptidase from the hyperthermophilic Archaeon *Pyrococcus furiosus*: cloning and overexpression in *Escherichia coli* of the gene, and its application to protein sequence analysis. *Journal of Biochemistry (Tokyo)* 1998 124: 778-783.

Turk, B; Krizaj, I; Kralj, B; Dolenc, I; Popovic, T; Bieth, JG; Turk, V. Bovine stefin C, a new member of the stefin family. *Journal of Biological Chemistry* 1993 268: 7323-7329.

Vaas, P-R. Molecular characterisation of a recombinant human neuropeptide-inactivating proteinase. PhD thesis. Dublin, Ireland: Dublin City University; 2005.

Yamada, M; Mori, M. Thyrotropin-releasing hormone-degrading enzyme in human serum is classified as type II of pyroglutamyl aminopeptidase: influence of thyroid status. *Proceedings of the Society for Experimental Biology and Medicine* 1990 194: 346-351.

Yoshimoto, T; Shimoda, T; Kitazono, T. Pyroglutamyl peptidase gene from *Bacillus amyloliquefaciens*: cloning, sequencing, expression and crystalisation of the expressed enyme. *Journal of Biochemistry (Tokyo)* 1993 113: 67-73.

In: Biotechnology: Research, Technology and Applications ISBN 978-1-60456-901-8
Editor: Felix W. Richter © 2008 Nova Science Publishers, Inc.

Chapter 6

Biofuels Production by Cell-Free Synthetic Enzymatic Technology

Y. H. Percival Zhang[*,1,2,3], *Xinhao Ye*[1] *and Yiran Wang*[1]
[1]Biological Systems Engineering Department
[2]Institute for Critical Technology and Applied Science (ICTAS)
Virginia Polytechnic Institute and State University
210-A Seitz Hall, Blacksburg, Virginia 24061, USA
[3]DOE Bioenergy Science Center (BESC), Oak Ridge, TN, USA

Abstract

Biomass is the only renewable resource that can provide a sufficient fraction of both future transportation fuels and renewable materials at the same time. Synthetic biology is an emerging interdisciplinary area that combines science and engineering in order to design and build novel biological functions and systems. Different from *in vivo* synthetic biology, cell-free *in vitro* synthetic biology is a largely unexplored strategy. Cell-free synthetic enzymatic pathway engineering (SEPE) is to *in vitro* assemble a number of enzymes and coenzymes to implement complicated biotransformations that can mimic natural fermentation or achieve unnatural processes. Recently, a novel synthetic enzymatic pathway composed of 13 enzymes and a cofactor has been demonstrated to produce 12 molecules of hydrogen per molecule of glucose unit of starch and water (PLoS One, 2007, 2:e456). This new sugar-to-hydrogen technology promises to solve several obstacles to the hydrogen economy – cheap hydrogen production, high hydrogen storage density (14.8 H_2 mass%), and costly hydrogen infrastructure, and to eliminate safety concerns about mass utilization of hydrogen. Furthermore, the advantages and limitations of producing liquid biofuels -- ethanol and butanol -- from sugars by SEPE are discussed. The research and development of SESE require more efforts, especially in low-cost recombinant thermophilic enzyme building block manufacturing, efficient cofactor recycling, enzyme and cofactor stabilization, and so on.

---

[*] Email: biofuels@vt.edu; Tel: (540) 231-7414 [O]; Fax: (540) 231-3199.

Keywords: biofuels, biomass, metabolic engineering, synthetic biology, synthetic enzymatic pathway, the hydrogen economy.

I. Introduction

Transportation fuels have several special requirements, such as, high energy storage density (MJ/L of and MJ/kg of fuel) due to limited spaces in vehicles, high power density (kW/kg of system), a rapid response time for power output, affordable power systems, fuel stability, available or low-cost infrastructure for fuel storage and distribution, fast fuel refilling, safe utilization, and so on. Therefore, a combination of liquid fuels (e.g., gasoline and diesel) and internal combustion engine (ICE) is the dominant transportation approach that has replaced heavy and low-energy-efficiency coal-steam engine. It is expected that a transition from the liquid fuel-ICE to the clean and high-energy-efficiency hydrogen-fuel-cell-electricity system will take place eventually.

Synthetic biology is an emerging interdisciplinary area that combines science (biology and chemistry) and engineering in order to design and build novel biological functions and systems that function unnaturally or function much better than natural counterparts [11, 23]. It is also interpreted as the engineering-driven building of increasingly complicated biological entities (parts, devices, and systems) from simple and basic building blocks [28]. The applications of synthetic biology include production of new drugs, materials and energy, programmable bio-devices and control logic, nano-material assembly, molecular media devices, computation and signal processing, biosensors, etc.

Synthetic biology projects can be divided into two classes: *in vivo* and *in vitro* [25]. As compared to *in vivo* living biological entity-based synthetic biology, *in vitro* cell-free synthetic biology is a largely unexplored or ignored approach. But the benefits of *in vitro* synthetic biology should not be underestimated because basic knowledge of *in vitro* synthetic biology is so clear that we are able to assemble a new system much more easily than to modify a living system. Engineering flexibility *in vitro* is much greater, i.e., unshackled from cellular viability, complexity, physiology, and membrane/wall. Many biopolymer syntheses have already been implemented in cell-free systems, for example, synthesis of DNAs by PCR amplification, synthesis of RNAs by *in vitro* transcription, and synthesis of protein/ polypeptide by *in vitro* transcription/translation. Recently, a new *in vitro* synthetic biology technology called "synthetic enzymatic pathway engineering (SEPE)" has been proposed to break down a biopolymer (starch) to small molecules. This example is to assemble 13 natural enzymes to create an unnatural catabolic pathway that completely converts starch and water to hydrogen and carbon dioxide [$C_5H_{10}O_5$ (eq) + 7 H_2O (l) $\rightarrow$ 12 H_2 (g) + 6 CO_2 (g)] [62].

Synthetic enzymatic pathway engineering can be also regarded as "*in vitro* metabolic engineering". Table 1 presents the comparison between metabolic engineering (ME)-modified microbe fermentation and synthetic enzymatic pathway (SEP)-based biocatalysis. Since SEP-based biocatalysis has a clearly-designed biochemical pathway and does not produce any cellular mass or other unwanted byproducts, a yield of the desired product should be very close to a theoretical value if the reaction is complete or the inhibitive product can be removed *in situ*. Also, higher product purity is expected. Because no cell wall or cell membrane slows down the substrate/product transportation across the cell membrane, higher reaction rates may be obtained for SEP biocatalysis.

Table 1. Comparison of microbial fermentation and synthetic enzymatic pathway (SEP)-based biocatalysis

Feature	Microbial fermentation	SEP-based biocatalysis
Product yield	Low-modest	High (theoretical)
Product purity	Low-modest	High
Reaction rate	Low-modest	Low-likely high
Reaction condition	Narrow	Broad
Substrate/Product	Non-toxic to microbes	Non-toxic to enzymes
Reaction control	Difficult	Easy
Implementation rate	Slow	Fast
Implementation risks	Modest-high	Low
Production costs	Low	High – likely very low
Application	A large number	Few -- more

SEP-based biocatalysis may be conducted under broader reaction conditions because most enzymes can tolerate more severe reaction conditions, such as high temperatures, low pH, high organic solvent concentration, etc. A number of microbial fermentations are restricted by toxicity of products, especially for cellular membranes, while SEP-based biocatalysis can be conducted under broad product/substrate conditions. Enzyme-based biocatalysis can be controlled more easily than can fermentation because it has a short feedback control loop and a clear control mechanism. What is more important, ME-modified microbe construction needs a very long time, and the success rates are not high because of some uncertainties in introduction of multiple (foreign) genes, recombinant protein expression and folding, metabolic pathway control, gene regulation, and so on. *In vivo* metabolic engineering is like rational enzyme design, where limited knowledge of the protein sequence, crystalline structure, and catalytic mechanism cannot guarantee solving problems as we expect.

SEP-based biocatalysis has its own limitations. For example, it has few applications; it needs high-cost purified enzymes and cofactors; enzymes could denature. But such situations are anticipated to change rapidly if mass recombinant (thermophilic) enzyme production is well established; $NAD(P)^+$ can be efficiently recycled or stabilized; low cost building blocks (such as enzyme or multiple enzyme complexes) are widely available; low-cost enzyme purification technologies are developed [31]; and enzyme stability can be enhanced *via* immobilization or protein engineering or added stabilizers, etc. We estimate that a theoretically low protein cost by fermentation could be as low as $1.00/kg of enzyme with the assumption of a protein yield (0.30 kg protein/kg of glucose, $0.15/kg of glucose) and the mature enzyme production costs (50% for processing costs; 50% from sugar). Current cellulase production costs are approximately $2.00/kg of crude cellulase made from Novozyme and Genencor [63]. In the long term, markets will be the large driving force to decrease enzyme production costs close to a value of the theoretical enzyme production cost through intensive R and D.

In this book chapter, biofuels are defined as transportation fuels made from biomass. Three examples of biofuels production from biomass sugars through cell-free synthetic enzymatic technology are presented and discussed.

II. Hydrogen Production

Abundant, clean, and carbon-neutral hydrogen is widely believed to be an ultimate mobile energy carrier replacing gasoline, diesel, and ethanol because a high energy conversion efficiency (~50-70%) can be achieved *via* fuel cells without pollutants produced [29]. Hydrogen production from less costly abundant biomass is a shortcut for producing low-cost hydrogen without net carbon emissions [1, 4, 20, 21, 27, 33, 43, 49].

Figure 1 shows the synthetic enzymatic pathway comprised of 14 reversible enzymatic reactions by 13 enzymes: (i) a chain-shortening phosphorylation reaction for producing G-1-P catalyzed by starch phosphorylase (Eq. 1); (ii) conversion of glucose-1-phosphate (G-1-P) to glucose-6-phosphate (G-6-P) catalyzed by phosphoglucomutase (Eq. 2); (iii) a pentose phosphate pathway containing 10 enzymes for producing 12 NADPH per G-6-P (Eq. 3); and (iv) hydrogen generation from NADPH catalyzed by hydrogenase (Eq. 4). The enzyme names, enzyme catalogue numbers, and their catalytic reactions are presented in Table 2.

$$(C_6H_{10}O_5)_n + H_2O + P_i \leftrightarrow (C_6H_{10}O_5)_{n-1} + \text{G-1-P} \qquad [1]$$

$$\text{G-1-P} \leftrightarrow \text{G-6-P} \qquad [2]$$

$$\text{G-6-P} + 12\ NADP^+ + 6\ H_2O \leftrightarrow 12\ NADPH + 12\ H^+ + 6\ CO_2 + P_i \qquad [3]$$

$$12\ NADPH + 12\ H^+ \leftrightarrow 12\ H_2 + 12\ NADP^+ \qquad [4]$$

The combination of Equations 1-4 leads to Equation 5:

$$C_6H_{10}O_5\ (aq) + 7\ H_2O\ (l) \rightarrow 12\ H_2\ (g) + 6\ CO_2\ (g) \qquad [5]$$

The proof-of-concept experiment has been conducted by assembling 12 mesophilic enzymes from commercial animal, plant, bacterial, and yeast sources, plus an archaeal hyperthermophilic hydrogenase. Hydrogen and CO_2 are produced (Figure 2). Clearly, CO_2 is evolved before H_2, in a good agreement with the mechanism that CO_2 is released by 6-phosphogluconate dehydrogenase before NADPH accumulates beyond the critical value for hydrogen generation by hydrogenase. The overall reaction (Eq. 5) is a spontaneous endothermic process (ΔG^o = -48.9 kJ/mol and ΔH^o = 596 kJ/mol) (62). The gaseous reaction products (H_2 and CO_2) are removed from the aqueous reaction solution, whose reaction temperatures are below water boiling temperature, suggesting a complete conversion. As compared to other sugar-to-hydrogen approaches [1, 4, 20, 21, 27, 33, 43, 49], the special features of this new technology -- such as the highest hydrogen yield, no impure product, and modest reaction conditions -- make it more appealing for transportation applications [61].

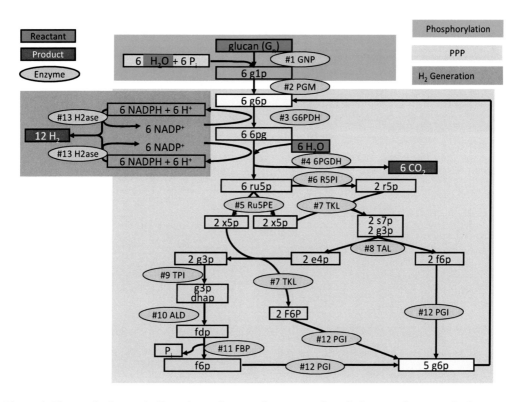

Figure 1. The synthetic metabolic pathway for complete conversion of glucan and water to hydrogen and carbon dioxide. PPP, pentose phosphate pathway. The enzymes are: #1 GNP, glucan phosphorylase, EC 2.4.1.1; #2 PGM, phosphoglucomutase, EC 5.4.2.2; #3 G6PDH, G-6-P dehydrogenase, EC 1.1.1.49; #4 6PGDH, 6-phosphogluconate dehydrogenase, EC 1.1.1.44; #5 R5PI, phosphoribose isomerase, EC 5.3.1.6; #6 Ru5PE, ribulose 5-phosphate epimerase, EC 5.1.3.1; #7 TKL, transketolase, EC 2.2.1.1; #8 TAL, transaldolase, EC 2.2.1.2; #9 TPI, triose phosphate isomerase, EC 5.3.1.1; #10 ALD, aldolase, EC 4.1.2.13; #11 FBP, fructose-1, 6-bisphosphatase, EC 3.1.3.11; #12 PGI, phosphoglucose isomerase, EC 5.3.1.9; and #13 H2ase, hydrogenase, EC 1.12.1.3. The metabolites and chemicals are: g1p, glucose-1-phosphate; g6p, glucose-6-phosphate; 6pg, 6-phosphogluconate; ru5p, ribulose-5-phosphate; x5p, xylulose-5-phosphate; r5p, ribose-5-phosphate; s7p, sedoheptulose-7-phosphate; g3p, glyceraldehyde-3-phosphate; e4p, erythrose-4-phosphate; dhap, dihydroxacetone phosphate; fdp, fructose-1,6-diphosphate; f6p, fructose-6-phosphate; and P_i, inorganic phosphate.

Table 2. The enzymes used for hydrogen production from starch and water, and their reaction mechanisms, sources, and amounts used in the reaction

No.	Enzyme Name	E. C.	Reaction
1	starch phosphorylase (GNP)	2.4.1.1	$g_n + P_i + H_2O \rightarrow g_{n-1} + g1p$
2	phosphoglucomutase (PGM)	5.4.2.2	$g1p \rightarrow g6p$
3	glucose-6-phosphate dehydrogenase (G6PDH)	1.1.1.49	$g6p + NADP^+ \rightarrow 6pg + NADPH$

Table 2. (Continued)

No.	Enzyme Name	E. C.	Reaction
4	6-phosphogluconic dehydrogenase (6PGDH)	1.1.1.44	$6pg + H_2O + NADP^+ \rightarrow ru5p + NADPH + CO_2$
5	ribulose-5-phosphate 3-epimerase (Ru5PE)	5.1.3.1	$ru5p \rightarrow x5p$
6	ribose 5-phosphate isomerase (R5PI)	5.3.1.6	$ru5p \rightarrow r5p$
7	transketolase (TKL)	2.2.1.1	$x5p + r5p \rightarrow s7p + g3p$ $x5p + e4p \rightarrow f6p + g3p$
8	transaldolase (TAL)	2.2.1.2	$s7p + g3p \rightarrow f6p + e4p$
9	triose-phosphate isomerase (TPI)	5.3.1.1	$g3p \rightarrow dhap$
10	aldolase (ALD)	4.1.2.13	$g3p + dhap \rightarrow fdp$
11	fructose-1,6-bisphosphate (FBP)	3.1.3.11	$fdp + H_2O \rightarrow f6p + Pi$
12	phosphoglucose isomerase (PGI)	5.3.1.9	$f6p \rightarrow g6p$
13	*P. furiosus* hydrogenase I (H2ase)	1.12.1.3	$NADPH + H^+ \rightarrow NADP^+ + H_2$

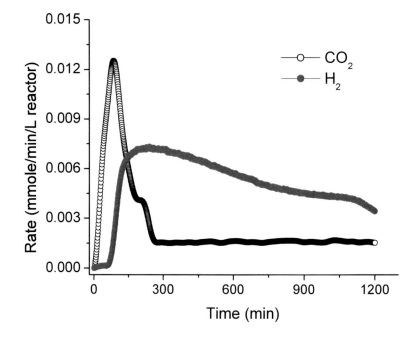

Figure 2. Experimental data for cell-free enzymatic hydrogen production from starch and water. The reaction mixture contains 2 mM starch, 10 units of GNP, 10 units of PGM, 1 unit each of the enzymes of the pentose phosphate cycle, 0.5 mM thiamine pyrophosphate, 2 mM NADP, and ~70 units of *P. furiosus* hydrogenase in 2.0 ml of 0.1M HEPES buffer (pH 7.5) containing 10 mM $MgCl_2$, 0.5 mM $MnCl_2$, and 4 mM P_i at 30 °C. Production of H_2 and CO_2 was measured in a continuous flow of helium at a flow rate of 50 mL/min and calibrated with an inline electrolysis cell and Faraday's law of electrochemical equivalence connected in tandem with the hydrogen detection system.

III. Ethanol Production

Ethanol is an attractive liquid fuel because it can be blended with gasoline at various percentages for internal combustion engines. Ethanol has been regarded as a partial solution to meeting the world's transportation fuel demand and reducing greenhouse gas emissions [26, 42, 48]. Ethanol is usually produced through microbial fermentation by the yeast *Saccharomyces cerevisiae* or the facultative bacterium *Zymomonas mobilis* from soluble sugars derived from sucrose-containing plants and starchy crops [8, 26, 50]. Ethanol yields from hexose are as high as 90-97% of the theoretical yield (0.51 g ethanol per g of glucose or 2 mole ethanol per mole of glucose), and final ethanol concentrations are up to 10-12% (w/v) [36, 44]. Large-scale ethanol production from lignocellulosic biomass will come true within next several years [41, 42].

Ethanol production by cell-free enzymatic technology could overcome the weaknesses of microbial fermentation, such as product inhibition and its associated unsteady states and oscillation, low product yield, and difficult process control [2, 8]. Cell-free ethanol production mediated by a number of enzymes was discovered by Eduard Buchner in 1897 [16]. This achievement led Buchner to the Nobel prize in 1907 and also initiated the study of enzyme biochemistry and metabolism. However, ATP imbalance (two net ATP produced per glucose) (Figure 3) results in incomplete conversion of sugars to ethanol by cell extracts (45, 56, 59). Scopes and his colleagues demonstrated the feasibility of high-rate ethanol production by a reconstituted yeast glycolytic enzyme system with supplement of ATPase, which can break the accumulated ATP to ADP and phosphate [56]. The cell-free system is capable of totally converting 180 g/L glucose to 90 g/L ethanol within 8 hours, very close to the theoretical yield [56]. Alternatively, arsenate, uncoupling ATP synthesis of phosphoglycerate kinase, can replace costly ATPase for high-rate ethanol production [56]. Recently, model simulation of the reconstituted enzymatic ethanol-producing system suggests the feasibility of high-speed ethanol production [2].

IV. Butanol Production

Butanol, a four-carbon liquid alcohol, has several advantages over ethanol, such as higher energy contents (MJ/L or MJ/kg), lower water absorption, better blending ability, and immediate usage in conventional combustion engines without any modification [22]. Butanol fermentation was firstly reported by Pasteur in 1861. Chaim Weizmann, a German chemist, isolates a microorganism, *Clostridium acetobutylicum*, which produces acetone, butanol, and ethanol (ABE). ABE fermentation was once the second largest industrial fermentation, followed by ethanol fermentation by yeast, in the 20th century. Currently, there is reviving interest in production of butanol as an alternative transportation fuel.

Typical ABE fermentation is a two-step fermentation. Butyric, propionic, lactic and acetic acids are produced by *C. acetobutylicum* in the first acidogenesis step. When the culture pH drops below a critical value due to acid accumulation, butanol, acetone, isopropanol, and ethanol are produced in the second solventogenesis step. ABE fermentation by *Clostridium* sp. is complicated and difficult to control, probably due to a lack of understanding of the global regulation of butanol production and the physiology of the solventogenic *Clostridia* [6, 24, 53].

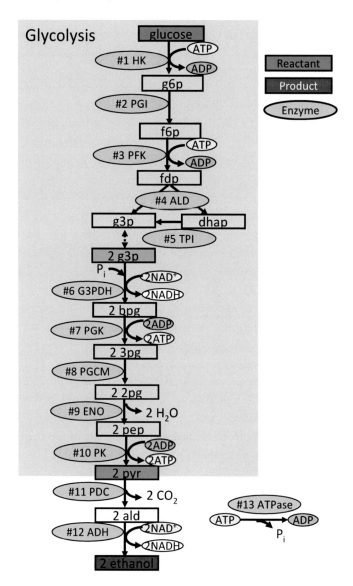

Figure 3. Cell-free enzymatic ethanol production from glucose. The enzymes are: #1 HK, hexokinase, EC 2.7.1.1; #2 PGI, phosphoglucose isomerase, EC 5.3.1.9; #3 PFK, phosphofructokinase I, EC 2.7.1.11; #4 ALD, adolase, EC 4.1.2.13; #5 TPI, triose phosphate isomerase, EC 5.3.1.1; #6 G3PDH, glyceraldehydes 3-phosphate dehydrogenase, EC 1.2.1.12; #7 PGK, phosphoglycerate kinase, EC 2.7.2.3; #8 PGCM, phosphoglycerate mutase, EC 5.4.2.1; #9 ENO, enolase, EC 4.2.1.11; #10 PK, pyruvate kinase, EC 2.7.1.40; #11 PDC, pyruvate dehydrogenase, EC 4.1.1.1; #12 ADH, alcohol dehydrogenase, EC 1.1.1.1; and #13, ATPase, EC 3.6.1.3. The abbreviated names of metabolites are: g6p, glucose-6-phosphate; f6p, fructose 6-phosphate; fdp, fructose 1,6-diphosphate; g3p, glyceraldehydes 3-phosphate; dhap, dihydroxacetone phosphate; bpg, 1,3-bisphosphoglycerate; 3pg, 3-phosphoglycerate; 2pg, 2-phosphoglycerate; pep, phosphoenolpyruvate; pyr, pyruvate; and ald, acetaldehyde.

Butanol is a stronger inhibitor of microbial metabolism and membrane integrity than is ethanol. Hydrophobicity of butanol can destabilize the phospholipids component of the cell membrane [14, 55]. An increase in membrane fluidity results in disruption of membrane functions. For example, butanol can inhibit membrane-bound ATPase activity, stop the uptake of sugars and amino acids, and abolish the membrane pH gradient [14, 37]. In addition, butanol can lead to autolytic degradation of solvent-producing bacterial cells by release of autolysin [3, 7]. One percent of butanol significantly inhibits cell growth and the fermentation process; butanol concentration in conventional ABE fermentations is usually lower than 1.3% [32, 37]; and 1.6% of butanol completely inhibits cell growth and terminates the fermentation [37]. In order to mitigate butanol inhibition, *in situ* product removal technologies, such as gas stripping, pervaporation, and extraction, have been investigated [24, 37, 46]. Recently, a hyper-butanol producer *Clostridium beijerinckii* BA101 has been found to produce up to 33 g/L butanol [46].

The theoretical yield of butanol is 1 mole of butanol per mole of glucose or 0.411 gram of butanol per gram of glucose (Figure 4). But practical butanol yields in most ABE fermentations are fairly low (~36.5% of theoretical yield) and rarely exceed 80%. In contrast, practical ethanol yields are at least 90% or as high as 95-98% of theoretical yield [57]. Recently, a new process has been developed using continuous immobilized cultures of *Clostridium tyrobutyricum* and *C. acetobutylicum* to produce butanol, where *C. tyrobutyricum* is responsible for producing hydrogen and butyric acid from glucose and *C. acetobutylicum* is responsible for converting butyric acid and hydrogen to butanol [47]. The combinatory two-step process drastically decreases formation of acetic acid, lactic acid, propionic acids, acetone, isopropanol, and ethanol and produces only butanol and carbon dioxide. The reported butanol yields could be as high as 0.47 g/g carbon source (i.e. 114% as calculated from the amount of glucose contained in corn) [64]. This result needs further confirmation and it could be attributed to consumption of other sugars.

Butanol production by cell-free synthetic enzyme technology could have several distinctive advantages: (1) a theoretically high butanol yield due to neither byproduct formation nor biomass synthesis, (2) possibly a high butanol titer because most enzymes can tolerate higher solvent levels than their parental microbes, (3) much easier process control because 2-step conversion is not necessary, and (4) no risk of phage contamination.

Figure 4 presents the synthetic cell-free enzymatic pathway for butanol production and the corresponding reactions and enzymes. Only 18 enzymes are involved for butanol conversion from glucose, far fewer than the enzymes need for cellular metabolism, plus an enzyme ATPase, which keeps ATP balanced. Glucose is metabolized *via* glycolysis with the conversion of 1 mol of hexose to 2 mol of pyruvate, with a net production of 2 mol of ATP and 2 mol of NADH; the pyruvate resulting from glycolysis is cleaved by pyruvate ferredoxin oxidoreductase in the presence of coenzyme A (CoA) to yield carbon dioxide, acetyl-CoA, and reduced ferredoxin; the conversion of pyruvate to butanol actually consumes 4 moles of NADH or its equivalent. The produced NADPH comes from NADH-ferredoxin dehydrogenase and NADPH-ferredoxin dehydrogenase recycling [37, 38, 52]. Similar to cell-free ethanol fermentation, the accumulating ATP (2 ATP per glucose) would slow down butanol production. It is expected that addition of ATPase or arsenate would enable sustainable butanol conversion.

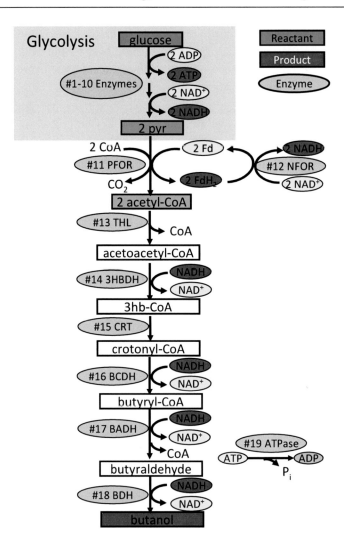

Figure 4. Cell-free enzymatic butanol production from glucose. The first ten steps are the same as for the glycolysis of ethanol fermentation (in Figure 3). The other enzymes (from #11- #19) are: #11 PFOR, pyruvate ferrodoxin oxidoreductase, EC 1.2.7.1; #12 NFOR, NADH ferrodoxin oxidoreductase, EC 1.18.1.3; #13 THL, thiolase (or AtoB, acetyle CoA acetyltransferase), EC 2.3.1.9; #14 3HBDH, hydroxybutyryl CoA hydrogenase, EC 1.1.1.157; #15 CRT, crotonase, EC 4.2.1.17; #16 BCDH, butyryl-CoA dehydrogenase, EC 1.1.1.157; #17 BADH, butyraldehyde dehydrogenase, EC 1.2.1.-; #18 BDH, butanol dehydrogenase, EC 1.1.1.-; and #19 ATPase, EC 3.6.1.3. The abbreviated names of metabolites are: Fd, ferredoxin; fdh2, reduced ferredoxin; and 3hb-CoA, hydroxybutyryl CoA.

Enzymes could tolerate higher levels of organic solvents than can the microorganisms' hydrophobic membrane. For example, fungal cellulase loses only 20% of its activity at 200 g/L ethanol and 20% of its activity in the presence of 100 g/L butanol [30]. A large number of enzyme-mediated catalyses have been conducted in the presence of organic solvents or of non-aqueous phase [34, 40, 51]. In addition, protein engineering has been used to improve enzyme properties through a slight change in one or several amino acids of the protein in higher levels of the organic solvent [5]. For example, the activity of subtilisin E was improved by 10-fold upon mutagenesis of only two amino acids in 20% (v/v) dimethylformamide solution [18, 19]. The specific activity of cytochrome P450 BM-3 monooxygenase has been improved by 6-fold in 25 % (v/v) dimethyl sulfoxide [58].

V. Perspectives

One of the simplest synthetic biology examples is to assemble enzymes to implement an unnatural or a natural process. Methods and techniques to utilize the (nearly-)purified enzyme components for the reconstituted pathway for biofuels production have received relatively scant attention as compared to microbial fermentation, despite the fact that microorganisms represent extraordinarily complex and typically labile systems, in which only a small fraction of cellular enzymes are responsible for the formation of the final product. Developing methodologies exploiting only relevant key biofuels-producing enzymes leads to a much superior system for high conversion efficiency and stability since each component can simply be individually optimized *in vitro*. It is anticipated that cell-free synthetic enzymatic pathway engineering (SEPE) may be applied to many areas because of its unique advantages: nearly theoretical product yields, board reaction conditions (e.g., high temperature and low pH), possibly high reaction rate, easy implementation, and operation and control (Table 1).

Nevertheless, two main technical obstacles to wide application of cell-free enzymatic technology are (i) relatively slow reaction rates and (ii) high costs of purified enzymes and cofactors. These challenges can be overcome *via* optimization of enzyme components by kinetic modeling, metabolic control flux analysis, experimental validation, and higher reaction temperatures by substitution of thermophilic or even hyperthermophilic enzymes. Recently, a mathematical model simulating cell-free ethanol conversion clearly infers the potential of increasing ethanol production rate by several fold through enzyme component optimization [2]. Ye and Zhang [60] have presented the feasibility of increasing the hydrogen production rate by at least several hundred folds *via* enzyme ratio optimization and higher enzyme loading. The high cost of enzymes can be relieved by a combination of thermostable enzyme high-throughput discovery, thermostability improvement by protein engineering [17, 54], enzyme immobilization [12, 13, 15], high level expression of recombinant protein [10], low-cost protein purification [9, 31], etc. Moreover, efficient costly NAD(P)H recycling in enzymatic reactions has been implemented for more than 125,000 times, or even 650,000 times [35, 39].

One of the most important lessons for the modern world is to decrease product costs through standardization and mass production, for example, cellular phone, personal computer, railroad gauges, screw threads, internet addresses, gasoline formulations, and so on. We envision that renewable liquid biofuels (cellulosic ethanol and butanol) will replace significant fractions of fossil fuels within 30 years. In the long term, hydrogen-fuel cell systems will play more and more important roles in the transportation section by replacing internal combustion engines, because carbohydrates from biomass can be easily converted to hydrogen by enzymatic technology and carbohydrates can be used as a high-hydrogen-density carrier (14.8 H_2-mass%) [62]. With technology development and integration with polymer electrolyte membrane fuel cells, future mobile appliances are believed to store solid sugar, produce hydrogen from sugar and water *via* cell-free enzymatic reaction, and then to generate electricity by hydrogen fuel cells at the same compact place [61].

Acknowledegment

This work was supported by the USDA, DOE BioEnergy Science Center, the Institute for Critical Technology and Applied Science (ICTAS) of Virginia Tech, the Air Force Office of Scientific Research (FA9550-08-1-0145), and DuPont Young Professor Award.

References

[1]   Adams, M. W. W., and E. I. Stiefel. 1998. Biological hydrogen production: Not so elementary. *Science* 282:1842-1843.

[2]   Allain, E. J. 2007. Cell-free ethanol production: the future of fuel ethanol? *J. Chem. Technol. Biotechnol.* 82:117-120.

[3]   Allcock, E. R., S. J. Reid, D. T. Jones, and D. R. Woods. 1981. Autolytic Activity and an Autolysis-Deficient Mutant of *Clostridium acetobutylicum*. *Appl. Environ. Microbiol.* 42:929-935.

[4]   Antal, M. J., S. G. Allen, D. Schulman, X. Xu, and R. J. Divilio. 2000. Biomass gasification in supercritical water. *Ind. Eng. Chem. Res.* 39:4040-4053.

[5]   Arnold, F. H. 1990. Engineering enzymes for non-aqueous solvents. *Trends Biotechnol.* 8:244-249.

[6]   Atsumi, S., A. F. Cann, M. R. Connor, C. R. Shen, K. M. Smith, M. P. Brynildsen, K. J. Chou, T. Hanai, and J. C. Liao. 2007. Metabolic engineering of *Escherichia coli* for 1-butanol production. *Met. Eng.* Epub:doi:10.1016/j.ymben.2007.08.003

[7]   Baer, S. H., H. P. Blaschek, and T. L. Smith. 1987. Effect of Butanol Challenge and Temperature on Lipid Composition and Membrane Fluidity of Butanol-Tolerant *Clostridium acetobutylicum*. *Appl. Environ. Microbiol.* 53:2854-2861.

[8]   Bai, F. W., W. A. Anderson, and M. Moo-Young. 2008. Ethanol fermentation technologies from sugar and starch feedstocks. *Biotechnol. Adv.* 26:89-105.

[9]   Banki, M. R., L. Feng, and D. W. Wood. 2005. Simple bioseparations using self-cleaving elastin-like polypeptide tags. *Nat. Methods* 2:659-662.

[10]   Barnard, G., G. Henderson, S. Srinivasan, and T. Gerngross. 2004. High level recombinant protein expression in *Ralstonia eutropha* using T7 RNA polymerase based amplification. *Protein Expr. Purif.* 38:264-271.

[11]   Benner, S. A., and A. M. Sismour. 2005. Synthetic biology. *Nat. Rev. Genetics* 6:533-543.

[12]   Bilal, E.-Z., H. Jia, and P. Wang. 2004. Enabling multienzyme biocatalysis using nanoporous materials. *Biotechnol. Bioeng* 87:178-183.

[13]   Bolivar, J. M., L. Wilson, S. A. Ferrarotti, J. M. Guisan, R. Fernandez-Lafuente, and C. Mateo. 2006. Improvement of the stability of alcohol dehydrogenase by covalent immobilization on glyoxyl-agarose. *J. Biotechnol.* 125:85-94.

[14] Bowles, L. K., and W. L. Ellefson. 1985. Effects of butanol on *Clostridium acetobutylicum. Appl. Environ. Microbiol.* 50:1165-1170.

[15] Brandi, P., A. D'Annibale, C. Galli, P. Gentili, and A. S. N. Pontes. 2006. In search for practical advantages from the immobilisation of an enzyme: the case of laccase. *J. Mol. Cataly B: Enzymatic* 41:61-69.

[16] Buchner, E. 1897. Alkoholische Gahrung ohne Hefezellen (Vorlaufige Mittheilung). *Ber. Chem. Ges.* 30:117-124.

[17] Chautard, H., E. Blas-Galindo, T. Menguy, L. Grand'Moursel, F. Cava, J. Berenguer, and M. Delcourt. 2007. An activity-independent selection system of thermostable protein variants. *Nat. Methods* 4:919-921.

[18] Chen, K. Q., and F. H. Arnold. 1991. Enzyme engineering for nonaqueous solvents: random mutagenesis to enhance activity of subtilisin E in polar organic media. *Biotechnology* 9:1073-1077.

[19] Chen, K. Q., A. C. Robinson, M. E. Van Dam, P. Martinez, C. Economou, and F. H. Arnold. 1991. Enzyme engineering for nonaqueous solvents. II. Additive effects of mutations on the stability and activity of subtilisin E in polar organic media. *Biotechnol. Prog.* 7:125-129.

[20] Cortright, R. D., R. R. Davda, and J. A. Dumesic. 2002. Hydrogen from catalytic reforming of biomass-derived hydrocarbons in liquid water. *Nature* 418:964-967.

[21] Deluga, G. A., J. R. Salge, L. D. Schmidt, and X. E. Verykios. 2004. Renewable hydrogen from ethanol by autothermal reforming. Science 303:993-997.

[22] Durre, P. 2007. Biobutanol: an attractive biofuel. *Biotechnol. J.* 2:1525-1534.

[23] Endy, D. 2005. Foundations for engineering biology. *Nature* 438:449-453.

[24] Ezeji, T., N. Qureshi, and H. P. Blaschek. 2007. Butanol production from agricultural residues: Impact of degradation products on *Clostridium beijerinckii* growth and butanol fermentation. *Biotechnol. Bioeng* 97:1460-1469.

[25] Forster, A. C., and G. M. Church. 2007. Synthetic biology projects *in vitro. Genome* Res. 17:1-6.

[26] Goldemberg, J. 2007. Ethanol for a Sustainable Energy Future. *Science* 315:808-810.

[27] Hallenbeck, P. C., and J. R. Benemann. 2002. Biological hydrogen production; fundamentals and limiting processes. *Int. J. Hydrogen Energy* 27:1185-1193.

[28] Heinemann, M., and S. Panke. 2006. Synthetic biology--putting engineering into biology. *Bioinformatics* 22:2790-2799.

[29] Hoffert, M. I., K. Caldeira, G. Benford, D. R. Criswell, C. Green, H. Herzog, A. K. Jain, H. S. Kheshgi, K. S. Lackner, J. S. Lewis, H. D. Lightfoot, W. Manheimer, J. C. Mankins, M. E. Mauel, L. J. Perkins, M. E. Schlesinger, T. A. Volk, and T. M. Wigley. 2002. Advanced technology paths to global climate stability: energy for a greenhouse planet. *Science* 298:981-987.

[30] Holtzapple, M. T., M. Cognata, Y. Shu, and C. Hendrickson. 1990. Inhibition of *Trichoderma reesei* cellulase by sugars and solvents. *Biotechnol. Bioeng.* 36:275-287.

[31] Hong, J., Y. Wang, Ye X, and Z. Y.-H. P. 2008. Simple protein purification through affinity adsorption on regenerated amorphous cellulose followed by intein self-cleavage. *J. Chromatogr. A* 1194:150-154.

[32] Huang, W. C., D. E. Ramey, and S. T. Yang. 2004. Continuous production of butanol by *Clostridium acetobutylicum* immobilized in a fibrous bed bioreactor. *Appl. Biochem. Biotechnol.* 113-116:887-898.

[33] Huber, G. W., J. W. Shabaker, and J. A. Dumesic. 2003. Raney Ni-Sn catalyst for H_2 production from biomass-derived hydrocarbons. *Science* 300:2075-2077.

[34] Hudson, E., R. Eppler, and D. Clark. 2005. Biocatalysis in semi-aqueous and nearly anhydrous conditions. *Curr. Opin. Biotechnol.* 16:637-643.

[35] Hummel, W., H. Schutte, E. Schmidt, C. Wandrey, and M. R. Kula. 1987. Isolation of L-Phenylalanine Dehydrogenase from *Rhodococcus* Sp M4 and Its Application for the Production of L-Phenylalanine. *Appl. Microbiol. Biotechnol.* 26:409-416.

[36] Jeffries, T. W. 2005. Ethanol fermentation on the move. *Nat. Biotechnol.* 23:40-41.

[37] Jones, D. T., and D. R. Woods. 1986. Acetone-butanol fermentation revisited. *Microbiol. Rev.* 50:484-524.

[38] Jungermann, K., E. Rupprecht, C. Ohrloff, R. Thauer, and K. Decker. 1971. Regulation of the reduced nicotinamide adenine dinucleotide-ferredoxin reductase system in *Clostridium kluyveri. J. Biol. Chem.* 246:960-963.

[39] Kragl, U., D. Vasicracki, and C. Wandrey. 1993. Continuous-Processes with Soluble Enzymes. *Indian J. Chemy Sec.* B 32:103-117.

[40] Lee, M., and J. Dordick. 2002. Enzyme activation for nonaqueous media. *Curr. Opin. Biotechnol.* 13:376-384.

[41] Lin, Y., and S. Tanaka. 2006. Ethanol fermentation from biomass resources: current state and prospects. *Appl. Microbiol. Biotechnol.* 69:627-642.

[42] Lynd, L. R., M. S. Laser, D. Bransby, B. E. Dale, B. Davison, R. Hamilton, M. Himmel, M. Keller, J. D. McMillan, J. Sheehan, and C. E. Wyman. 2008. How biotech can transform biofuels. *Nat. Biotechnol.* 26:169-172.

[43] Matsumura, Y., T. Minowa, B. Potic, S. R. A. Kersten, W. Prins, W. P. M. van Swaaij, B. van de Beld, D. C. Elliott, G. G. Neuenschwander, A. Kruse, and M. Jerry Antal Jr. 2005. Biomass gasification in near- and super-critical water: Status and prospects. *Biomass Bioenergy* 29:269-292.

[44] Mohagheghi, A. E., K.; Chou, Y-C.; Zhang, M. 2002. Cofermentation of Glucose, Xylose,and Arabinose by Genomic DNA-IntegratedXylose/Arabinose Fermenting Strainof Zymomonas mobilis AX101. *Appl. Biochem. Biotechnol.* 100:885-898.

[45] Pye, E. K. 1969. Biochemical mechanisms underlying the metabolic osillations in yeast. *Can. J. Botany* 47:271-285.

[46] Qureshi, N., and H. P. Blaschek. 2001. Recent advances in ABE fermentation: hyper-butanol producing *Clostridium beijerinckii* BA101. *J. Ind Microbiol. Biotechnol.* 27:287-291.

[47] Ramey, D., and S. Yang. 2004. Production of Butyric Acid and Butanol from Biomass. U.S. Department of Energy, Morgantown, WV Contract No.: DE-F-G02-00ER86106.

[48] Richter, Burton, and L. D. Schmidt. 2004. Using Ethanol as an Energy Source. *Science* 305:340.

[49] Salge, J. R., B. J. Dreyer, P. J. Dauenhauer, and L. D. Schmidt. 2006. Renewable hydrogen from nonvolatile fuels by reactive flash volatilization. *Science* 314:801-804.

[50] Sanderson, K. 2006. US biofuels: A field in ferment. *Nature* 444:673-676.

[51] Serdakowski, A., and J. Dordick. 2008. Enzyme activation for organic solvents made easy. *Trends Biotechnol.* 26:48-54.

[52] Thauer, R. K., E. Rupprecht, C. Ohrloff, K. Jungermann, and K. Decker. 1971. Regulation of the reduced nicotinamide adenine dinucleotide phosphate-ferredoxin reductase system in *Clostridium kluyveri. J. Biol. Chem.* 246:954-959.

[53] Tomas, C. A., N. E. Welker, and E. T. Papoutsakis. 2003. Overexpression of groESL in *Clostridium acetobutylicum* results in increased solvent production and tolerance, prolonged metabolism, and changes in the cell's transcriptional program. *Appl. Environ. Microbiol.* 69:4951-4965.

[54] Unsworth, L. D., J. van der Oost, and S. Koutsopoulos. 2007. Hyperthermophilic enzymes -- stability, activity and implementation strategies for high temperature applications. *FEBS Journal* 274:4044-4056.

[55] Vollherbst-Schneck, K., J. A. Sands, and B. S. Montenecourt. 1984. Effect of butanol on lipid composition and fluidity of *Clostridium acetobutylicum* ATCC 824. *Appl. Environ. Microbiol.* 47:193-194.

[56] Welch, P., and R. K. Scopes. 1985. Studies on cell-free metabolism: Ethanol production by a yeast glycolytic system reconstituted from purified enzymes. *J. Biotechnol.* 2:257-273.

[57] Wheals, A. E., L. C. Basso, D. M. G. Alves, and H. V. Amorim. 1999. Fuel ethanol after 25 years. *Trends Biotechnol.* 17:482-487.

[58] Wong, T. S., F. H. Arnold, and U. Schwaneberg. 2004. Laboratory evolution of cytochrome p450 BM-3 monooxygenase for organic cosolvents. *Biotechnol. Bioeng* 85:351-358.

[59] Xu, J., and K. B. Taylor. 1993. Characterization of Ethanol Production from Xylose and Xylitol by a Cell-Free *Pachysolen tannophilus* System. *Appl. Environ. Microbiol.* 59:231-235.

[60] Ye, X., and Y. H. P. Zhang. 2007. Presented at the AIChE's 2007 *Annual Meeting Salt Lake City, Utah.*

[61] Zhang, Y.-H. P. 2008. Reviving the carbohydrate economy *via* multi-product biorefineries. *J. Ind. Microbiol. Biotechnol.* 35:367-375

[62] Zhang, Y.-H. P., B. R. Evans, J. R. Mielenz, R. C. Hopkins, and M. W. W. Adams. 2007. High-yield hydrogen production from starch and water by a synthetic enzymatic pathway. PLoS One 2:e456.

[63] Zhang, Y.-H. P., M. Himmel, and J. R. Mielenz. 2006. Outlook for cellulase improvement: Screening and selection strategies. *Biotechnol. Adv.* 24:452-481.

[64] Zhu, Y., Z. Wu, and S. T. Yang. 2002. Butyric acid production from acid hydrolysate of corn fibre by *Clostridium tyrobutyricum* in a fibrous-bed bioreactor. *Proc. Biochem.* 38:657-666.

In: Biotechnology: Research, Technology and Applications   ISBN 978-1-60456-901-8
Editor: Felix W. Richter   © 2008 Nova Science Publishers, Inc.

Chapter 7

Genetic and Environmental Regulation and Artificial Metabolic Manipulation of Artemisinin Biosynthesis

Qing-Ping Zeng[1]*, Rui-Yi Yang[1,] Li-Ling Feng[2] and Xue-Qin Yang[2]

[1]Laboratory of Biotechnology, Tropical Medicine Institute
[2]Artemisinin Research Center; Guangzhou University of Chinese Medicine
Guangzhou 510405 China

Abstract

Artemisia annua is currently sole herbaceous biomass for industrial manufacture of artemisinin, an antimalarial sesquiterpene lactone with the unique endoperoxide architecture. Due to presence in trace amount, artemisinin has been targeted for *in planta* overproduction by genetic modification of *A. annua*. Beneficial from such pursuits, the overall enzymatic cascades involving artemisinin biogenesis have been elucidated and a dozen of critical artemisinin responsible genes identified. Consequently, transgenic *A. annua* plants with enhanced artemisinin production are available although substantial and profound potentials in artemisinin accumulation expected. Alternatively, due to conservation of the entire terpene pathways among higher plants and eukaryotic or even prokaryotic microbes, re-establishment of extended or diverted pathways toward *de novo* microbial artemisinin production has been eagerly attempted in genetically tractable microbes. In such aspect, a suit of downstream pathway genes specific for artemisinin biogenesis have been transplanted from *A. annua* into *Sacchromyce cerevisiae* and *Escherichia coli*, in which a series of incredible amounts of artemisinin precursors manufactured. The next-step goal is to further accelerate forward the total artemisinin biosynthesis through biotransformation of the artemisinin precursor(s) either *in vivo* or *in vitro*. For this purpose, the putative oxidant sink molecule capable of quenching the reactive oxygen species (ROS), in particular, the singlet oxygen ($^1O_2$), must be produced,

---

* Corresponding author, E-mail: qpzeng@gzhtcm.edu.cn

in a large scale, in genetically modified microbes or transgenic *A. annua* plants. Whether dihydroartemisinic acid or artemisinic acid is such a $^1O_2$-scavenging direct intermediate has not been convinced, but conversion from dihydroartemisinic acid or artemisinic acid to artemisinin recognized as a bottleneck for artemisinin biosynthesis and versatile strategies aiming at breaking the rate-limited step enthusiastically pursued in *A. annua*, for example, by utilization of the primary abiotic or biotic stress signals or secondary stress signal transducers. These achievements should benefit our future intervention with the homeostatic tempo-spatial regulation mode of genetic background-based and environment-dependent artemisinin accumulations. This article introduces, from the genomics, transcriptomics, proteomics and metabolomics, the updated literatures describing the relationship between artemisinin biosynthetic gene overexpression and subsequent artemisinin overproduction as well. It should shed light on further elucidation of the intrinsic rule and mechanism underlying that artemisinin biochemical synthesis is fine-tuned by the genetic and environmental regulators, and should also urge the researchers all over the world more intensively investigating the intriguing *A. annua* plant that has implications in the medicinal and aromatic industries.

Keywords: *Artemisinin annua*, Artemisinin, Environmental induction, Gene expression, Microbial engineering, Signal transduction, Transgenic plant.

1. Introduction

Artemisia annua, the Latin nomenclature of Qinghao in Chinese and annual wormwood or sweet wormwood in English, is a well-known medicinal herbage for chemically extracting antimalarial artemisinin, from which a series of therapeutically efficient artemisinin derivatives, including artemether, arteether, artesunate and artelinate were artificially semi-synthesized (Figure 1).

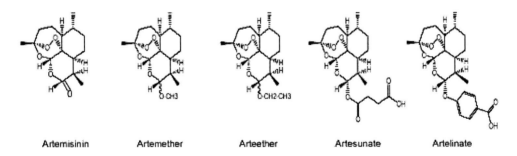

| Artemisinin | Artemether | Arteether | Artesunate | Artelinate |

Figure 1. The structure of naturally isolated artemisinin and artificially semi-synthesized artemisinin derivatives.

Nowadays, these improved artemisinin analogs have been recommended by the World Health Organization (WHO) as the essential component for artemisinin-based combination therapies (ACTs) in order to combat the multi-drug resistant malaria occurring in malarial endemic districts (WHO, 2001). For such consideration, three kinds of artemisinin derivetives, artesunate and artemether have been described in the *International pharmacopoeia* (WHO, 2003) and listed in the *WHO Model List* of *Essential Medicines* (WHO, 2005).

Based on the thousand-year experiences and practices in ancient China for treatment of fever probably caused by malaria, Chinese scientists pioneered extensive search for antimalarial compound(s) in *A. annua* and first purified the antimalarial monomer, artemisinin, a molecule known as qinghaosu or arteannuin as an original nomenclature, at that time, and subsequently identified artemisinin as an intriguing type of the sesquiterpene lactone with a unique endoperoxide bridge (Liu et al. 1979). During that period, artemisinin was analyzed by the nuclear magnetic resonance (NMR), mass spectroscopy (MS) and X-ray crystal diffraction, from which the molecular weight (282.3), chemical formula ($C_{15}H_{22}O_5$) and configuration of artemisinin were successfully revealed.

As a natural secondary metabolite, artemisinin seems to serve as a phytoalexin that confers resistance to the plant itself against pathogenic microorganisms (Stoessl et al. 1976) and attraction to protective insects (Kappers et al. 2005) as other volatile terpenes do. Moreover, it has been recently suggested that artemisinin is most likely as a scavenger of the reactive oxygen species (ROS) to protect cells from damage caused by oxidative stresses (Wallaart et al. 2001).

In broad definition, the term "stress" indicates all extreme environmental conditions that induce generation of ROS. The natural stress circumstances are classified as the biotic stresses (disease and pest ingression and mechanical wounding, etc.) and abiotic stresses (cold, heat, irradiation, anoxia, drought, saline and alkaline, etc (Xiong et al. 2002). Although oxidative stress-induced overproduction of plant secondary metabolites has been documented in *Oryza sativa* (Nojiri et al. 1996; Tamogami et al. 1997), *Catharanthus roseus* (Menke et al. 1999), *Arabidopsis thaliana* (Brader et al. 2001), and *Cupressus lusitanica* (Zhao and Sakai, 2001), investigation regarding the relationship between oxidative stress and artemisinin biosynthesis are just thriving, and available data from these researches are fragmentary and incomplete. Nevertheless, the entire physiochemical process involving generation, reception and transduction of the oxidative stress signals, activation of the transcription factors and their binding to promoters, and induced expression of target genes has been initiated and may soon lead to validation for the hypothesis of oxidative stress induction as a rate-limited step in artemisinin biosynthesis (Zeng et al. 2008a).

This article prospectively summarize the cutting-edge technical frontier and the updated academic viewpoints that intend to attract and encourage global researchers dedicating their endeavors to exploration on the molecular mechanism of artemisinin and other valuable terpenes, and to stimulate their interests and enthusiasms in the understanding of oxidative stress-induced up-regulation of the artemisinin responsible genes as well as other plant secondary metabolism implicated in agriculture, medicine and industry.

2. An Overview on Species Character and Genetic Background of *A. annua*

2.1. Genome Organization and Phylogeny of *Artemisia* Genus

Torrell and Vallès (2001) estimated the genome size of *Artemisia* genera among five subgenera that compose 21 species and three subspecies and summarized their life forms, DNA amounts, ploidy states and karyotypes (Table 1).

Table 1. DNA content and karyological characters of *Artemisia* populations

Taxon	Life cycle	DNA content (pg)	Ploidy level (2n)	Chromosome length (µm)
Subgenus *Seriphidham*		8.80		
A. *fragrans*	P	5.35	18, 2x	24.84
A. *caerulescens*	P	6.66	18, 2x	30.17
A. *herba-alba* subsp. *valentina*	P	6.57	18, 2x	36.16
A. *herba-alba* subsp. *herba-alba*	P	12.48	36, 4x	56.45
A. *barrelieri*	P	12.96	36, 4x	65.36
Subgenus *Artemisia*		7.15		
A. *annua*	A	3.50	18, 2x	19.58
A. *tournefortiana*	A/B	6.69	18, 2x	
A. *vulgaris*	P	6.08	16, 2x	29.51
A. *vulgaris*	P	9.74	34, 4x	44.27
A. *chamaemelofolia*	P	6.04	18, 2x	27.77
A. *molinieri*	P	5.96	18, 2x	26.45
A. *lucentica*	P	7.68	16, 2x	27.33
A. *Judaica*	P	11.52	16, 2x	43.44
Subgenus *Absinthium*		11.26		
A. *absinthium*	P	8.52	18, 2x	38.46
A. *thuscula*	P	10.52	18, 2x	38.33
A. *numelliformis*	P	12.41	34, 4x	63.38
A. *splendens*	P	13.59	32, 4x	58.64
Subgenus *Dracunculus*		13.67		
A. *campestris*	P	5.87	18, 2x	25.29
A. *campestris*	P	11.00	36, 4x	47.18
A. *monosperma*	P	11.02	36, 4x	64.61
A. *crithmifolia*	P	15.60	54, 6x	81.92
A.*dracunculus*	P	23.22	90, 10x	-
Subgenus *Tridentae*		16.91		
A. *tridentatae* subsp. *spiciformis*	P	8.18	18, 2x	-
A. *cana*	P	25.65	72, 8x	-

Note: A: annual; B: biennial; P: perennial.

Although multiple ploidy levels of 2x, 4x, 6x, 8x and 10x are presented in *Artemisia*, two basic chromosome numbers, x = 9 and x = 8, were proposed for the genus, in which the latter chromosome sets may be derived from the former chromosome sets by a possible descendent chromosome fusion-based dysploidy mechanism (Vallès and Siljak-Yakovlev 1997). Later, Kreitschitz and Vallès (2003) counted the chromosome numbers of five *Artemisia* species pooled from Poland and identified the diploid *A. annua* (2n=18), tetraploid *A. absinthium* and *A. abrotanum* (2n=36), and decaploid *A. dracunculus* (2n=90) in Polish *Artemisia* species.

The first report on the cytological description of *A. annua* chromosomes appeared in 1928, when Weinedel-Libebau recognized the basic chromosome numbers (x=9) and somatic diploid (2n=18) among this species. Subsequently, more investigators provided evidence supporting that *A. annua* plants possess diploid chromosomes although the plants under investigation distribute in distinct geological areas in the World, including Hungary (Polya 1949), Japan (Suzuka 1950; Arano 1964), Bulgaria (Kuzamanov et al. 1986), Spain (Valles 1987), the south part of far east region in former Soviet Union (Vokova and Boyco 1986) and the inner Mongolia district of northern China (Fu 1991).

Wang et al. (1999) counted the chromosomes and analyzed the karyotypes of five species growing in the northeast provinces of China and belonging to *Abrotanum* sector of *Artemisia* genus, *A. tanacetifolia*, *A. annua*, *A. adamsii*, *A. palustris* and *A. aurata*. Their investigation indicated that these species are diploid with the common chromosome numbers (2n=18). The somatic chromosome morphology and karyotype of these five *Artemisia* species are illustrated in Figure 2.

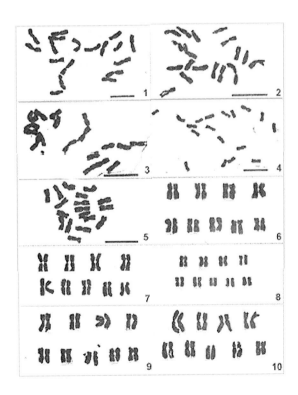

Figure 2. The morphology of somatic chromosome and karyotype in the five species of Artemisia. 1 and 7. A. tanacetifolia; 3 and 9. A. annua; 2 and 10. A. adamsii; 4 and 8. A. palustris; 5 and 6. A. aurata. Bar=10 μm. Adopted from Fu et al. (1991).

The tribe Anthemideae (Asteraceae) was classified into one subtribe Artemisiinae and other outgroup genera. The subtribe Artemisiinae includes the *Artemisia* subgenera (*Artemisia, Absinthium, Dracunculus*) and *Seriphidium* (new world *Seriphidium* and old world *Seriphidium*), which altogether comprise the *Artemisia* group. Watson et al. (2002) described the molecular phylogeny of subtribe Artemisiinae by comparison of the internal transcribed spacers (ITS) of nuclear ribosomal DNA (rDNA). The resulting evolutionary tree is comprised of three major clades that correspond to one radiate genus and two clades of *Artemisia* species. The sequences of ITS1 and ITS2 of rDNA were also analyzed by Vallès et al. (2003). Among 44 *Artemisia* species that represent five classical subgenera, the established eight main clades definitely supported the monophyly of *Artemisia* genus.

2.2. Morphological and Ecophysiological Traits of *A. annua*

A. annua, belonging to the Asteraceae (Compositae) family, is a species of aromatic annual herbal plant that originally distributes in the high altitude steppe region of north China (40_ N, 109_ E and 1000 to 1500 m above the sea level) (Wang 1961). However, the herbal plant now cosmopolitanly grows wild in the World and thrives in Asian, European and American countries (Lin and Lin 1991). Although the tropical Africa seems not suitable to cultivate the plant, large-scale plantation of *A. annua* has been in trials in Kenya and Tanzania.

As a short-day plant with a photoperiod between 12 and 16 h (13.5 h in average and 1000 h in total annual light hours), *A. annua* is very sensitive to short-day light stimulation, by which the branched plants will bloom after two weeks of such light induction. *A. annua* is reproduced by seeds that germinate on the temperature above 7°C and grow at 13.5-17.5°C with the total annual temperature needs of 3500-5000°C (Paniego et al. 1993). The whole plant of *A. annua* compose a single stem (2-6 mm in diameter) with alternative branches reaching 30-100 cm (more than 200 cm for cultivars) in height and deeply dissected leaves ranging from 3–5 cm in length and 2-4 cm in width. The aerial surface of leaves and flowers are covered with multi-celled biseriate glandular trichomes that accumulate mono- and sesquiterpenes (Duke et al. 1987). The glandular trichomes are comprised of ten cells that differentiate into five cell pairs with each cell pair apparently having different functions as deduced from the ultrastructural difference (Duke and Paul 1993). By the light microscopy, the basal, stalk and apical cell pairs are colorless, whereas the two subapical cell pairs are green (Figure 3).

Duke et al. (1994) concluded that artemisinin is sequestered in glandular trichomes of *A. annua* because flowers and leaves that store artemisinin and essential oils own abundant glandular trichomes, while those in a special *A. annua* biotype without or with scarce glands contains no or low artemisinin. Artemisinin is present in *A. annua* in a small amount range from 0.01 to 0.8 % of dry weight (Abdin et al. 2003). The time-course determination of artemisinin content at the different developmental stages revealed a positive correlation between plant growth stages and artemisinin yields, which was attributed to gradual expansion of the leaf surface and progressive increase of the biomass weight during plant growth (Singh et al. 1988). Furthermore, artemisinin content of inflorescence in the bud stage is not higher than in leaves, but it is four- to 11-fold higher in flowers at full bloom than in leaves (Ferreira et al. 1995). Some researchers indicated that artemisinin content is highest just before flowering (Acton and Klayman 1985; Liersh et al. 1986; ElSohly et al. 1990), but others observed that a peak artemisinin yield is reached during full flowering (Pras et al. 1991; Morales et al. 1993; Ferreira et al. 1995).

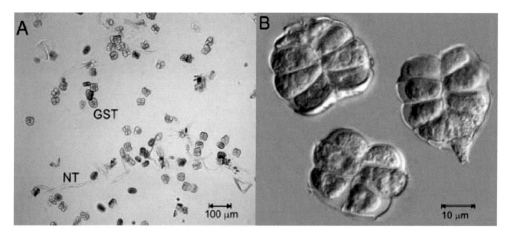

Figure 3. The microscopic morphology of glandular trichomes isolated from the floral bud of *A. annua*. (A) low magnification micrographic view; (B) high magnification phase contrast micrographic view. GST, glandular secretory trichome; NT, nonglandular trichome. Adopted from Covello et al. (2007).

The most abundant type of sesquiterpenes in *A. annua* is artemisinic acid, which occurs in an eight- to tenfold higher concentration (Roth and Acton 1989; Jung et al. 1990) and followed by arteannuin B (Klayman 1993). The fluctuation of artemisinin and the related sesquiterpene intermediates, such as artemisinic acid, arteannuin B, and artemisitene, in *A. annua* during the vegetation period was studied in Vietnamese varieties (Woerdenbag et al. 1994). The highest levels of artemisinin (0.86% dry weight), artemisinic acid (0.16% dry weight) and arteannuin B (0.08% dry weight) were measured in the leaves of five- month-old plants that possess highest leaf weight.

Recently, Lommen et al. (2007) found that the concentration levels of total artemisinin plus its precursors that include dihydroartemisinic acid, dihydroartemisinic aldehyde, dihydroartemisinic alcohol, artemisinic acid, artemisinic aldehyde, and artemisinic alcohol are higher in green leaves than in dead leaves in the younger stage, but are comparable at the final harvest stage. This means that conversion of the precursors to artemisinin is more advanced in dead leaves than in green leaves, which leads to the concentration of artemisinin being higher in dead leaves than in green leaves.

2.3. Chemical Constituents of *A. annua* Implicated in Medical and Fragrant Industries

The extensive searches for active compounds in *A. annua* have led to isolation of more than 150 kinds of secondary metabolites, including terpenoids, flavonoids, coumarins, steroids, phenolics, purines, lipids and aliphatic compounds (Bhakuni et al. 2001). The reported olefinic monoterpenes, sesquiterpenes, diterpenes and triterpenes are alphabetically listed in Table 2.

Table 2. The identified terpenoids and isolated terpene intermediates in *A.annua*

Terpenoids	Terpenoids
abscisic acid	chrysanthenone
abscisic acid methyl ester	cineol, 1-4
amorpha-4, 11-diene	cineol, 1-8
amyrenone, α	copaene, α
amyrin, α	copaene, β
amyrin, β	cymene, para
amyrin acetate, β	decan-2-one
annuic acid, *nor*	elemene, β
annulide	eudesma-4(15)-11-diene, 5-α-hydroxy
annulide, iso	farnesene, β
artemisia alcohol	farnesene, β trans
artemisia ketone	farnesene, *trans*-β
artemisia ketone, iso	fenchol
artemisinic acid	fenchone
artemisinic acid, dihydro	friedelan-3-beta-ol
artemisinic acid hydroperoxide, dihydro	friedelin
artemisinic alcohol	germacrene A
artemisinic alcohol, dihydro	germacrene D
artemisinic aldehyde	germacrene, bicyclo
artemisinic aldehyde, dihydro	guaiene, α
arteannuin A(Qinghaosu I)	hentriacontan-1-ol-triacontanoate
arteannuin B (Qinghaosu II)	hepta-3-trans-5-diene-2-one, 6-methyl
arteannuin B, deoxy (Qinghaosu III)	hept-2-ene, bicyclo (3, 1, 1), 3-7-7-trimethyl
arteannuin B, dihydro	hex-2-en-al
arteannuin B, dihydrodeoxy	hexcis-3-en-1-ol
arteannuin B, dihydro-*epi*-deoxy	hexacosan-1-ol
arteannuin B, *epi*-deoxy	hexadecanoic acid ethyl ether
arteannuin C	hexan-1-ol acetate
arteannuin D (Qinghaosu IV)	hexan-1-ol, 2-ethyl
arteannuin E (Qinghaosu V)	humulene, α
arteannuin F	limonene
arteannuin G	linalool
arteannuin H	linalool acetate
arteannuin I	longipinene
arteannuin J	menthen-4-ol, para
arteannuin K	menthol
arteannuin L	menthol, 2-hydroxy
arteannuin M	muurola-4,11-diene

Table 2. (Continued).

Terpenoids	Terpenoids
arteannuin N	myrcene
arteannuin O	myrcene, α-hydroperoxide
artemisin	myrcene, β-hydroperoxide
artemisinic acid methyl ester	myrtenal
artemisinic acid, α-epoxy	myrtenol
artemisinic acid, dihydro	nerolidol
artemisinic acid, 6, 7-dehydro	nortaylorione
artemisinic acid, dihydro, methyl ester	octan-1-ol
artemisinin (qinghaosu)	oleanolic acid
artemisinin, dehydro	phytene-1-2-diol
artemisinin, deoxy	phytol, *trans*
artemisinol	pinene
artemisitene	pinene, α
bisabolene, β	pinene, β
borneol	pinocamphone
borneol, acetate	pinocarveol, trans
cadina-4(15)-11-dien-9-one	sabiene
cadina-4(7)-11-dien-12-al	sabiene, cis-hydrate
cadinane, *seco*	selina-4,11-diene
cadinane, dihydroxy	selinene, β
cadinene, γ	taraxasterone
cadin-4-en-11-ol, 3-iso-butyryl	taraxerol acetate
cadin-4-ene, 3-α-7-α-dehydroxy	terpinen-4-ol
camphene	terpinene, α
camphene hydrate	terpinene, γ
camphor	terpineol, α
cineole, 1, 8	thujene, α
caryophyllene, α	thujone
caryophyllene, β	thujone, α
caryophyllene, oxide	thujone, iso
caryophyllene, trans	Trycyclene
cedrol	Ylangene

The content of essential oil in *A. annua* is generally as 0.02-0.49% (on a fresh weight basis) or 0.04-1.9% (on a dry weight basis), but depends on its geographical origin. Woerdenbag et al. (1993) detected 4.0% and 1.4% essential oil (V/W) from Chinese and Vietnamese varieties, respectively. The Chinese varieties-originated essential oil includes *artemisia* ketone (63.9%), *artemisia* alcohol (7.5%), myrcene (5.1%), *a*-guainene (4.7%), and camphor (3.3%), while the Vietnamese varieties-derived essential oil contains camphor (21.8%), germacrene D (18.3%), *a*-caryophyllene (5.6%), trans-*a*-farnesene (3.8%), and 1,8-cineole (3.1%), but no *artemisia* ketone. Woerdenbag et al.(1994) measured the maximum oil content (containing 55% of monoterpenes) prior to the flowering period in Vietnamese varieties. Hethelyi et al. (1995) analyzed Hungarian varieties-isolated essential oil content (0.48-0.81%) from the flowering shoots, and found that the essential oil mainly consists of *artemisia* ketone and *artemisia* alcohol that vary from 33% to 75% and from 15% to 56%, respectively. *A. annua* plants grown in India contains *artemisia* ketone (58.8%), camphor (15.8%), 1, 8-cineole (10.2%), and germacrene D (2.4%) (Bhakuni et al. 2001). Until now, more than 70 constituents have been investigated and identified in *A. annua* (Li et al. 2006).

Artemisinin is a potent antimalarial agent and can clean the chloroquine- and quinine-resistant *Plasmodium falciparum* strains, also known as multi-drug resistant strains. Administration of artemisinin will allow 90% of malarial patients getting rid of death within 48 h, although the disease may re-occur in a short intermittent phase. Therefore, WHO has recommended the ACTs for malaria therapy, which include other long-term antimalarial drugs in addition to artemisinin. Furthermore, artemisinin is able to kill other parasites, such as *Schistosoma japonicum*, *Clonorchis sinensis*, *Theileria annulatan* and *Toxoplasma gondii*. Although the neurotoxicity occurs in experimental animals upon high doses of artemisinin, no significant clinic toxicity appears in patients. In similar, high-dose of artemisinin induces fatal resorption in animals, but does not show any mutagenic and teratogenic effects on the pregnant women infected by severe malaria or uncomplicated malaria.

In the cytotoxicity assays of artemisinin and derivatives to Ehrlich ascite tumor cells, Woedenbag et al. (1993) estimated the 50% inhibitory concentration (IC_{50}) of artemisinin, artesunate, artemether and arteether as 12.2-29.8 μM, artemisitene as 6.8 μM, and the dimmers of dihydroartemisinin as 1.4 μM. Zheng et al. (1994) and Jung (1997) determined the significant cytotoxicity of artemisinin and semi-synthetic analogs on L-1210, P-388, A-549, HT-29, MCF-7 and KB tumor cells. Beekman et al. (1997a; 1997b) also detected the stereochemistry-dependent cytotoxicity of artemisinin and analogs.

Artemisinin even exhibits activities of anti-arrhythmia (Wang et al. 1998) and anti-hepatitis B virus (Romero et al. 2005), while artemisinic acid shows antibacterial activity (Roth and Acton 1989). Artemisinin and dihydroartemisinin play marked suppression effects on the humeral responses in mice at the high dosage, but do not alter the delayed-type hypersensitivity response to sheep erythrocytes (Tavlik et al. 1990). In a therapy trial, 56 patients with systemic lupus erythematosus (with the types of DLE 16, SCLE 10 and SLE 30) were treated with intravenous injection of artesunate (60 mg once a day, 15 days of a course, and two to four course), during which the therapy efficacies of 94%, 90% and 80% were estimated for each type of systemic lupus erythematosus. *Artemisia* ketone plays versatile roles of anti-inflammation, angiotensin converting enzyme inhibition, cytokinin-like and antitumor. Polymethoxyflavones, casticin (Yang et al. 1988), artemetin, chrysosplenetin, chrysosplenol-D and circilineol (Cubukcu et al. 1990) possess the weaker activity against *P. falciparum*. The coumarin scopoletin has anti- inflammatory activity (Huang et al. 1993) and the flavonoids fisetin and patuletin-3,7- dirhamnoside are non-peptide angiotensin converting enzyme inhibitors (Lin et al. 1994;).

Furthermore, Duke et al. (1987) and Chen et al. (1987) found that artemisinin exhibits the growth inhibitory activity to plant growth and accounted reduction of root growth in lettuce for about 50% at 33 μM of artemisinin. Bagchi et al. (1997) also observed the plant growth regulatory activity of artemisinin and artesunate. In addition, Shukla et al. (1992)

described the plant growth regulatory activity of abscisic acid, abscisic acid methyl ester and bis(1-hydroxy-2-methylpropyl) phthalate. All these results demonstrated that artemisinin and other terpene metabolites presented in *A. annua* may be used as potential herbicides in agriculture.

3. Elucidation of *In Vivo* Artemisinin Biosynthetic Pathway

3.1. Unconfirmed Artemisinin Biosynthetic Pathway Deduced from Intermediates

Although the complete pathway for artemisinin biosynthesis has not been established, most biochemical intermediates have been identified and some enzyme catalytic steps elucidated *in vitro* and *in vivo*. Akhila et al. (1987) proposed an artemisinin biosynthetic pathway that starts from mevalonate, in which inclusion of the intermediate candidates, isopentenylpyrophosphate (IPP), farnesylpyrophosphate (FPP), germacrane skeleton, dihydrocostunolide, cadinanolide and arteannuin B were suggested. Later, these authors detected generation of artemisinic acid from mevalonate (Akhila et al. 1990). In *A. annua,* artemisinic acid is 8-10 times abundant as artemisinin, so artemisinic acid was suggested as another biogenetic precursor of artemisinin (El-Feraly et al. 1986; Roth et al. 1987; Jung et al. 1990; Sangwan et al. 1993).

By the radioactive isotope-labeled precursor feeding, Wang et al. (1988) converted ^{3}H-labeled artemisinic acid (C-15) to arteannuin B and artemisinin although via separate pathways, so they tentatively concluded that artemisinic acid may be a common precursor of arteannuin B and artemisinin. This result was later supported by Sangwan et al. (1993), who confirmed *in vitro* and *in vivo* transformation of artemisinic acid to arteannuin B and artemisinin. Other confirming experimental data were then compiled by several research groups. Kudakasseril et al. (1987) and Martinez and Staba (1988) converted IPP to arteannuin B and artemisinin. Nair and Basile (1993) and Roth and Acton (1989) converted arteannuin B into artemisinin. Bharel et al. (1998) accompleshed *in vitro* biotransformation of artemisinic acid, artemisinin B and dihydroartemisinin B to artemisinin. However, in a similar experiment, Wang et al. (1993) found that artemisinin B is not a precursor to artemisinin.

On the other hand, as an endoperoxide closely related to artemisinin, artemisitene was ever isolated and characterized from *A. annua* (Acton et al. 1985). Artemisitene presents at all stages of development with amounts ranging from 0.002% to 0.09% dry mass, and the ratio of artemisitene to artemisinin increases from 1:10 in the early growth stage to 1:1 when flowers develop. Therefore, artemisitene was also reasonably considered as a candidate precursor to artemisinin. Afterwards, Kim and Kim (1992) transformed dihydroartemisinic acid into artemisinin by the cell-free extracts from teratoma cells but not from leaves or calli of *A. annua.* Li et al. (1994) synthesized [15-^{14}C]-labeled artemisinin in the supernatants prepared from the tender leaves of ripe *A. annua* with addition of [15-^{14}C] dihydroartemisinic acid as a starting compound.

Abdin et al. (2000) isolated two kinds of proteins from *A. annua* leaf cell extracts and confirmed their involvement in conversion from artemisinic acid to artemisinin. An enzyme that catalyzes artemisinin B to generate artemisinin in *A. annua* leaves was partially purified by Dingra et al. (2000) and then completely purified by Dhingra and Narasu (2001). Unfortunately, the target genes encoding those above enzymes have not been cloned even past so many years. Following isolation of amorpha-4, 11-diene synthase (ADS), amorpha-4, 11-diene was verified as the committed product from the cyclization reaction of FPP (Bouwmeester et al. 1999). Later, Wallaart et al. (2000) found that increased artemisinin content is concomitant with decreased dihydroartemisinic acid content in *A. annua*. Meanwhile, those plants with increased artemisinin content exhibit a higher dihydroartemisinic acid level, but a lower artemisinic acid level.

This result led them to conclude that dihydroartemisinic acid rather than artemisinic acid is an immediate precursor of artemisinin, and that conversion from dihydroartemisinic acid to the corresponding hydroperoxide may represent a rate-limiting step during artemisinin biosynthesis. Bertea et al. (2005) detected from *A. annua* not only relevant intermediates including (dihydro)artemisinic alcohol, aldehyde and acid in addition to artemisinic acid, but also multiple enzymatic activities presumably derived from unidentified cytochrome P450 enzyme(s) that at least involve amorphadiene hydroxylase, artemisinic alcohol dehydrogenase and artemisinic aldehyde dehydrogenase. As expected by above presume, a multi-functional enzyme, cytochrome P450 monooxygenase (CYP71AV1), was experimentally identified by Teoh et al. (2006) from *A. annua* and further confirmed by the Keasling Laboratory through expression of the recombinant *CYP71AV1* gene in microbes (Ro et al. 2007; Chang et al. 2007).

The cytosolic isoprene or terpene metabolic pathway in plants, now known as the mevalonate (MVA) pathway that is involved in artemisinin biosynthesis, can be divided into the upstream common stage that presents in all plants and the downstream specific stage that only sequesters in *A. annua*. It has been known that the linear terpene backbone is synthesized in the upstream route from acetyl coenzyme A to FPP, and the circular terpene precursor is generated during the downstream phase from FPP to amorpha-4,11-diene, artemisinic alcohol, aldehyde, acid and artemisinin in *A. annua*. However, the exact chemical process from amorpha-4,11-diene to artemisinin acturally remains suggestive and awaits further experimental elucidation.

Based on all available data, Bertea et al. (2005) have suggested an integrative and bidirectional biosynthetic pathway from amorpha-4,11-diene to artemisinin, in which one direction is via the currently detected intermediates *in vivo*, (dihydro) artemisinic alcohol, aldehyde, and acid; another direction is with inclusion of the previously identified arteannuin B, artemisitene and artemisinic acid (Figure 4).

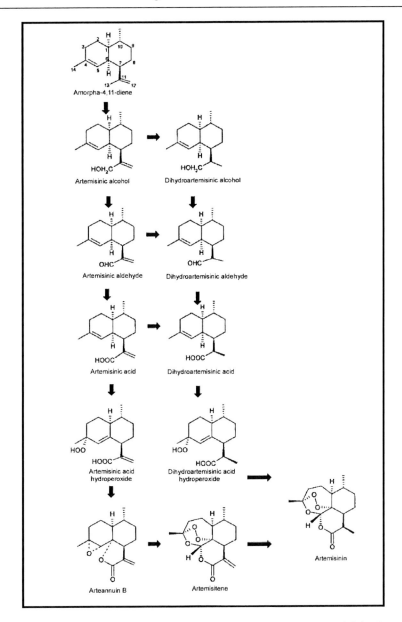

Figure 4. The proposed biosynthetic pathway from amorpha-4,11-diene to artemisinin. Some of the enzymatic or non-enzymatic steps shown by the dash arrows and intermediates indicated in the figure are not yet clearly identified. Adopted from Zeng et al. (2008a).

3.2. Possibly Concomitant or Exclusive Presence of Enzymatic and Non-Enzymatic Reactions

The possibility regarding concomitant presence of the bidirectional pathways toward artemisinin either from artemisinic acid or dihydroartemisinic acid can not be completely eliminated, but the detailed mechanism and subsequent fate (continuous conversion or as a final product?) are uncertain. Brown and Sy (2004) fed the intact *A. annua* plant with isotope-labeled dihydroartemisinic acid and detected 16 kinds of 12-carboxyamorphane and cadinane sesquiterpenes that include a small proportion of labeled artemisinin, suggesting that dihydroartemisinic acid is converted to artemisinin. Furthermore, they also confirmed that the committed product of dihydroartemisinic acid is an allylic hydroperoxide that originates from a non-enzymatic catalysis by the molecular oxygen rather than from an enzymatic step. This observation led them to conclude that the main 'metabolic route' for dihydroartemisinic acid in *A. annua* involves a spontaneous autooxidation mechanism.

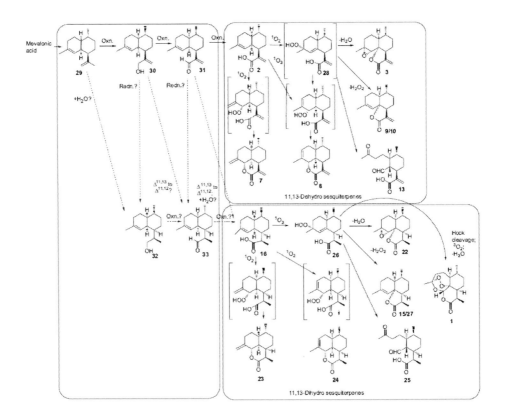

Figure 5. A unified biosynthetic pathway deduced from isotope-labeling data of 11,13-unsaturated or saturated sesquiterpenes in A. annua. 1. artemisinin; 2. artemisinic acid; 3. arteannuin B; 7. annulide; 8. isoannulide; 9. epi-deoxyarteannuin B; 10. deoxyarteannuin B; 13. seco-cadinane; 15. dihydro-epi-deoxyarteannuin B; 16. dihydroartemisinic acid; 22. dihydroarteannuin B; 23. arteannuin I; 24. arteannuin J; 25. seco-cadinane; 26. allylic hydroperoxide of dihydroartemisinic acid; 27. dihydro-deoxyarteannuin B; 28. allylic hydroperoxide of artemisinic acid; 29. amorpha-4,11-diene; 30. artemisinic alcohol; 31. artemisinic aldehyde; 32. dihydro artemisinic alcohol; 33. dihydroartemisinic aldehyde. Oxn. oxidation; Redn. Reduction. Adapted from Brown and Sy (2007).

If this conclusion is neatly reaching the reality, we are able to reasonably infer that the non-enzymatic pathway directing dihydroartemisinic acid to artemisinin is likely presented exclusively in *A. annua*. Brown and Sy (2007) fed isotope-labeled artemisinic acid to *Artemisia annua* and isolated seven labeled sesquiterpene metabolites, comprising of arteannuin B, annulide, isoannulide, *epi*-deoxyarteannuin B, deoxyarteannuin B, *seco*-cadinane and artemisinic acid methyl ester, but not including artemisinin. The fact that all of the sesquiterpenes retain their unsaturation at the 11,13-position and do not convert into any 11,13-dihydro metabolites allowed them to draw a conclusion that artemisinic acid-derived amorphane and cadinane sesquiterpenes do not convert to artemisinin (Figure 5).

Nevertheless, the authors also acknowledged that although any significant 'direct' pathway from artemisinic acid to artemisinin appear to be ruled out, but it is still possible that the 'indirect' route via arteannuin B may be presented. In addition, due to a short-term feeding course, there may be not sufficient time to accumulate a detectable amount of labeled artemisinin.

This conclusion is conceivable if the conversion rate from arteannuin B to artemisinin is much slower than the observed conversion rate from artemisinic acid to arteannuin B. Therefore, this issue may be resolved, in the future, by using the isotope-labeled arteannuin B as a feeding precursor. From such conclusive result, we are able to figure out although two parallel pathways toward artemisinic acid and dihydroartemisinic acid are concomitantly presented in *A. annua*, their subsequent fatalities are totally distinct, i.e. while dihydroartemisinic acid can convert to artemisinin, artemisinic acid cannot.

3.3. Experimental Evidence on Dihydroartemisinic Acid as a ROS Scavenger and Artemisinin as a ROS Pool

The installation of endoperoxide bridge structure in artemisinin, either enzymatically or non-enzymatically, is still not conclusive. Wallaart et al. (1999a; 1999b) found a novel intermediate dihydroartemisinic acid hydroperoxide that was previously undetected in *A. annua*, and suggested that dihydroartemisinic acid hydroperoxide may be derived from oxidation of dihydroartemisinic acid by $^1O_2$. From this point, they further proposed that dihydroartemisinic acid is first converted to dihydroartemisinic acid hydroperoxide via a light-involved and $^1O_2$-catalyzed reaction, and then the resultant dihydro- artemisinic acid hydroperoxide is autooxidized to artemisinin in air. Such deduced reaction mechanism underlying the non-enzymatic conversion from dihydroartemisinic acid to artemisinin has been previously validated by Sy and Brown (2002), who demonstrated that dihydroartemisinic acid can be slowly autooxidized into artemisinin through two steps, the first step involves light but the second step completes in the dark.

Since Knox and Dodge (1985) concluded that $^1O_2$ can be generated and emitted from the plant cell as exposure to CO_2 shortage, freezing, strong irradiation and supplement with photosynthesis inhibitors, more reports have indicated that artemisinin production is strongly influenced by the climatological conditions (Chen and Zhang 1987; Martinez and Staba 1988; Ferreira et al. 1995). The post-harvest drying process of *A. annua* is beneficial to boosting artemisinin accumulation (Laughlin 1993). Environmental stresses, such as extreme light, temperature, water and salt, significantly alter the artemisinin yield (Weathers et al. 1994). Irfan et al. (2005) reported that high concentration of salts and heavy metals augment the intracellular osmotic pressure and lead to high efficient conversion to artemisinin. Lommen et al. (2007) found that the artemisinin content in dead leaves is higher than young leaves in *A. annua,* thereby potentiating the previous postulation of post-harvest drying effects on artemisinin accumulation. Yin et al. (2008) established a positive correlation of chilling stress to overexpression of artemisinin biosynthetic genes and to overproduction of artemisinin.

It seems that dihydroartemisinic acid may act as a scavenger of ROS capable of specifically quenching $^1O_2$ in cytosol, like the carotenoids in chloroplast, for protecting the mesophyll cells from oxidative stress-mediated damage. Nevertheless, what is the real ecological implications of artemisinin and how the cytosolic dihydroartemisinic acid cooperates with the plastidic carotenoids in dealing with the harmful $^1O_2$ needs more detailed investigations.

3.4. Crosstalk and Flux Exchange between Cytosolic and Plastidic Terpene Pathways for Artemisinin Biosynthesis

It has been know that in higher plants, two independent pathways that locate in the separate intracellular compartments are involved in terpene synthesis: the cytosolic MVA pathway and the plastidic non-MVA pathway, also called 2-C-methyl-D-erythritol-4-phosphate (MEP) or 1-deoxy-D-xylulose-5-phosphate (DXP) pathway (Newman et al. 1999). The MVA pathway provides the biogenic precursors of sesquiterpenes, triterpenes (including sterols and brassinosteroids), polyterpenes (e.g. dolichol), polyprenols and the phytohormone cytokinin, whereas the MEP/DXP pathway is involved in generation of monoterpenes, diterpenes (e.g. phtoenol), photosynthesis-related terpenoids such as carotenoids, plastoquinone, phylloquinones and the side chains of chlorophylls, and phytohormones including abscisic acid (ABA) and gibberellins. In the cytosol, farnesyl pyrophosphate (FPP) is as a common precursor for all terpene synthesis; In plastids, geranylpyrophosphate (GPP) is a precursor of monoterpenes, while geranylgeranyl pyrophosphate (GGPP) is a precursor of diterpenes, tetraterpene and plastoquinones (Mahmoud and Croteau 2002). In addition, except for the cytosolic and plastidic pathways, there is an incomplete mitochondrial pathway that uniquely involves ubiquinone biosynthesis (Figure 6).

As seen from Figure 6, the shuttle intermediate among different compartments for carbon flux exchange is IPP, which can be transported into/out of plastids through the membrane-located transporters. Whether the IPP pool stored in *A. annua* plastids contributes a much proportion of carbon source to artemisinin biosynthesis carrying out in the cytosol is yet to be elucidated.

Although the subcellular compartmentations allow two distinct pathways to operate independently, there is increasing evidence that they well cooperate in metabolite biosynthesis. For example, sesquiterpene labeling and quantitative ^{13}C- nuclear magnetic resonance spectroscopy showed that the chamomile sesquiterpene is composed of two C5 terpenoid units formed via the MEP/DXP pathway with a third unit being derived from both the MVA and MEP/DXP pathways (Adam and Zapp 1998). In *Arabidopsis thaliana*, the MEP/DXP pathway can compensate for the reduced flux through the inhibited MVA pathway and *vice versa* (Laule et al. 2003). Using a tobacco Bright Yellow-2 cell suspension system, Hemmerlin et al. (2003) also investigated the cross-talk between such two subcellular pathways by incorporation of the labeled 1-deoxy-D-xylulose into the intact plants. Their results indicated that the sterols normally derived from MVA pathway can be also synthesized via the MEP/DXP pathway in presence of an inhibitor of HMG-CoA reductase (HMGR), mevinolin (MEV), and that growth inhibition caused by an inhibitor of DXP reductoisomerase (DXR), fosmidomycin (FSM), can be partially overcome by the MVA pathway (Hemmerlin et al. 2003).

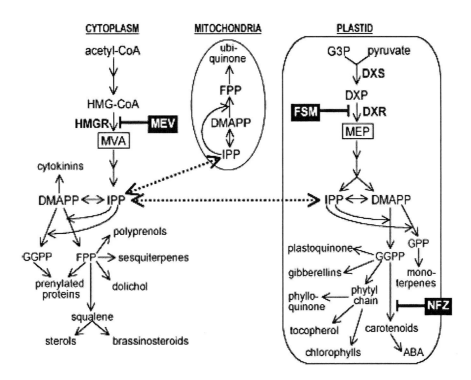

Figure 6. Isoprenoid biosynthetic pathways in distinct subsellular compartments of plant cells. ABA: abscisic acid; DMAPP: dimethylallyl diphosphate; DXP: deoxyxylulose 5-phosphate; DXR: DXP reductoisomerase; DXS: DXP synthase; FPP: farnesyl pyrophosphate; FSM: fosmidomycin; GGPP: geranylgeranyl pyrophosphate; HMG-CoA: 3-hydroxy-3-lmethylglutaryl coenzyme A; HMGR: HMG-CoA reductase; G3P: glyceraldehyde 3-phosphate; GPP: geranyl pyrophosphate; IPP: isopentenyl pyrophosphate; MEP: methylerythritol 4-phosphate; MEV: mevinolin; MVA: mevalonic acid; NFZ: norflurazon. Adopted from Rodriguez- Concepcion et al. (2004).

Evidence of cross-talk between pathways has also been documented in other plant species (Croteau et al. 2000). For instance, interaction of both pathways on biosynthesis of monoterpenes and sesquiterpenes in lima beans (Piel et al. 1998), of gibberellins in *Arabidopsis* (Kasahara et al. 2002), and of sesquiterpene germacrene D in *Solidago Canadensis* (Steliopoulis et al. 2002). All these results indicated that MVA-derived precursors seems to be imported into the plastids, and correspondingly, MEP/DXP- derived precursors can also be exported to the cytosol. However, Dudareva et al. (2005) convinced that the MEP/DXP pathway provides IPP for both plastidic monoterpene and cytosolic sesquiterpene biosynthesis in the epidermis of snapdragon petals, but the trafficking of IPP occurs unidirectionally from plastids to the cytosol.

Recent experimental results provided much support to presence of the exporter system on the plastid envelope membranes that enable terpenoid intermediates cross out of plastids in *Arabidopsis thaliana*, spinach, kale, and Indian mustard. The isolated chloroplast envelopes can transport IPP and GPP with higher efficiencies, but transport FPP and DMAPP with lower rates. Such conclusion implied that there are specific transporters embedded in the plastid bilayer membranes (Bick and Lange 2003). Flugge et al. (2005) confirmed that plastidic phosphate translocator family members are capable of exporting the terpene intermediates across the plastid envelopes.

Despite there is ample experimental data regarding the metabolic cross-talk between two pathways, little is known about the regulation manner of flux exchange. After treating *A. thaliana* with MEV or FSM, researchers found that the dramatic inhibitor-mediated changes in the end-product levels are not reflected in the altered gene expression levels of biosynthetic enzymes (Laule et al. 2003), thereby indicating that the post- transcriptional event plays a major role in the temperal regulation of the pathways. Some novel regulatory factors comprising exoribonuclease polyribonucleotide phosphorylase (Sauret-Gueto et al. 2006) and pentatricopeptide repeat protein (Kobayashi et al. 2007) were thought to regulate the normal accumulation of enzyme quantities by a post- transcriptional regulation mode. Moreover, phytochromes may also regulate biosynthesis and exchange of terpenoid precursors through light perception and signal transduction (Rodriguez-Concepcion et al. 2004).

Adam et al. (1998) and Steliopoulis et al. (2002) showed that the sesquiterpene-directed IPP carbon comes from both the MEP/DXP and MVA pathways in chamomile and goldenrod, which are members of the Compositae family and closely related to *A. annua*. Now the question is whether the MEP/DXP pathway contributes to artemisinin biosynthesis that normally take place via the MVA pathway. By treating *A. annua* plants with the MVA pathway inhibitor MEV or/and MEP/DXP pathway inhibitor FSM, Towler and Weathers (2007) found that artemisinin production is significantly reduced in presence of each inhibitor, thereby proving that both pathways are involved in artemisinin biosynthesis. Our experimental data also demonstrated that both the pathway inhibitors, lovastatin (LVT) and FSM promote transcription of genes that encode the plastidic HMGR and DXR in *A. annua*, in which block of the MVA pathway with 5 µM of LVT led to twofold elevation of *hmgr* and *dxr* mRNA levels, and block of the MEP/DXP pathway with 100 µM of FSM led to threefold elevation of *hmgr* and *dxr* mRNA levels (data not shown). These data indicated a possible metabolic cross-talk between the cytosolic and plastidic pathways in regard of terpene biosynthesis in *A. annua*.

4. Genetic and Environmental Regulation Mechanisms on Artemisinin Biosynthesis

4.1. Artemisinic Biosynthetic Genes and Novel EST Cloning from *A. annua*

Since 1995, a dozen of genes related to artemisinin biosynthesis have been cloned from *A. annua*, and their complete or partial mRNA sequences accessed in GenBank (Table 3).

At present, functional genomic investigations regarding artemisinin biosynthesis in *A. annua* have been emphasized in several laboratories (Bertea et al. 2006; Covello et al. 2007; Zeng et al. 2008b), which would, in the near future, lead to discovery of a large batch of useful genes encoding structural and regulatory proteins involved in the anabolism of artemisinin and other valuable secondary metabolites. While two plastidic enzymes, DXS and DXR, in the MEP/DXP pathway may be indirectly responsible for artemisinin biosynthesis, other cytoplasmic enzymes including ADS, CYP71AV1 and CPR are particularly relevant to such a process.

Bouwmeester et al. (1999) first reported the partial purification of natural ADS enzyme from *A. annua*. This enzyme has a pH optimum around 6.5-7.0, a K_m of 0.6 µM for FPP, and a molecular weight of 56 kDa. The *ADS* gene were later identified and cloned by several groups independently (Mercke et al. 2000; Chang et al. 2000; Wallaart et al. 2001).

Table 3. Artemisinin biosynthetic genes and other related *A. annua* genes accessed in GenBank

Gene	Enzyme/location/sequence type	GenBank accession	Submitter
HMGR /HMG	3-hydroxy-3-methylglutaryl coenzyme A (HMG-CoA)		
	reductase, cytosol	AAU14624	Kang et al. 1995
	mRNA, complete cds	AAU14625	Kang et al. 1995
	mRNA, complete cds	AF142473	Chen et al. 1999
	mRNA, partial cds	AF156854	Chen et al. 1999
	mRNA, complete cds		
FPPS/FPS	farnesyl diphosphate (FPP) synthase, cytosol		
	mRNA, complete cds	AF136602	Chen et al. 1999
	mRNA, complete cds	AF149257	Liu et al. 1999
	mRNA, complete cds	AF112881	Chen et al. 2000
DXS	1-deoxy-D-xylulose-5-phosphate synthase (DXS), plastid	AF182286	Wobbe et al. 2000
	mRNA, complete cds		
DXR	1-deoxy-D-xylulose-5-phosphate reductoisomerase		
	(DXR), plastid	AF182286	Wobbe et al. 2000
	mRNA, partial cds	AF182287	Wobbe et al. 2000
	mRNA, complete cds		
ADS	amorpha-4,11-diene synthase (ADS), cytosol		
	mRNA, complete cds	AF138959	Mercke et al. 2000
	mRNA, complete cds	AF327527	Liu et al. 2001
	mRNA, complete cds	AY006482	Wallaart et al.
	mRNA, complete cds	AJ251751	2001
	mRNA, complete cds	DQ241826	Chang et al. 2005
	mRNA, complete cds	EF197888	Huang et al. 2005
			Kong et al. 2007

Table 3. (Continued).

Gene	Enzyme/location/sequence type	GenBank accession	Submitter
CYP71AV1	cytochrome P450 monooxygenase (CYP71AV1), endoplasmic reticulum		
	mRNA, complete cds	DQ268763	Ro et al. 2006
	mRNA, complete cds	DQ315671	Teoh et al. 2006
	mRNA, complete cds	DQ453967	Olsson et al. 2006
	mRNA, complete cds	DQ872632	Yin et al. 2006
	mRNA, complete cds	DQ667171	Kong et al. 2006
	mRNA, complete cds	EF197889	Kong et al. 2007
CPR	cytochrome P450 reductase (CPR), cytosol		
	mRNA, complete cds	DQ104642	Ro et al. 2006
	mRNA, complete cds	DQ318192	Yin et al. 2006
	mRNA, complete cds	EF197890	Kong et al. 2007
SS/SQS	squalene synthase (SS), endoplasmic reticulum		
	mRNA, complete cds	AF181557	Wobbe et al. 1999
	mRNA, complete cds	AF405310	Liu et al. 2001
	mRNA, complete cds	AY445505	Zeng et al. 2003

The recombinant ADS prepared from *E. coli* by Mercke et al. (2000) has a broad pH optimum between 7.5-9.0, a K_m of 0.9 μM for FPP at pH 7.5, and a molecular mass of 63.9 kDa, which is higher than the native plant ADS identified by Bouwmeester et al. (1999). Expression of *ADS* gene in *E. coli* leads to generation of a number of terpene precursors, predominantly amorpha-4,11-diene (91.2%, w/w). Picaud et al. (2005) reported that the recombinant ADS expressed in *E. coli* gives rise to amorpha-4,11-diene as a major product, but 15 sesquiterpenes in total are simultaneously produced.

As a huge plant gene family, *CYP* gene family exhibits tremendous sequence diversity among their member genes (Schuler and Werck-Reichhart 2003). Teoh et al. (2006) isolated a cDNA clone, designated as *CYP71AV1*, from an *A. annua* glandular trichome library, and identified *CYP71AV1* gene belonging to *CYP71D* subfamily that encodes a lot of plant terpene hydroxylases. They assayed the yeast microsomal membrane-bound recombinant CYP71AV1 with a variety of substrates in NADPH involvement. When amorpha-4,11-diene was supplemented, artemisinic alcohol synthesized in an NADPH- dependent manner. Their experiment suggested that CYP71AV1 is a multifunctional enzyme that converts amorpha-4,11-diene to artemisinic acid through three sequential oxidation steps.

Alternatively, by analyzing the data pooling in a database of expressed sequence tags (ESTs) that focuses on *Lactuca*, *Helianthus* and other Compositae plants (http://www. cgpdb. ucdavis. edu), Ro et al. (2006) identified two major *CYP* subfamilies, *CYP71* and *CYP82*. Using degenerate primers designed based on the conserved sequence of *CYP71* subfamily, a full-length cDNA that encodes an open reading frame of 495 amino acids, i.e. *CYP71AV1*, was isolated from *A. annua*. When *CYP71AV1* was co-expressed with *ADS* in *S. cerevisiae*, amorpha-4,11-diene is promptly oxidized to artemisinic acid in the engineered yeast cells.

On the other hand, Bertea et al. (2006) constructed a cDNA library starting from the total RNA isolated from the glandular trichomes of *A. annua*. About 900 of randomly selected clones were partially sequenced and analyzed for sequence homologies using the BLAST algorithm. Fragment assembly identified a total of 459 contigs and 900 ESTs and then assigned functions based on the highest similarity. The enzyme types encoded by ESTs are listed in Table 4.

Table 4. Selection of isoprene biosynthetic ESTs identified from *A. annua*

EST identification	No. of hits
deoxyxylulose 5-phosphate synthase	3
isopentenyl diphosphate isomerase	1
geranyl diphosphate synthase (small subunit)	1
pinene synthase	3
limonene synthase	2
linalool synthase	8
farnesyl diphosphate synthase	1
amorpha-4,11-diene synthase	1
germacrene A synthase	2
other sesquiterpene synthases	4

In our recent work, homology of newly isolated sequences with the accessed genes in GenBank were browsed by the online BLAST software, thereby conferring homology- based functional annotations of these sequences. Among 75 accessed *A. annua* sequences in a format of either CoreNucleotide or ESTs in GenBank, four full-length cDNAs are highly homologous to the known *A. annua* genes, other 71 ESTs do not have sequence records in *A. annua*, but in which 34 ESTs are homologous to other plant genes, including 24 known protein-coding sequences and 10 unknown protein-coding sequences, other 27 ESTs do not have sequence records in any plants. Table 5 list all *A. annua* cDNAs and ESTs with annotated function. Besides, ten sequences classified as gene fragments encoding unknown proteins homologous to other plant genes, and 37 sequences recognized as gene fragments encoding unknown proteins without homology to any plant genes are also listed.

Table 5. Functional annotation of *A. annua* cDNAs and ESTs by homology comparison

GenBank accession No.	Homology comparison-based functional annotation
AY445506	*A. annua* squalene synthase, SS
DQ241826	*A. annua* amorpha-4, 11-diene synthase, ADS
DQ838799	unknown protein without homology to any plant genes
DQ872632	*A. annua* cytochrome P450 monooxygenase, CYP
DQ984181	*A. annua* cytochrome P450 reductase, CPR
DQ838800	vacular processing enzyme-1b
DQ838801	membrane protein
EF050423	structural constituent of ribosome
EF050424	chitinase, CHI
EF050425	unknown protein without homology to any plant genes
EF050426	unknown protein without homology to any plant genes
EF050427	ribulose-1,5-bisphosphate carboxylase/oxygenase small subunit, RuBPC/O
EF050428	unknown protein without homology to any plant genes
EF050429	cytosolic NADP-malic enzyme
EF379388	unknown protein without homology to any plant genes
EF494771	unknown protein without homology to any plant genes
EF494772	unknown protein without homology to any plant genes
EF494773	unknown protein without homology to any plant genes
EF549580	40S ribosomal protein S9
EF549581	15.9kDa subunit of RNA polymerase II
EF549582	calmodulin, CaM
EF549583	histone H4-like protein
EF549584	light-harvesting chlorophyll a/b-binding protein, LHCBP
EF549585	ubiquitin-conjugating enzyme, UCE
EF660343	eukaryotic translational factor TIF3B1
ES494773	unknown protein homologous to other plant genes
ES582125	drought/low temperature and salt responsive protein, D/LTSRP
ES582126	unknown protein homologous to other plant genes
ES582127	unknown protein without homology to any plant genes
ES582128	hydroxyl praline-rich protein
ES582129	RNA-binding glycine-rich protein, RGP
ES582130	unknown protein homologous to other plant genes
ES582131	thioredoxin
ES582132	acyl-ACP thioesterase FATA1
ES582133	unknown protein homologous to other plant genes
ES582134	unknown protein homologous to other plant genes
ES582135	unknown protein homologous to other plant genes
ES582136	unknown protein without homology to any plant genes
ES582137	unknown protein without homology to any plant genes
ES582138	unknown protein without homology to any plant genes

Table 5. (Continued).

GenBank accession No.	Homology comparison-based functional annotation
ES582139	unknown protein without homology to any plant genes
ES582140	unknown protein homologous to other plant genes
ES582141	unknown protein without homology to any plant genes
ES582142	unknown protein homologous to other plant genes
ES582143	DICER-like 2/3 spliceform 2
ES582144	unknown protein homologous to other plant genes
ES582145	auxin-repressed/dormancy-associated protein, AR/DAP
ES582146	unknown protein without homology to any plant genes
ES582147	unknown protein without homology to any plant genes
ES582148	unknown protein without homology to any plant genes
ES582149	unknown protein without homology to any plant genes
ES582150	unknown protein without homology to any plant genes
ES582151	unknown protein homologous to other plant genes
ES582152	cytosolic malate dehydrogenase
ES582153	40S ribosomal protein S30-like protein
ES582154	secretory peroxidase
ES582155	unknown protein homologous to other plant genes
ES880929	unknown protein without homology to any plant genes
ES880930	unknown protein without homology to any plant genes
ES880931	unknown protein without homology to any plant genes
ES880932	unknown protein without homology to any plant genes
ES880933	unknown protein without homology to any plant genes
ES880934	unknown protein without homology to any plant genes
ES880935	unknown protein without homology to any plant genes
ES880936	unknown protein without homology to any plant genes
ES880937	unknown protein without homology to any plant genes
ES880938	unknown protein without homology to any plant genes
ES880939	unknown protein without homology to any plant genes
EV780877	unknown protein without homology to any plant genes
EV780878	unknown protein without homology to any plant genes
EV780879	unknown protein without homology to any plant genes
EV780880	unknown protein without homology to any plant genes
EV780881	unknown protein without homology to any plant genes
EV780882	unknown protein without homology to any plant genes
EV780883	unknown protein without homology to any plant genes

From the known functions served by above *A. annua* genes, one can find out a lot of genes coding for the environment inducible or stress responsive proteins. The diverse genes include those for metabolic engineering-directed breeding for boosting artemisinin production (*ADS*, *CYP71AV1*, *CPR*, and *SS*), those for breeding toward the highly effective photosynthesis (*RuBPC/O*, and *LHCBP*, etc.), and those for breeding on the disease and pest insect resistance (*CHI, D/LTSRP, AR/DAP*, and *CaM*, etc.). Moreover, all novel genes with unknown protein-encoding function or no sequencing records await further identifications by the so-called 'gain/lost-of-function' approaches, such as site-directed gene mutagenesis, homologous recombination-mediated gene knockout, and anti-sense inhibition or microRNA interference.

4.2. Temporal and Spatial Expression Patterns of Artemisinin Biosynthetic Genes

Teoh et al. (2006) proved that *CYP71AV1* gene is expressed in *A. annua* at a maximum level in glandular trichomes, a moderate level in leaves, and a minimum level in roots, which is just in accordance with the distribution pattern of artemisinin, i.e. glandular trichomes give rise to the highest artemisinin content, leaves display a lower level of artemisinin yield, and artemisinin is presented in trace amount in roots. Our recent immunoquantitative assay of organ-specific distribution of CYP71AV1 showed that abundance of the tested enzym is highest in leaves, moderate in stems and lowest in roots (unpublished data). From artemisinin determination results throughout the vegetative stage of *A. annua*, it was known that top leaves (later initiated) exhibit generally higher artemisinin content, but bottom leaves (earlier initiated) show lower artemisinin content (Liersch et al. 1986; Ferreira et al. 1995). The previous follow-up determination of artemisinin showed that a highest artemisinin content was often achieved just before flowering (Morales et al. 1993) or under flowering (Gupta et al. 2002; Laughlin 1995). However, the highest artemisinin content was determined in dry leaves exposed to the post-harvest maturation (Lommen et al. 2006) and even in dead leaves experienced the programmed cell death (apoptosis) (Lommen et al. 2007).

Weathers et al. (2006) cited their unpublished experimental results that *ADS* mRNA is ubiquitously present in all tissues including roots, stems, leaves, and flowers in mature *A. annua* plants. However, considering the fact that artemisinin is only accumulated in the glandular trichomes on leaves and flowers, they explained that *ADS* mRNA may not be translated into protein in all tissues (translationally regulated) or ADS may not be active in all tissues (post-translationally regulated), and addressed that the most probable situation is that amorpha-4,11-diene may be synthesized in all tissues, but then transported to leaves and flowers for further artemisinin biosynthesis, or amorpha-4,11- diene synthesized in roots and stems may be used to produce compounds other than artemisinin. These suggestions need verification by quantitative assay of *ADS* mRNA in different tissues and *in situ* immunoquantitative determination of ADS concentration or enzymatic detection of ADS activity in all tissues.

4.3. Quantification on Environmental Stress-Induced Overexpression of Artemisinin Biosynthetic Genes

It is obviously from the artemisinin biosynthetic pathway that dihydroartemisinic acid as a precursor for artemisinin biosynthesis experiences the hydroperoxide intermediate stage. This means that artemisinin biosynthesis may involve induction of artemisinin biosynthetic genes evoked by environmental stresses. However, current investigations regarding stress-induced gene expression in *A. annua* were restricted to a few stress factors, e.g. extensive light illumination (Souret et al. 2002; 2003). Yin et al. (2008) treated *in vitro* cultural *A. annua* plants by cold, heat and ultraviolet light, and then quantified the transcripts of three artemisinin biosynthetic genes, *ADS*, *CYP71AV1* and *CPR*, which demonstrated that *ADS* and *CYP71AV1* genes are markedly up-regulated, while *CPR* gene keeps stable expression either prior to or post the treatment. The real-time fluorescent quantification data further revealed that as exposure to chilling stress, the copy numbers of *ADS* and *CYP71AV1* mRNAs in *A. annua* plants were accounted as eleven- and sevenfold elevations as the control plants, respectively. Nevertheless, artemisinin content in those *A. annua* plants exposed to chilling stress does not increase in proportion with elevated levels of *ADS* and *CYP71AV1* mRNAs, implying that artemisinin biosynthesis may be modulated by more than one step regulations, which determine the conversion efficiency from artemisinic acid to artemisinin.

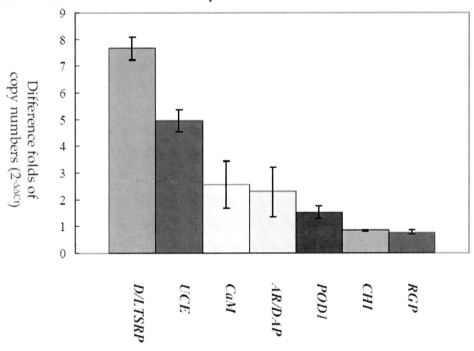

Figure 7. Regulation of chilling stress-induced expression in A. annua. AR/DAP: Auxin-repressed/dormancy- associated protein; CaM: calmodulin gene; CHI: chitinase gene; D/LTSRP: drought/low temperature and salt responsive protein gene; POD1: peroxidase 1 gene; RGP: RNA-binding glycine-rich protein gene; UCE: ubiquitin-conjugating enzyme gene. Adopted from Zeng et al. (2008b).

Wallaart et al. (1999) ascertained that only those *A. annua* shoots treated by drought (30% relative humidity) and light stress (6000 lux) can be amplified from *ADS* mRNA, but no product is available from amplification of non-treated *A. annua* shoots. In similar, as we detected the amplification products by gel electrophoresis, the amplicons from *ADS* and *CYP71AV1* mRNAs were only available from the chilling stress-exposed *A. annua* plants, while no amplicon was detected from the untreated *A. annua* plants (Yin et al. 2008). The results let us to envisage that expression of *ADS* and *CYP71AV1* genes in *A. annua* may be controlled by environmental stresses, at least by low temperature in this situation.

To experimentally verify the function of our newly isolated *A. annua* ESTs, the responsive pattern of seven novel ESTs to low temperature was chosen as an evaluation criteron during quantitatively evaluating their chilling stress-induced overexpression levels by the real-time fluorescent quantitative polymerase chain reaction. The result showed that upon standing at 4°C for 48 h, the expression levels of five ESTs (*D/LTSRP*, *UCE*, *CaM*, *AR/DAP*, and *POD1*) were significantly up-regulated, while those of other two ESTs (*CHI* and *RGP*) were not predominantly fluctuated (Figure 7).

4.4. Artemisinin Biosynthetic Gene Transcription-Based Identification of Chilling Stress Signal Transduction

It has been known that low temperature-induced gene expression is essentially involved in the signal transduction pathway. Therefore, we treated cultural *A. annua* plants with Ca^{2+} channel inhibitor $LaCl_3$ and Ca^{2+} chelator ethylene glycol tetra-acetic acid (EGTA) to assess their effects on chilling-mediated signal transduction (Zeng et al. 2008b). When supplemented with $LaCl_3$ or EGTA, chilling-induced expression of *ADS* and *CYP71AV1* genes in *A. annua* was suppressed to defferent degrees, in which the former one is more potently attanuated than the later one. As $LaCl_3$ or EGTA was depleted, chilling-induced expression of *CYP71AV1* gene recovered immediately, while that of *ADS* gene retrieved more slowly. In contrast, either with or without $LaCl_3$ or EGTA, expression of *CPR* gene was not affected by the treatment. Moreover, calmodulin gene (*CaM*) was up-regulated by 2.5 folds upon chilling exposure. Lin et al. (2004) also observed the elevated CaM content and enhanced antioxidant enzyme activity in *Populus tomentosa* during cold- acclimation. We thus inferred that cyclization from FPP to form amorpha-4,11-diene might be regulated at the transcription level, seemingly involving activation binding of transcription factor(s) with *ADS* and *CYP71AV1* promoters through the Ca^{2+}-CaM signal transduction pathway. At present, there has no report demonstrating interaction of the specific transcription factor(s) with these *A. annua* promoters. Nevertheless, the promoter sequence of *ADS* gene has been accessed in GenBank, which should earge investigation of the stress-inducible binding of transcription factors to the promoter. Wang et al. (2001; 2002) discovered that the oligosaccharide elicitor derived from *Colletotrichum* sp. can trigger the signaling cascade involving rapid Ca^{2+} accumulation, plasma membrane NAD(P)H oxidase activation and ROS release.

4.5. Stress-Responsive and Other *Cis*-Regulatory Elements in ADS Promoter

The isoprene intermediate FPP is a common precursor of sterols and sesquiterpenes. The enzymes that initiate these two branching biosynthetic pathways, squalene synthase and sesquiterpene synthases, compete for FPP. In plants, different structural types of sesquiterpenes are synthesized from FPP upon catalysis by distinct sesquiterpene synthases. At least four sesquiterpene synthases have been identified in *A. annua* (Figure 8).

What is the regulation mechanism to normally cooperate with expression of all these sesquiterpene synthase gene? When are they expressed to serve the diversed cellular function? We still do not have ideas to account for them. Although the cDNA encoding ADS was isolated by cDNA library screening (Merck et al. 2000) and degenerated reverse transcription-polymerase chain reaction (Wallaart et al. 2001), its regulated expression manner is uncertain. The indirect evidence regarding regulation of *ADS* gene expression only came from amplification of *ADS* mRNA, in which no amplicon was visible on the gel when RNA was amplified from non-stressed *A. annua* plants. This is consistent with the characteristics of terpene synthases themselves, of course, which occur in very low intracellular concentrations in plant tissues.

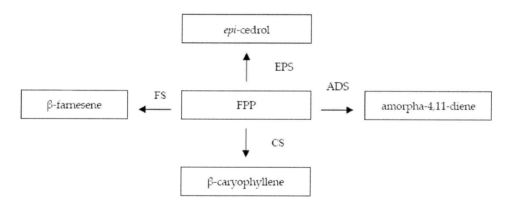

Figure 8. Sesquiterpene synthases isolated from A. annua. ADS: amorpha- 4,11-diene synthase; CS: β-caryophyllene synthase; EPS: epi-cedrol synthase; FPP: farnesyl perophosphate; FS: β-farnesene. Adopted from Weathers et al. (2006).

However, specific transcription was exclusively induced in plants pre-exposed to drought (30% relative humidity) and photo (6000 lx) stresses (Wallaart et al. 2001). The transcript level of *ADS* gene was also found to be elevated by cold, heat and ultraviolet light exposure of *A. annua* plants (Yin et al. 2008). The real-time quantitative assay showed that *A. annua* plants under a transient chilling exposure resulted in elevenfold increase in *ADS* transcript level. These results indicated that *ADS* gene may be an inducible one capable of highly responding to stresses. Therefore, a hypothesis has been proposed that dihydroartemisinic acid may act as an antioxidant by quenching $^1O_2$ (Wallaart et al. 1999a) and generated dihydroartemisinic acid hydroperoxide consequently converted to artemisinin in response to $^1O_2$ emission (Wallaart et al. 1999b).

Several groups independently isolated and sequenced *ADS* promoter with the accessed GenBank accession No: DQ448294, DQ448295, DQ448296, DQ448297 and AY528931. Our online sequence analysis (http://www.dna.affrc.go.jp/PLACE/signalscan.html) revealed that there are a lot of potential *cis*-regulatory elements in *ADS* promoter that responses to drought, cold, salt, heat shock, mechanic wound, pathogens and phytohormones (Table 6).

Table 6. The cis-element of *ADS* promoter in *A. annua*

Cis-element	Induction	Core sequence	Location (Strand)	Reference
ABRELATERD1	dehydration	ACGTG	1064(-), 1928(-)	Simpson et al. 2003
ABRERATCAL	Ca^{2+}	MACGYGB M=C/A Y=C/T B=T/C/G	2159(+), 1415(-), 1927(-)	Kaplan et al. 2006
CRT/DRE/CBF2	cold	GTCGAC	1141(+), 1141(-)	Xue 2003
CBFHv	cold	RYCGAC R=A/G Y=C/T	59(-)	Xue 2002; Svensson et al. 2006
CCAATbox1	heat shock	CCAAT	251(+), 806(+), 844(+), 1109(+), 1507(-), 1860(-), 2068(-)	Rieping et al. 1992
GT1CONSENSUS	light	GRWAAW R=A/G W=A/T	663(+), 746(+), 752(+), 1264(+), 1317(+), 1402(+), 1488(+), 1489(+), 2230(+), 2464(+), 148(-), 283(-),694(-), 775(-), 1291(-), 1507(-), 1902(-), 149(-), 584(-), 1047(-), 2216(-), 2238(-), 2305(-)	Terzaghi et al. 1995
GT1core	light	GGTTAA	640(+), 1676(+), 642(-)	Terzaghi et al. 1995
I-box	light	GATAA	1402(+), 284(-), 585(-), 1598(-), 1903(-)	Terzaghi et al. 1995
GT1GMSCAM4	salt and pathogen	GAAAAA	663(+), 752(+), 1489(+)	Park et al. 2004
W-boxATNPR1	salicylic acid and pathogen	TTGAC	799(+), 2146(+), 316(-), 1503(-), 2295(-)	Yu et al. 2001; Chen et al. 2002

Table 6. (Continued).

Cis-element	Induction	Core sequence	Location (Strand)	Reference
WRKY710S	gibberellins, abscisic acid and pathogen	TGAC	244(+), 381(+), 800(+), 848(+), 941(+), 1122(+), 1195(+), 1250(+), 1666(+), 1955(+), 1977(+), 2147(+), 170(-), 316(-), 688(-), 1503(-), 1755(-), 1996(-), 2295(-)	Zhang et al. 2004; Xie et al. 2005
W-boxHVISO1	sugar	TGACT	244(+), 1195(+), 1955(+), 169(-), 315(-), 687(-), 1995(-)	Sun et al. 2003
W-boxNTERF3	wound	TGACY Y=C/T	244(+), 848(+), 941 (+), 1195(+), 1955(+), 2147 (+), 169(-), 315(-), 687(-), 1995 (-), 1502 (-), 1754 (-)	Nishiuchi et al. 2004

The future work should be identification of these *cis*-element-bound transcription factors and their *trans*-action upon environmental stresses. In this regard, the comparative genomics, transcriptomics, proteomics and metabolomics among plants with the closed evolutionary relations would be helpful and reasonable. Furthermore, direct isolation of the stress-corresponding transcription factors with the gel retardation assay and DNA fingerprinting.

5. Primary and Secondary Stress Signal-Induced Artemisinin Overproduction

5.1. Primary stress signal induction

Liu et al. (1997) supplemented cell homogenate of *Aspergillus oryzae* to the cultural hairy roots of *A. annua* and detected increase of artemisinin content up to 550 mg/L. Wang et al. (2000) investigated the stimulating effects of fungal elicitors on artemisinin biosynthesis in *A. annua* hairy root cells. After adding chitosan (150 mg/L) to *A. annua* hairy root cells and keeping for six days, Putalun et al. (2007) determined artemisinin content for sixfold increase to 1.8 µg/mg dry weight, and observed a similar effect as yeast extracts (2 mg/mL) were supplemented instead. Kapoor et al. (2007) applied phosphate fertilizer to field-grown *A. annua* plants with two fungi, *Glomus macrocarpum* and *Glomus fasciculatum*, and quantified the tremendously elevated artemisinin level together with the significant high density of glandular trichomes on *A. annua* leaves.

In cultural cells and cell-free extracts of *A. annua*, supplement with sterol inhibitors, miconazol and naphtiphine, promotes incorporation of a large quantity of ^{14}C-labeled IPP into artemisinin (Kudakasseril et al. 1987). In tobacco suspension cells, supplement of a fungal elicitor leads a sharp dropped activity level of squalene synthase accompanying with a large magnitude of enhancement of sesquiterpene phytoalexin biosynthesis (Vogeli et al. 1998), which demonstrated that specific induction that is mediated by the fungal elicitors may be undergone by 'influx' into the sesquiterpene biosynthetic branch and 'efflux' from sterol biosynthetic branch.

In addition, the types of monosaccharide and disaccharide (Wang et al. 2007) and microelement deficiency (Ferreira et al. 2007) can affect artemisinin biosynthesis at a certain extent upon stress signal transduction.

5.2. Secondary Stress Signal Induction

Salicylic acid (SA), jasmonic acid (JA) and methyl jasmonate (MJ) are ubiquitous signal molecules in plant cells and known as the 'second messengers', which can transduce most of the extreme environmental stimuli to initiate the cellular responsive mechanics against the oxidative stresses. It was previously reported that MJ is a senescence- promoting inducer in *Artemisia absinthium* (Ueda et al. 1980), and MJ emitted by *Artemisia tridentate* can induce approximated tomato plants to express proteinase inhibitor gene and exhibit protection reaction (Farmer et al. 1990). The recent investigation by Afitlhile et al. (2005) showed that the plants belonging to *Artemisia* accumulate higher levels of JA and MJ than other plants, and accumulated JA is higher up to eightfold in field-grown *Artemisia* plants than in greenhouse-cultivated *Artemisia* plants, which can be divided into the high level group (30 nmol/g dry weight), moderate level group (10-20 nmol/g dry weight) and low level group (10 nmol/g dry weight). Wallaart et al. (2000) also measured various content levels in artemisinin and its precursors in *A. annua* plants, thus suggesting that different chemotypes may be mixed in *A. annua*. Nevertheless, it is unknown if these chemotypes directly lead to variation of JA and MJ levels in plants. In addition, Lommen et al. (2007) determined highest artemisinin content in senescent and dead leaves, but whether this finding is indeed related to MJ-promoted senescence and MJ-induced artemisinin overproduction needs elucidations.

Signal molecules that induce accumulation of plant secondary metabolites have more reports (Woerdenbag et al. 1993; O'Donnell et al. 1996; Walker et al. 2002). Recently, Baldi et al. (2007) evaluated the effects of SA, MJ, gibberellic acid (GA$_3$) and CaCl$_2$ on artemisinin biosynthesis in *A. annua* cultural cells. Consequently, they found that 20-50 mg/L SA, 5-10 mg/L MJ, and 10 mg/L GA$_3$ can significantly increase artemisinin content, the highest magnitude of such increase is sixfold as the control. The previous research also showed that GA$_3$-treated *A. annua* plants can significantly increase their artemisinin content (Fulzele et al. 1995; Paniego and Giulietti 1996). Zhang et al. (2005) demonstrated that exogenous GA$_3$ can divert the carbon flux to artemisinin by feed-back inhibition of GA$_3$ biosynthesis, thereby suggesting presence of a regulon at the conversion step from artemisinic acid to artemisinin.

To investigate the relationship between artemisinin accumulation and blooming, Wang et al. (2004; 2007) transformed *A. thaliana* flowering-promoting factor 1 gene (*FPF1*) and flowering gene (*CO*) into *A. annua*. Although transgenic plants bloom early for 20 days and 14 day, respectively, their artemisinin content does not increase significantly. Therefore, they concluded that flowering is not a prerequisite for enhanced artemisinin production in general.

Weathers et al. (2005) found that *A. annua* hairy roots in the cultural medium supplemented with cytokinin 2-isoprenyl adenine give rise to substantially increased artemisinin content. After introducing *ipt* gene into *A. annua*, Geng et al. (2001) detected two- to threefold elevated cytokinin level and 30-70% boosted artemisinin content.

5.3. Combined Applications of Primary and Secondary Stress Signals

Nitric oxide (NO) is a novel signal molecule (Delledonne et al. 1998) that enhance production of taxol in *Taxus* (Wang et al. 2004) and of catharanthine in *Catharanthus* (Xu et al. 2005), but not of artemisinin in *Artemisia* (Zheng et al. 2007). However, NO can potentiate fungal elicitor-induced overproduction of ginseng saponin (Hu et al. 2003), taxol (Xu et al. 2004), hypericin (Xu et al. 2005), puerarin (Xu et al. 2006), and artemisinin (Zheng et al. 2007). These investigations also indicated that NO-mediated fungal elicitor induction of secondary metabolites is an essential outcome from the burst of ROS *en route* in JA-dependent signal transduction pathway. Zheng et al. (2007) demonstrated that artemisinin content in 20 day-old hairy roots elevates from 7 mg/g dry weight to 13 mg/g dry weight upon application of the oligosaccharide elicitor for four days, and the combined treatments of the oligosaccharide elicitor with sodium nitroprusside (SNP) lead to increase of artemisinin content even higher from 12-22 mg/g dry weight.

6. Artificial Metabolic Manipulation: Artemisinin Biosynthetic Gene Transfer into *A. annua* and Other Plants

6.1. Upstream Pathway Gene Transfer

Chen et al. (1999) employed *Agrobacterium rhizogenes* to produce transgenic *A. annua* hairy roots that express FPP synthase gene (*FPS*) driven by 35S promoter of cauliflower mosaic virus (CaMV). A number of resulting roots produce artemisinin up to 2-3 mg/g dry weight, 3-4 times that of the control roots. When *A. tumefaciens* was used instead, regenerated transgenic plants yield artemisinin at 8–10 mg/g dry weight (Chen et al. 2000). Both studies showed that although manipulation of *FPS* increase artemisinin content, the yields are only comparable to or slightly higher than wild-type plants, suggesting that the multiple downstream pathways toward so many kinds of desired or unexpected products may make it complicated in obtaining a singular target metabolite.

Davidovich-Rikanati et al. (2007) modified the flavor and aroma of tomatoes by overexpressing the *Ocimum basilicum* geraniol synthase gene under the control of the tomato ripening–specific promoter from polygalacturonase gene. A majority of untrained taste panelists preferred the transgenic fruits over controls. Monoterpene accumulation was at the expense of reduced lycopene accumulation. Similar approaches may be applicable for carotenoid-accumulating fruits and flowers in other plant species.

6.2. Antisense-Based Genetic Modification

Wang et al. (2001) identified a cytochrome P450 hydroxylase gene (CYPH) specific to the glandular trichomes and used both antisense and sense co-suppression strategies to investigate impact of such genetic modification on the plant behavior. As a result, CYPH-suppressed transgenic tobacco plants demonstrate a ≥41% decrease in the predominant exudate component, cembratriene-diol (CBT-diol), and a ≥19-fold increase in its precursor, cembratricne-ol (CBT-ol). Thus, CBT-ol level is raised from 0.2 to ≥4.3% of leaf dry weight. Exudates from antisense-expressing plants exhibit higher aphidicidal activity, and transgenic plants with exudates containing high concentration of CBT-ol show a greatly diminished aphid colonization response. Their results demonstrated the feasibility of modifying the natural product composition and aphid-interactive properties of gland exudates using metabolic engineering.

We recently introduced *A. annua* anti-sense squalene synthase gene (*asSS*) into *A. annua* to attempt enhancing artemisinin production by suppressing sterol biosynthesis. The elevated *SS* mRNA level and dropped sterol content in transgenic plants were quantified. The determination results of artemisinin content showed that in transgenic plants artemisinin yield reaches 1.66 mg/g dry weight (T47) and 1.26 mg/g dry weight (T81), while in untreated transgenic plants artemisinin content is 1.23 mg/g dry weight (T47) and 0.93 mg/g dry weight (T81). As comparison, artemisinin content in the control plant (WT) is only 4.5 mg/g dry weight.

6.3. Nuclei-Coded and Plastid-Targeted Artemisinin Precursor Biosynthetic Enzymes

Wu et al. (2006) engineered high level terpene production in tobacco plants by diverting carbon flow from cytosolic or plastidic IPP through expression of an avian *FPS* gene in the cytosol but targeting to plastids. The strategy used in the present study increased amorpha-4,11-diene content more than 1000-fold up to 25 μg/g fresh weight (Figure 9), and seems to be suitable generating high levels of other cytosolic or plastidic terpenes for scientific research, industrial production or therapeutic applications.

Terpenes represent over a $1-billion market value to the flavor and fragrance industry, but the market for pharmaceutics and medicine industry may be astonishingly attractive. Whereas tremendous progress has been made in engineering microbial platforms for terpene production, plant systems also addressed due to inexpensive carbon feedstocks, low processing costs and a greater elaboration potential. This is especially the case for those terpenes requiring decoration with carbohydrate and/or aryl substituents, and introduction of peroxide functionalities.

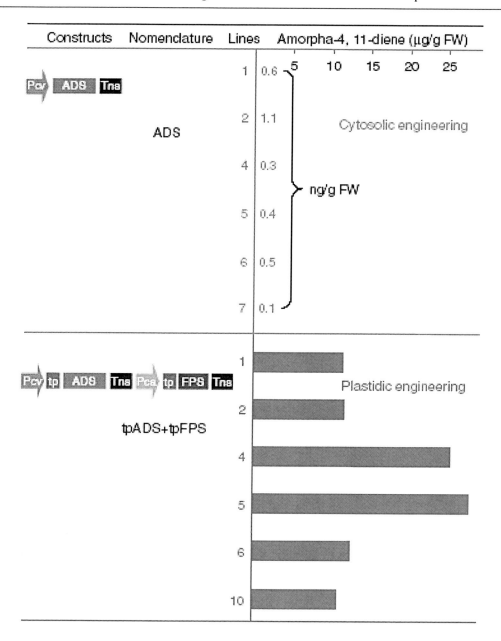

Figure 9. Engineering production platforms for amorpha-4,11-diene by diverting carbon flux from the cytosolic MVA pathway and the plastidic MEP/DXP pathway. ADS: amorpha-4,11-diene synthase; FPS: farnesyl phosphate synthase; FW: fresh weight; Pca, 35S cauliflower mosaic viral promoter; Pcv, cassava mosaic viral promoter; Tns: nopaline synthase gene (NOS) terminator; tp: plastid targeting signal sequence. Adopted from Wu et al. (2006).

7. Artificial Metabolic Manipulation: Re-Establishment of Artemisinin Biosynthetic Pathway in Microorganisms

7.1. Production of Artemisinin Precursors by Genetically Modified Yeast

As a species of eukaryotic microbes, yeast offers promise for overexpression of plant genes, especially those encoding the protein containing membrane-bound domains (DeJong et al. 2006). One of the pioneer examples is heterologous expression of *Taxus* taxane-10β-hydroxylase gene in yeast (Schoendorf et al. 2001). Another successful attempt is functional expression of *A. annua epi*-cedrol synthase gene (*ECS*) in yeast (Jackson et al. 2003). Adequate attachment of the membrane-anchored cytochrome P450 is crucial for high-level production of artemisinin in yeast. Moreover, the high rate of isoprene metabolism in yeast must be, of course, taken into account for which an ideally engineered yeast strain should be one with an increased capacity in FPP production and a decreased level of sterol accumulation.

The Brodelius group compared the effects of two different formats of *A. annua ADS* gene expression in yeast on amorpha-4,11-diene production. For doing this, *ADS* gene was either introduced into yeast as episomal plasmids or inserted into the yeast genome by homologous recombination. The plasmid-harboring and genome-integrated yeast strains with the functionally expressed *ADS* gene exhibit sixfold increase of amorpha-4, 11-diene (600 µg/L versus 100 µg/L) in 16-day batch cultivations (Lindahl et al. 2006), demonstrating that sesquiterpene production in yeast may be positively correlated with the gene dosage although the insufficient FPP pool is also a possible limiting factor.

Production of artemisinic acid in the engineered yeast strain was achieved by multiple gene transfer as illustrated in Figure 10. Firstly, *HMGR* and *ADS* genes were transferred into yeast for increasing the FPP pool and catalyzing amorpha-4,11-diene production, respectively. Further augmented levels of FPP and amorpha-4,11-diene were realized through down-regulation of *SS* gene by a methionine-repressible promoter. Subsequently, *CYP71AV1* and *CPR* genes were introduced into the engineered yeast strain to enable the three-step oxidation of amorpha-4,11-diene to generate artemisinic acid.

When evaluated step-by-step, expression of *ADS* gene alone only resulted in a small quantity of amorpha-4,11-diene (4.4 mg/L), while co-expression of a truncated *HMGR12* gene (*tHMGR*) improved amorpha-4,11-diene yield by about fivefold. Introduction of a methionine repressible promoter down-regulated the sterol biosynthetic gene, *ERG9*, and led to an additional twofold increase of amorpha-4,11-diene. Down-regulation of *ERG9* gene and overexpression of *upc2-1*, a dominant mutated gene encoding the Upc2p transcription factor capable of regulating sterol biosynthesis, elevated amorpha-4,11-diene level to 105 mg/L. Integration of an additional copy of *HMGR* gene into the chromosome enhanced amorpha-4, 11-diene production to 149 mg/L. Overexpression of *FPPS* (*ERG20*) gene decreased cell density and, in turn, increased amorpha-4,11-diene content by about additional 10% (w/w). All these multi-gene manipulations consequently led to a total enhancement in amorpha-4,11-diene output up to 153 mg/L, an elevation of nearly 500-fold of what was shown previously (Jackson et al. 2003).

In the next step, two genes, *CYP71AV1* and *CPR*, directly responsible for artemisinin biosynthesis were introduced into the engineered yeast cells to make it possible for conversion from amorpha-4,11-diene to artemisinic acid. In a shake flask, approximately 32 mg/L of artemisinic acid was measured, but only negligible artemisinic alcohol and artemisinic aldehyde were detected.

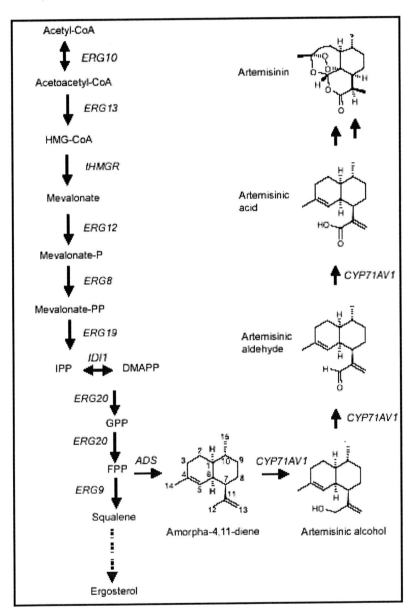

Figure 10. Engineered artemisinic acid biosynthetic pathway in S. cerevisiae. ERG: yeast MVA pathway genes; ADS: A. annua amorpha-4,11-diene synthase gene; A. annua CYP71AV1: cytochrome P450 monooxygenase gene; IDI: IPP isomerase gene; tHMGR: truncated HMG-CoA reductase gene. A. annua CPR gene is not indicated. Adopted from Zeng et al. (2008a).

The engineered yeast cells expressing *ECS* gene converted their endogenous FPP to *epi*-cedrol up to 90 µg/L. Further optimization by introducing *upc2-1* gene increased the availability of FPP and yielded 370 µg/L of *epi*-cedrol, a sixfold increase compared with what were reported in *E. coli* (Cane et al. 1993; Martin et al. 2001). The engineered yeast cells can produce comparable levels of artemisinic acid as *A. annua* on a biomass basis but in a much shorter time (4-5 days for yeast versus several months for *A. annua*). As such viewpoint, artemisinic acid productivity of the engineered yeast is nearly two orders of magnitude greater than *A. annua*, demonstrating an advantage in yeast over in plants for large-scale production of plant terpenes. Moreover, secretion of artemisinic acid to the medium by yeast cells simplifies fractionation and purification of the desired product, representing another potential cost-saving benefit in industrial processing.

7.2. Production of Artemisinin Precursors by Genetically Modified Bacteria

The classical bacterial species, *E. coli*, remains an amenable host for industrial fermentative production of plant secondary metabolites due to its speedy growth rate and easy genetic manipulation. However, difficultis in assembly of cytochrome P450 enzymes and other membrane-bound proteins renders employment of such prokaryotic platform for isoprene production extremely challenging.

More efforts on engineering *E. coli* for isoprene production often underwent by manipulation of the native DXP/MEP pathway. However, the metabolic flux in this endogenous pathway is subject to strict intracellular regulatory control. Therefore, to rebuild an artemisinin biosynthetic pathway in *E. coli*, Martin et al. (2003) bypassed the DXP/MEP pathway by introducing entire MVA pathway genes from yeast. These genes were organized into two artificial operons, in which *MevT* operon is responsible for conversion from acetyl-CoA to mevalonate, and *MBIS* operon is involving generation of FPP from mevalonate. Reconstruction of the MVA pathway in *E. coli*, however, resulted in severe growth inhibition due to the excessive accumulation of FPP, IPP and/or DMAPP. Co-expression of a codon optimized *ADS* gene alleviated the intermediate toxicity and finally yielded amorpha-4,11-diene with a titer of 280-480 mg/L in a fed-batch bioreactor (Newman et al. 2006).

While expression of *MevT* operon led to HMG-CoA toxicity in *E. coli*, co-expression of *HMGR* gene prevented the harmful buildup of HMG-CoA and increased mevalonate production for almost twofold (Pitera et al. 2007). Manipulation of the intergenic *MevT* operon regions strongly impacted the balancing expression pattern of individual genes and consequently resulted in sevenfold increase in the mevalonate titer (Pfleger et al. 2006). Enhancement of gene stability by chromosomal integration of the MVA pathway genes also greatly enhanced mevalonate production (Yuan et al. 2006). These efforts demonstrated that the carbon source can be effectively channeled to the exogenous isoprene pathway in the engineered *E. coli* strain even though the flux of native isoprene biosynthesis remains a low level.

The recombinant sesquiterpene synthases appear to lack the specificity that their native equivalents should have. For example, the engineered bacteria-expressed *Abies grandis* δ-selinene synthase and γ-humulene synthase give rise to 34 and 52 sesquiterpenes, respectively (Steele and Crock 1998). Similarly, the recombinant *A. annua* ADS enzyme generates 16 sesquiterpenes. However, the fidelity of recombinant ADS-catalyzed reaction in yeast can be considerably improved in presence of the divalent metal irons, Mn^{2+} or Co^{2+} (Picaud et al. 2005). The byproduct (δ-cadinene) of peppermint (*E*)-β-farnesene synthase can be eliminated as Mn^{2+} presents (Crock et al. 1997). Two recombinant maize sesquiterpene synthases enable production of at least 23 sesquiterpenes, but their product spectra shift toward (*E*)-β-farnesene with Mn^{2+} supplement (Kollner et al. 2004).

Functional expression of the plant-isolated *CYP* genes in *E. coli* represents a challenge in proper domain folding, membrane insertion, cofactor incorporation, post-translational modification and essential factor interactions (Carter et al. 2003). By exploiting some of plant-derived *CYP* genes, Chang et al. (2007) demonstrated that *E. coli* can be genetically engineered to produce high levels of complex sesquiterpenes. Although expression of the native *CYP71AV1* gene only led to the undetectable CYP activity, codon optimization coupled with transmembrane domain modification resulted in generation of some intermediates such as artemisinic alcohol at a relative low concentration (0.18-0.45 mg/L). Replacement of *Candida tropicalis CPR* gene (*ctCPR*) with *A. annua CPR* gene (*aaCPRct*) increased artemisinic alcohol for 12-fold to 5.6 mg/L. Furthermore, choice of the plasmid vectors and *E. coli* strains can affect the artemisinic acid titer. Use of pCWori plasmid and *E. coli* strain DH1 further increased the oxidized product of amorpha-4, 11-diene by more than 1000-fold to 553 mg/L.

7.3. Engineering Microbes as Factories for Total Artemisinin Production

Why expression of *CYP71AV1* gene in engineered microbes only led to generation of artemisinic acid rather than artemisinin? We argue that the catalytic reaction of CYP71AV1 enzyme may be changed in microbial circumstances where it is only able to catalyze conversion from amorpha-4,11-diene to artemisinic acid; or CYP71AV1 enzyme is actually another type of cytochrome P450 enzyme (CYP) that naturally produces artemisinic acid. In other words, the genuine CYP gene *bona fide* responsible for generation of dihydroartemisinic acid may not be isolated until now. In fact, *CYP71AV1* gene that was first cloned by the Covello group (Teoh et al. 2006) and by the Keasling group (Ro et al. 2006) is both from amplification of the *A. annua* CYP by a pair of degenerate primers designed according to the bioinformatic analysis data. Therefore, Covello et al. (2007) have suggested that other genes encoding the putative double bond reductase and possibly corresponding enzymes presumably allow bioengineering toward dihydroartemisinic acid production (and possibly artemisinin) and avoid the chemical steps required to convert artemisinic acid to dihydroartemisinic acid.

There have two alternative options for us: the first one is pursuing to find out the original CYP gene responsible for dihydroartemisinic acid biosynthesis; the second one is to modify the currently isolated CYP gene to evolved it into a novel CYP gene that encodes an enzyme mimicking the artemisinin biosynthetic reaction. For the former strategy, one can browse the CYP gene website (http://drnelson.utmem.edu/CytochromeP450.html) and check out the candidate CYP genes among more than 1000 accessed CYP genes to help designing hybridization probes for further cDNA library screening. CYP represents a huge and complicated gene family whose members show a wide range of homology from 1-100%, so it must be a great challenge for the sequence-dependent CYP gene isolation; For the later strategy, it will be promise to use the presently isolated CYP genes as the start templates for their artificial evolution by the error-prone PCR or DNA shuffling procesure.

The site-directed artificial evolution of enzymes is the most efficient strategy that well simulate the natural evolution, because the artificial evolution can be promptly accomplished *in vitro*, thereby enabling the evolutionary process shortening from more than million years to a couple of years or even several months (Stemmer 1994). For example, the stereo-specificity of a kind of S-transaminase has increased from 65% to 80-94% only screening approximately 10 000 random mutants (Matchem and Bowen 1996). It may be a preferential choice to employ DNA shuffling method among the homologous sequences for CYP gene mutagenesis, because the benefit mutation can be quickly accumulated within the desired segments.

Conclusion

Terpenes are synthesized by the complicated reactions associated with two independent anabolistic pathways that operate in plants, i.e. the MVA pathway in the cytosol and the MEP/DXP pathway in plastids. Their interactions provide an array of terpenes that regulate plant growth and development, and mediate plant-environment interactions. Intensive attempts to manipulate terpene anabolism have been carried out in many plant species for generating valuable metabolites that may meet the industrial and commercial needs, for evaluating contribution of the specific classes of terpenes to the life spans, and for annotating the function of putative terpene biosynthetic genes.

Engineering of terpene metabolism in plants is an attractive goal due to their elaborate biosynthetic potential and obvious economic benefit conferred by the photosynthesis- driven energy and biomass production system. The appreciable successes by distinct strategies dealing with genetic modification of *A. annua*, such as the upstream pathway manipulation, antisense-based modification and compartment protein targeting, etc. have been deeply impressed although much spaces for improvement and optimization left yet.

Overexploitation of the natural products as the high-valued fine chemicals has stirred interests in establishing the alternative production platform to facilitate preservation of the endangered species. Recent efforts have succeeded to engineer *E. coli* with a combination of the MVA pathway genes and the MEP/DXP pathway genes. At the same time, *S. cerevisiae* has also been developed as an artemisinin production platform by which a high-level artemisinic acid was synthesized upon overexpression of *A. annua* genes.

Biosynthesis and accumulation of artemisinin is genetically and environmentally dependent in *A. annua* and can be modulated through two ways of interventions-genetic and environmental. From the genetic view, not only the chemical types with highly concentration of secondary messengers such as JA and MJ are presented in *Artemisia*, but also those with significantly different content of artemisinin and its precursors have been found in *A. annua* although it is unclear whether the former is one of the causes leading to the latter. From the environmental view, because generation of the secondary signals is obviously the reliable outcome of environmental induction, we can maximum the effects of environmental factors on terpene accumulation in the future endeavors toward artemisinin overproduction. In theory, a new *A. annua* plant lines with artemisinin overproduction should be readily created through the metabolic engineering. We may breed the transgenic plants, for example, capable of smoothly redirecting the carbon flux from the plastidic terpenoids or cytosolic steroids to artemisinin production in combination with the environmental induction. Otherwise, artemisinin is certainly enhanced only in a limited range as posed by environmental stimuli.

Acknowledgement

We thank Lu-Lu Yin, Xiao-Ling Xu, Xiao-Mei Zeng, Wen-Jie Lu, Li-Xiang Zeng, Xiao-Xia Guo, Ping-Zu Zhang for their helpful assistance and constructive discussion. This work is partially supported by the Natural Science Foundation of China (NSFC) with the Project Nos. 30672614 to Li-Ling Feng and 30271591 to Qing-Ping Zeng and by Guangdong Provincial Scientific Development Project of China (GPSDPC) under the approved Project No. 2007B031404008 to Qing-Ping Zeng.

References

[1]   Abdin, M.Z., Israr, M., Srivastava, P.S. Jain, S.K. (2000). In vitro production of artemisinin, a novel antimalarial compound from *Artemisia annua* L. *J. Med. Arom. Plant Sci.*, 22-4a: 378-384.

[2]   Acton, N., Klayman, D.L. (1985). Artemisitene, a new sesquiterpene lactone endoperoxide from *Artemisia annua*. *Plant Med.* 5: 441–442.

[3]   Acton, N., Klayman, D.L., Rollmann, I.J. (1985). Reductive electrochemical HPLC assay for artemisinin (qinghaosu). *Plant Med.* 5: 445–446.

[4]   Adam, K.P., Zapp, J. (1998). Biosynthesis of the isoprene units of chamomile sesquiterpenes. *Phytochemistry* 48: 953-959.

[5]   Afitlhile, M., Fukushige, H., McCraken, C., Hildebrand, D. (2005). Allen oxide synthase and hydroperoxide lyase product accumulation in *Artemisia* species. *Plant Sci.* 169: 139-146.

[6]   Akhila, A., Thakur, R.S., Popli, S.P. (1987). Biosynthesis of artemisinin in *Artemisia annua*. *Phytochemistry* 16: 1927–1930.

[7]   Akhila, A., Kumkum, R., Thakur, R.S. (1990). Biosynthesis of artemisinic acid in *Artemisia annua*. *Phytochemistry* 29: 2129–2132.

[8]   Arano, H. (1964). Cytotaxonomic studies in subfamily Carduoideae of Japanese Compositae.XI. The karyotype analysis in some species of *Artemisia*. *Kromosomo* 57-59: 1883-1888.

[9]   Bagchi, G. D., Jain, D. C. and Kumar, S. (1997). Arteether: a potent plant growth inhibitor from. *Artemisia annua*. *Phytochemistry* 45: 1131-1134.

[10] Baldi, A., Dixit, V.K. (2007). Yield enhancement strategies for artemisinin production by suspension cultures of *Artemisia annua*. *Bioresour Technol* doi: 10.1016/j.biotech.2007.06.061.

[11] Beekman, A.C., Barentsen, A.R.W., Woerdenbag, H.J., van Uden, W., Pras, N., El-Feraly, F.S., Galal, A.M. (1997). Stereochemistry-dependent cytotoxicity of some artemisinin derivatives. *J. Nat. Prod.* 60: 325-327.

[12] Beekman, A.C., Wierenga, P., Woerdenbag, H.J., van Uden, W., Pras, N., Konings, A., El-Feraly, F.S., Galal, A.M., Wikstrom, H.V. (1998). TI - Artemisinin-derived sesquiterpene lactones as potential antitumour compounds: cytotoxic action against bone marrow and tumour cells. *Plant Med.* 64: 615-619.

[13] Brown, G.D., Sy, L.K. (2004). *In vitro* transformations of dihydroartemisinic acid in *Artemisia annua* plants. *Tetrahedron* 60: 1139-1159.

[14] Bertea, C.M., Freije, J.R., van der Woude, H., Verstappen FW, Perk L, Marquez V, de Kraker JW, Posthumus MA, Jansen BJ, de Groot A, Franssen MC, Bouwmeester HJ (2005). Identification of intermediates and enzymes involved in the early steps of artemisinin biosynthesis in *Artemisia annua*. *Planta Med* 71: 40-47.

[15] Bertea, C.M., Voster, A., Verstappen, F.W., Maffei, M., Beekwilder, J., Bouwmeester, H.J. (2006). Isoprenoid biosynthesis in *Artemisia annua*: cloning and heterologous expression of a germacrene A synthase from a glandular trichome cDNA library. *Arch Biochem. Biophys.* 448: 3-12.

[16] Bharel, S., Gulati, A., Abdin, M.Z., Srivastava, P.S., Vishwakarma, R.A., Jain, S.K. (1998). Enzymatic synthesis of artemisinin from natural and synthetic precursors. *J. Nat. Prod.* 61: 633-636.

[17] Bhakuni, R.S., Jain, D.C., Sharma, R.P., Kumar, S. (2001). Secondary metabolites of *Artemisia annua* and their biological activity. *Curr. Sci.* 80: 35-48.

[18] Bick, J.A., Lange, B.M. (2003). Metabolic cross talk between the cytosolic and plastidic pathways of isoprenoid biosynthesis: unidirectional transport of intermediates the chloroplast envelope membrane. *Arch. Biochem. Biophys.* 415: 146-154.

[19] Brader, G., Tas, E., Palva, E.T. (2001). Jusmonate-dependent induction of indole glucosinolates in *Arabidopsis* by culture filtrates of the nonspecific pathogen *Erwinia carotovora*. *Plant Physiol.* 126: 849-860.

[20] Bouwmeester, H.J., Wallaart, T.E., Janssen, M.H.A., van Loo, B., Jansen, B.J., Posthumus, M.A., Schmidt, C.O., de Kraker, J.W., Knig, W.A., Franssen, M.C. (1999). Amorpha-4, 11-diene synthase catalyses the first probable step in artemisinin biosynthesis. *Phytochemistry* 52: 843-854.

[21] Brown, G.D., Sy, L.K. (2004). *In vivo* transformations of dihydroartemisinic acid in *Artemisia annua* plants. *Tetrahedron* 60: 1139-1159.

[22] Cane, D.E. (1981). Biosynthesis of sesquiterpene. In: *Biosynthesis of isoprenoid compounds*. Porter JW, Spurgeon SL (eds), Vol 1 and 2, John Wiley and Sons, New York, pp. 283-374.

[23] Cane, D.E., Wu, Z., Oliver, J.S., Hohn, T.M. (1993). Overproduction of soluble trichodiene synthase from Fusarium sporotrichioides in *Escherichia coli*. *Arch. Biochem. Biophys.* 300: 416-422.

[24] Chang, M.C.Y., Keasling, J.D. (2006). Production of isoprenoid pharmaceuticals by engineered microbes. *Nat. Chem. Biol.* 2: 674-681.

[25] Chang, M.C.Y., Eachus, R.A., Trieu, W., Ro, D.K., Keasling, J.D. (2007). Engineering Escherichia coli for production of functionalized terpenoids using plant P450s. *Nat. Chem. Biol.* 3: 274-277.

[26] Chang, Y.J., Song, S.H., Park, S.H., Kim, S.U. (2000). Amorpha-4, 11-diene synthase of *Artemisia annua*: cDNA isolation and bacterial expression of a terpene synthase involved in artemisinin biosynthesis. *Arch. Biochem. Biophys.* 383: 178-184.

[27] Charles, D.J., Simon, J,E,, Wood, K.V., Heinstein, P. (1990). Germplasm variation in artemisinin content of *Artemisia annua* using an alternative method of artemisinin analysis from crude plant extracts. *J. Nat. Prod.* 53: 157–160.

[28] Carter, O.A., Peters, R.J., Croteau, R. (2003). Monoterpen biosynthesis pathway construction in *Escherichia coli*. *Phytochemistry* 64: 425-433.

[29] Chen, A.X., Lou, Y.G., Mao, Y.B., Lu, S., Wang, L.J,, Chen, X.Y. (2007). Plant terpenoids: biosynthesis and ecological functions. *J. Integr. Plant Biol.* 49: 179-186.

[30] Chen, C., Chen, Z. (2002). Potentiation of developmentally regulated plant defense response by AtWRKY18, a pathogen-induced Arabidopsis transcription factor. *Plant Physiol.* 129: 706-716.

[31] Chen, F.T., Zhang, G.H. (1987). Studies on several physiological factors on artemisinin synthesis in *Artemisia annua* L. *Plant Physiol* 5: 26–30.

[32] Chen, P.K., Leather, G.R., Klayman, D.L. (1987). Allelopathic effect of artemisinin and its related compounds from *Artemisia annua*. *Plant Physiol* 68: 406.

[33] Connolly, J.D., Hill, R.A. (1991). *Dictionary of terpenoids*. Vol 1, Mono- and Sesquiterpenoids. Chapman and Hall, London.

[34] Covello, P.S., Teoh, K.H., Polichuk, D.R., Reed, D.W., Nowak, G. (2007). Functional genomics and the biosynthesis of artemisinin. *Phytochemistry* 68: 1864-1871.

[35] Crock, J., Wildung, M., Croteau, R. (1997). Isolation and bacterial expression of a sesquiterpene synthase cDNA clone from peppermint (*Mentha piperita* L.) that produces the aphid alarm pheromone (E)-beta-farnesene. *Proc. Natl. Acad. Sci. USA* 94: 12833-12838.

[36] Croteau, R., Kutchan, T.M., Lewis, N.G. (2000). Natural products (secondary metabolites). In: Buchanan B, Gruissem W, Jones R (eds) *Biochemistry and molecular biology of plants*. American Society of Plant Physiologists. Rockville MD, pp1250-1318.

[37] Cubukcu, B., Bray, D.H., Warhust, D.C., Mericli, A.H., Ozhalay, N., Sariyar, G. (1990). *In vitro* antimalarial activity of crude extracts and compounds from *Artemisia abrotanum* L. *Phytother Res.* 4: 203-204.

[38] Davidovich-Rikanati, R., Sitrit, Y., Tadmor, Y., Iijima, Y., Bilenko, N., Bar, E., Carmona, B., Fallik, E., Dudai, N., Simon, J.E., Pichersky, E., Lewinsohn, E. (2007). Enrichment of tomato flavor by diversion of the early plastidic terpenoid pathway. *Nat. Biotechnol.* 25: 899-901.

[39] DeJong JM, et al. (2006). Genetic engineering of taxol biosynthetic genes in *Saccharomyces cerevisiae*. *Biotechnol. Bioeng* 93: 212-224.

[40] Delledonne, M., Xia, Y., Dixon, R.A., Lamb, C. (1998). Nitric oxide functions as a signal in plant disease resistance. *Nature* 394, 585-588.

[41] Dhingra, V., Rajoli, C., Narasu, M.L. (2000). Partial purification of proteins involved in the bioconversion of arteannuin B to artemisinin. *Bioresour Technol* 73: 279-282.

[42] Dhingra, V., Narasu, M.L. (2001). Purification and characterization of an enzyme involved in biochemical transformation of arteannuin B to artemisinin from *Artemisia annua*. *Biochem. Biophys. Res. Commun.* 281: 558-561.

[43] Dudareva, N., Andersson, S., Orlova, I., Gatto, N., Reichelt, M., Rhodes, D., Boland, W., Gershenzon, J. (2005). The nonmevalonate pathway supports both monoterpene and sesquiterpene formation in snapdragon flowers. *Proc. Natl. Acad. Sci. USA* 102: 933-938.

[44] Duke, S.O., Vaughn, K.C., Croom, E.M., ElSohly, H.N. (1987). Artemisinin, a constituent of annual wormwood (*Artemisia annua)*, is a selective phytotoxin. *Weed Sci.* 35: 499–505.

[45] Duke, S.O., Paul, R.N. (1993). Development and fine structure of the glandular trichomes of *Artemisia annua* L. *Int. J. Plant Sci.* 154: 107–118.

[46] Duke, M.V., Paul, R.N., ElSohly, H.N., Sturtz, G., Duke, S.O. (1994). Localization of artemisinin and artemisitene in foliar tissues of glanded and glandless biotypes of *Artemisia annua* L. *Int. J. Plant Sci.* 155: 365–372.

[47] ElFeraly, F.S., AlMeshal, I.A., AlYahya, M.A., Hifnawy, M.S. (1986). On the possible role of qinghao acid in the biosynthesis of artemisinin. *Phytochemistry* 25: 2777–2778.

[48] ElFeraly, F.S., AlMeshal, I.A., Khalifa, S.I. (1989). *Epi*-deoxyarteannuin B and 6,7-dehydroartemisinic acid from *Artemisia annua*. *J. Nat. Prod* 52: 196–198.

[49] ElSohly, H.N., Croom, Jr E.M., ElFeraly, F.S., ElSherei, M.M. (1990). A large-scale extraction technique of artemisinin from *Artemisia annua*. *J. Nat. Prod.* 53: 1560–1564.

[50] Ferreira, J.F.S., Simon J.E., Janick, J. (1995a). Developmental studies of *Artemisia annua*: Flowering and artemisinin production under greenhouse and field conditions. *Plant Med.* 61: 167-170.

[51] Ferreira, J.F.S., Simon, J.E., Janick, J. (1995b). Relationship of artemisinin content of tissue cultured, greenhouse grown, and field grown plants of *Artemisia annua*. *Plant Med.* 61: 351–355.

[52] Ferreira, J.F.S. (2007). Nutrient deficiency in the production of artemisinin, dihydroartemisinic acid, and artemisinic acid in *Artemisia annua* L. *J. Agr. Food Chem.* 55: 1686-1694.

[53] Farmer, E.E., Ryan, C.A. (1990). Interplant communication: airborne methyl jasmonate induces synthesis of protein inhibitors in plant leaves. *Proc. Natl. Acad. Sci. USA* 87: 7713-7716.

[54] Flugge, U.I., Gao, W. (2005). Transport of isoprenoid intermediates across chloroplast envelope membranes. *Plant Biol* (Stuttg) 7:91-97.

[55] Fu, T.M. (1991). Study on five species of *Artemisia* plants in inner Mongolia. *Univ J Inner Mongolia* (Nat Sci Ed) 22: 422-427.

[56] Fulzele, D.P., Sipahimalani, A.T., Heble, M.R. (1995). Tissue culture of *Artemisia annua* L. plant cultures in bioreactor. *J. Biotechnol.* 40: 139-143.

[57] Geng, S., Ma, M., Ye, H.C., Liu, B.Y., Li, G.F., Chong, K. (2001). Effects of *ipt* gene expression on the physiological and chemical characteristics of *Artemisia annua* L. *Plant Sci.* 160: 691-698.

[58] Gupta, S.K., Singh, P., Bajpai, P., Ram, G., Singh, D., Gupta, M.M., Jain, D.C. (2002). Morphogenetic variation for artemisinin and volatile oil in *Artemisia annua*. *India Crop Prod* 16: 217-224.

[59] Helliwell, C.A., Poole, A., Peacock, W.J., Dennis, E.S. (1999). *Arabidopsis ent*-kaurene oxidase catalyzes three steps of gibberellin biosynthesis. *Plant Physiol* 119: 507-510.

[60] Helliwell, C.A., Chandler, P.M., Poole, A., Dennis, E.S., Peacock, W.J. (2001). The CYP88A cytochrome P450, *ent*-kaurenoic acid oxidase, catalyzes three steps of the gibberellin biosynthesis pathway. *Proc. Natl. Acad. Sci. USA* 98: 2065-2070.

[61] Hemmerlin, A., Hoeffler, J.F., Meyer, O., Tritsch, D., Kagan, I.A., Grosdemange-Billiard, C., Rohmer, M., Bach, T.J. (2003). Cross-talk between the cytosolic mevalonate and the plastidic methylerythritol phosphate pathways in Tobacco bright yellow-2 cells. *J. Biol. Chem.* 278: 26666-26676.

[62] Hethelyi, E. B., Cseko, I. B., Grosy, M., Mark, G. and Palinkas, J. J. (1995). Chemical composition of the *Artemisia annua* essential oils from Hungary. *J. Essential Oil Res* 7, 45-48.

[63] Hu, X.Y., Neill, S.J., Cai, W.M., Tang Z.C. (2003). Nitric oxide mediates elicitor-induced saponin synthesis in cell cultures of *Panax ginseng*. *Funct Plant Biol.* 30: 901-907.

[64] Huang, L., Liu, J.F. (1993). Studies on anti-inflammatory effects of *Artemisia annua*. *China J Chin Mat Med* 18, 44-48.

[65] Irfan, Q.M., Israr, M., Abdin, M.Z., Iqbal, M. (2005). Response of *Artemisia annua* L. to lead and salt-induced oxidative stress. *Environ. Exp. Bot.* 53: 185-193.

[66] Jackson, B.E., Hart-Wells, E.A., Matsuda, S.P.T. (2003). Metabolic engineering to produce sesquiterpenes in yeast. *Org. Lett.* 5: 1629-1632.

[67] Jung, M. (1997). Synthesis and cytotoxicity of novel artemisinin analogs. *Bioorg. Med. Chem. Lett.* 7: 1091-1094.

[68] Kaplan, B., Davydov, O., Knight, H., Galon, Y., Knight, M.R., Fluhr, R., Fromm, H. (2006). Rapid transcriptome changes induced by cytosolic Ca^{2+} transients reveal ABRE-related sequences as Ca^{2+}-responsive cis elements in Arabidopsis. *Plant Cell* 18: 2733-2748.

[69] Kapoor, R., Chaudhary, V., Bhatnaga, A.K. (2007). Effects of arbuscular mycorrhiza and phosphorus application on artemisinin concentration in *Artemisia annua* L. *Micorrhiza* 17: 581-587.

[70] Kappers, I.F., Aharoni, A., van Herpen, T.W.J.M., Luckerhoff, L.L.P., Dicke, M. Bouwmeester, H.J. (2005). Genetic engineering of terpenoid metabolism attracts bodyguards to *Arabidopsis*. *Science* 309: 2070-2072.

[71] Kasahara, H., Hanada, A., Kuzuyama, T., Takagi, M., Kamiya, Y., Yamaguchi, S. (2002). Contribution of the mevalonate and methylerythritol phosphate pathways to the biosynthesis of gibberellins in *Arabidopsis*. *J. Biol. Chem.* 277: 45188-45194.

[72] Khosla, C., Keasling, J.D. (2003). Metabolic engineering for drug discovery and development. *Nat. Rev. Drug Discov.* 2: 1019-1025.

[73] Kim, N.C., Kim, S.U. (1992). Biosynthesis of artemisinin from 11,12-dihydroarteannuic acid. *J. Korean Agric. Chem. Soc. Rev.* 35: 106–109.

[74] Klayman, D.L., Lin, A.J., Acton, N., Scovill, J.P., Hoch, J.M., Milhous, W.K., Theodarides, A.D., Dobek, A.S. (1984). Isolation of artemisinin(Qinghaosu) from *Artemisia annua* growing in the United States. *J. Nat. Prod.* 47: 715–717.

[75] Klayman, D.L. (1993). *Artemisia annua*: from weed to respectable antimalarial plant. In: Kinhorn AD, Balandrin MF (eds) *HumanMedicinal Agents from Plants*, pp. 242–255. American Chemical Society Symposium Series. ACS, Washington, DC.

[76] Knox, J.P., Dodge, A.D. (1985) Singlet oxygen and plants. *Phytochemistry* 24: 889-896.

[77] Kobayashi, K., Suzuki, M., Tang, J., Nagata, N., Ohyama, K., Seki, H., Kiuchi, R., Kaneko, Y., Nakazawa, M., Matsui, M., Matsumoto, S., Yoshida, S., Muranaka, T. (2007). Lovastatin insensitive 1, a novel pentatricopeptide repeat protein, is a potential regulatory factor of isoprenoid biosynthesis in Arabidopsis. *Plant Cell Physiol.* 48:322-331.

[78] Kollner, T.G., Schnee, C., Gershenzon, J., Degenhardt, J. (2004). The variability of sesquiterpenes emitted from two *Zea mays* cultivars is controlled by allelic variation of two terpene synthase genes encoding stereoselective multiple product enzymes. *Plant Cell* 16:1115-1131.

[79] Korenromp, E., Miller, J., Nahlen, B., Wardlaw, T., Young, M. (2005). World Malaria Report 2005. World Health Organization (WHO), Roll Back Malaria Partnership, Geneva, 2005.

[80] Kreitschitz, A., Vallès, J. (2003). New or rare data on chromosome numbers in several taxa of the genus *Artemisia* (Asteraceae) in Poland. *Folia Geobotanica*, 38: 333-343.

[81] Kudakasseril, G.J., Lam, L., Staba, J. (1987). Effect of sterol inhibitors on the incorporation of ^{14}C-isopentanyl phosphate into artemisinin by cell-free system from *Artemisia annua* cultures and plants. *Plant Med.* 53: 280-284.

[82] Kuzamanov, B.A., Georgieva, S.B., Nikolova, V.A. (1986). Chromosome numbers of Bulgarian flowering plants. I. Family Asteraceae. *Fitologija* (Sofia), , 31: 71-74.

[83] Lange, B.M., Rujan, T., Martin, W., Croteau, R. (2000). Isoprenoid biosynthesis: the evolution of two ancient and distinct pathways across genomes. *Proc. Natl. Acad. Sci. USA* 97: 13172-13177.

[84] Laughlin, J.C. 1993 Effect of agronomic practices on plant yield and anti-malarial constituents of *Artemisia annua* L. *Acta Hortic* 331: 53-61.

[85] Laughlin, J.C. (1995). The influence of distribution of antimalarial constituents in *Artemisia annua* L. on time and method of harvest. *Acta Hortic* 390: 67-73.

[86] Laughlin JC (2002). Post-harvest drying treatment effects on antimalarial constituents of *Artemisia annua* L. *Acta hortic* 576:315-320.

[87] Laule, O., Furholz, A., Chang, H.S., Zhu, T., Wang, X., Heifetz, P.B., Gruissem, W., Lange, B.M. (2003). Crosstalk between cytosolic and plastidic pathways of isoprenoid biosynthesis in *Arabidopsis thaliana*. *Proc. Natl. Acad. Sci. USA* 100: 6866-6871.

[88] Lee, P.C., Schmidt-Dannert, C. (2002). Metabolic engineering towards biotechnological production of carotenoids in microorganisms. *Appl. Micro Biotechnol* 60: 1-11.

[89] Li, Y., Yang, Z.X., Chen, Y.X., Zhang, X. (1994). Synthesis of (15_14C) labelled artemisinin. *Acta Pharmaceutica Sinica* 29: 713–716.

[90] Li, Y., Huang, H., Wu, Y.L. (2006). Qinghaosu artemisinin – a fantastic antimalarial drug from a traditional Chinese medicine. In: Liang, XT, Fang WS. (Eds.), *Medicinal Chemistry of Bioactive Natural Products.* John Wiley and Sons Inc, pp. 183–256.

[91] Liersh, R., Soicke, H., Stehr, C., T̈ullner, H.U. (1986). Formation of artemisinin in *Artemisia annua* during one vegetation period. *Plant Med* 52: 387–390.

[92] Lin, J.Y., Chen, T.S., Chen, C.S., Jpn-Kokai Tokkyo Koho Jp, Patent No. 06135830, 1994.

[93] Lin, S.Z., Zhang, Z.Y., Lin, Y.Z., Zhang, Q., Guo, H. (2004). The role of calcium and calmodulin in freezing-induced freezing resistance of *Populus tomentosa* cuttings. *Chinese J. Plant Physiol. Mol. Biol.* 30: 59-68.

[94] Lin, R., Lin, Y.R. (eds.). Flora of the People's Republic of China. Editorial Committee on Flora of the People's Republic of China of the Chinese Academy of Sciences. Beijing, Science Press, 1991. Vol. 76, No. 2.

[95] Lindahl, A.L., Olsson, M.E., Mercke, P., Tollbom, O., Schelin, J., Brodelius, M., Brodelius, P.E. (2006). Production of the artemisinin precursor amorpha-4, 11-diene by engineered *Saccharomyces cerevisiae. Biotechnol Lett* 28: 571-580.

[96] Liu, C.Z., Wang, Y.C., Ouyang, F., Ye, H.C., Li, G.F. (1997). Production of artemisinin by hairy root cultures of *Artemisia annua* L. *Biotechnol Lett* 19: 927-929.

[97] Liu J.M., Ni, M.Y., Fan, J.F., Tu, Y.Y., Wu, Z.H., Wu, Y.L., Chou, W.S. (1979). Structure and reaction of arteannuin. *Acta Chim. Sin* 37: 129-143.

[98] Lommen, W.J.M., Schenk, E., Bouwmeester, H.J., Verstappen, F.W.A. (2006). Trichome dynamics and artemisinin accumulation during development and senescence of *Artemisia annua* leaves. *Plant Med.* 72: 336-345.

[99] Lommen, W.J.M., Elzinga, S., Verstappen, F.W.A., Bouwmeester, H.J. (2007). Artemisinin and sesquiterpene precursors in dead and green leaves of *Artemisia annua* L. crop. *Plant Med* 73: 1133-1139.

[100] Lommen, W.J., Schenk, E., Bouwmeester, H.J., Verstappen, F.W. (2005). Trichome dynamics and artemisinin accumulation during development and senescence of *Artemisia annua* leaves. *Plant Med.* 72: 336-345.

[101] Mahmoud, S.S., Croteau, R.B. (2002). Strategies for transgenic manipulation of monoterpene biosynthesis in plants. *Trends Plant Sci.* 7: 366-373.

[102] Martin, V.J.J., Yoshikuni, Y., Keasling, J.D. (2001). The in vivo synthesis of plant sesquiterpenes by *Escherichia coli. Biotechnol Bioeng* 75: 497-503.

[103] Martin, V.J.J., Pitera, D.J., Withers, S.T., Newman, J.D., Keasling, J.D. (2003). Engineering a mevalonate pathway in *Escherichia coli* for production of terpenoids. *Nat. Biotechnol* 21: 796-802.

[104] Martinez, B.C., Staba, E.J. (1988). The production of artemisinin in *Artemisia annua* L. tissue cultures. *Adv. Cell Cult.* 6: 69–87.

[105] Matcham, G.W., Bowen, A.R.S. (1996). Biocatalysis for chiral intermediates: meeting commercial and technical challenges. *Chem. Today* 14: 20-24.

[106] Menke, F.L.H., Parchmann, S., Mueller, M.J., Kijne, J.W., Mcmelink, J. (1999). Involvement of the octadecanoid pathway and protein phosphorylation in fungal elicitor-induced expression of terpenoid indole alkaloid biosynthetic genes in *Catharanthus roseus. Plant Physiol.* 119: 1289-1296.

[107] Mercke, P., Bengtsson, M., Bouwmeester, H.J., Posthumus, M.A., Brodelius, P.E. (2000). Molecular cloning, expression, and characterization of amorpha-4, 11-diene synthase, a key enzyme of artemisinin biosynthesis in *Artemisia annua* L. *Arch Biochem. Biophys.* 381:173-180.

[108] Morales, M.M., Charles, D.J., Simon, J.E. (1993). Seasonal accumulation of artemisinin in *Artemisia annua* L. *Acta Hortic* 344: 416-420.

[109] Mutabingwa, T.K. (2005). Artemisinin-based combination therapies (ACTs): best hope for malaria treatment but inaccessible to the needy! *Acta Trop.* 95:305-315.

[110] Nair, M.S.R., Basile, D.V. (1999). Bioconversion of arteannuin B to artemisinin. *J. Nat. Prod.* 56: 1559-1566.

[111] Newman, J.D., Chappell, J. (1999). Isoprenoid biosynthesis in plants: carbon partitioning within the cytoplasmic pathway. *Crit. Rev. Biochem. Mol. Biol.* 34: 95-106.

[112] Newman, J.D., Marshall, J., Chang, M.C.Y., Nowroozi, F., Paradise, E., Pitera, D., Newman, K.L., Keasling, J.D. (2006). High-level production of amorpha-4,11-diene in a two phase partitioning bioreactor of metabolically engineered *Escherichia coli. Biotechnol Bioeng* 95:684-691.

[113] Newton, P., White, N. (1999). Malaria: new development in treatment and prevention. *Ann. Rev. Med.* 50: 179-192.

[114] Nishiuchi, T., Shinshi, H., Suzuki, K. (2004). Rapid and transient activation of transcription of the ERF3 gene by wounding in tobacco leaves: possible involvement of NtWRKYs and autorepression. *J. Biol. Chem.* 279: 55355-55361.

[115] Nojiri, H., Sugimori, M., Yamane, H., Nishimura, Y., Yamada, A., Shibuya, N., Kodama, O., Murofushi, N., Omori, T. (1996). Involvement of jasmonic acid in elicitor-induced phytoalexin production in suspension-cultured rice cells. *Plant Physiol.* 110: 387-392.

[116] O'Donnell, P.J., Calvert, C., Atzorn, R., Wasternack, C., Leyser, H.M.O., Bowles, D.J. (1996). Ethylene as a signal mediating the wound response of tomato plants. *Science* 274: 1914-1917.

[117] Paniego, N.B., Maligne, A.E., Giulietti, A.M. (1993). *Artemisia annua*: in vitro culture and the production of artemisinin. In: Bajaj YPS (ed) Medicinal and Aromatic Plants. Biotechnology in Agriculture and Forestry Vol 5, pp. 70–78. Springer-Verlag, Springer.

[118] Paniego, N.B., Giulietti, A.M. (1996). Artemisinin production by *Artemisia annua* L.-transformed organ culture. *Enzyme Microbiol. Technol.* 18: 526-530.

[119] Park, H.C., Kim, M.L., Kang, Y.H., Jeon, J.M., Yoo, J.H., Kim, M.C., Park, C.Y., Jeong, J.C., Moon, B.C., Lee, J.H., Yoon, H.W., Lee, S.H., Chung, W.S., Lim, C.O., Lee, S.Y., Hong, J.C., Cho, M.J. (2004). Pathogen- and NaCl-induced expression of the

SCaM-4 promoter is mediated in part by a GT-1 box that interacts with a GT-1-like transcription factor. *Plant Physiol.* 135: 2150-2161.

[120] Pfleger, B.F., Pitera, D.J., Smolke, C.D., Keasling, J.D. (2006). Combinatorial engineering of intergenic regions in operons tunes expression of multiple genes. *Nat. Biotechnol.* 24:1027-1032.

[121] Picaud, S., Olofsson, L., Brodelius, M., Brodelius, P.E. (2005). Expression, purification and characterization of recombinant amorpha-4,11-diene synthase from *Artemisia annua* L. *Arch. Biochem. Biophys.* 436: 215-226.

[122] Piel, J., Donath, J., Bandemer, K., Boland, W. (1998). Mevalonate-independent biosynthesis of terpenoid volatiles in plants - induced and constitutive emission of volatiles. *Angewandte Chemie* 37: 2478-2481.

[123] Pitera, D.J., Paddon, C., Newman, J.D., Keasling, J.D. (2007). Rebuilding a balanced heterologous mevalonate pathway for isoprenoid production in *Escherichia coli. Metab Eng* 9:193-207.

[124] Polya, L. (1949). Chromosome numbers of some Hungarian plants. *Acta Hungarica* 6: 12137.

[125] Pras, N., Visser, J.F., Batterman, S., Woerdenbag, H.J., Malingr´e, T.M., Lugt, C.B. (1991). Laboratory selection of *Artemisia annua* L. for high artemisinin yielding types. *Phytochem Anal* 2: 80–83.

[126] Putalun, W., Luealon, W., De-Eknamkul, W., Tanaka, H., Shoyama, Y. (2007). Improvement of artemisinin production by chitosan in hairy root cultures of *Artemisia annua* L. *Biotechnol. Lett* 29: 1143-1146.

[127] Rieping, M., Schoffl, F. (1992). Synergistic effect of upstream sequences, CCAAT box elements, and HSE sequences for enhanced expression of chimaeric heat shock genes in transgenic tobacco. *Mol. Gen. Genet* 231: 226-232.

[128] Ro, D.K., Paradise, E.M., Ouellet, M., Fisher KJ, Newman KL, Ndungu JM, Ho KA, Eachus RA, Ham TS, Kirby J, Chang MC, Withers ST, Shiba Y, Sarpong R, Keasling JD (2006). Production of the antimalarial drug precursor artemisinic acid in engineered yeast. *Nature* 440: 940-943.

[129] Rodriguez-Concepcion, M., Fores, O., Martinez-Garcia, J.F., Gonzalez, V., Phillips, M.A., Ferrer, A.,Boronat, A. (2004). Distinct light-mediated pathways regulate the biosynthesis and exchange of isoprenoid precursors during Arabidopsis seedling development. *Plant Cell* 16:144-156.

[130] Romero, M.R., Efferth, T., Serrano, M.A., Castano, B., Macias, R.I., Briz, O., Marin, J.J. (2005). Effect of artemisinin/artesunate as inhibitors of hepatitis B virus production in an in vitro replicative system. *Antiviral Research* 68:75–83.

[131] Roth, R.J., Acton, N.A. (1989). The isolation of Sesquiterpenes from *Artemisia annua*. *J. Chem. Educ.* 66: 349.

[132] Roth, R.J., Acton, N.A. (1989). A simple conversion of artemisinic acid into artemisinin. *J. Nat. Prod.* 52: 1183–1185.

[133] Sangwan, R.S., Agarwal, K., Luthra, R., Thakur, R.S., Sangwan, N.S. (1993). Biotransformation of arteannuic acid into arteannuin B and artemisinin in *Artemisia annua*. *Phytochemistry* 34:1301-1302.

[134] Sauret-Gueto, S., Botella-Pavia, P., Florres-Perez, U., Martinez-Garcia, J.F., San Roman, C., Leon, P., Boronat, A., Rodriguez-Concepcion, M. (2006). Plastid cues posttranscriptionally regulate the accumulation of key enzymes of the methylerythritol phosphate pathway in Arabidopsis. *Plant Physiol.* 141:75-84.

[135] Schuler, M.A., Werck-Reichhart, D. (2003). Functional genomics of P450s. *Ann Rev Plant Biol.* 54: 629-667.

[136] Schoendorf, A., Rithner, C.D., Williams, A.M., Croteau, R.B. (2001). Molecular cloning of a cytochrome P450 taxane 10β-hydroxylase cDNA from *Taxus* and functional expression in yeast. *Proc. Natl. Acad. Sci. USA* 98: 1501-1506.

[137] Shukla, A., Abad Farooqi, A. H., Shukla, Y. N., Sharma, S. (1992). Effect of triacontanol and chlormequat on growth, plant hormones and artemisinin yield in *Artemisia annua* L. *Plant Growth Regul* 11: 165.

[138] Simpson, S.D., Nakashima, K., Narusaka, Y., Seki, M., Shinozaki, K., Yamaguchi-Shinozaki, K. (2003). Two different novel cis-acting elements of erdl, a clpA homologous Arabidopsis gene function in induction by dehydration stress and dark-induced senescence. *Plant J.* 33: 259-270.

[139] Singh, A., Vishwakarma, R.A., Husain, A. (1988). Evaluation of *Artemisia annua* strains for higher artemisinin production. *Plant Med.* 54: 475–476.

[140] Souret, F.F., Weathers, P.J., Wobbe, K.K. (2002). The mevalonate independent pathway is expressed in transformed roots of *Artemisia annua* and regulated by light and culture age. *In Vitro Cell Dev. Biol. Plant.* 38: 581-588.

[141] Souret, F.F., Kim, Y., Wyslouzil, B.E., Wobbe, K.K., Weathers, P.J. (2003). Scale-up of *Artemisia annua* L. hairy root cultures produces complex patterns of terpenoid gene expression. *Biotechnol. Bioeng* 83: 653-667.

[142] Steele, C.L., Crock, J., Bohlmann, J., Croteau, R. (1998). Sesquiterpene synthases from grand fir (*Abies grandis*): Comparison of constitutive and wound-induced activities, and cDNA isolation, characterization, and bacterial expression of delta-selinene synthase and gamma-humulene synthase. *J. Biol. Chem.* 273: 2078-2089.

[143] Steliopoulis, P., Wust, M., Adam, K.P., Mosandl, A. (2002). Biosynthesis of the sesquiterpene germacrene D in *Solidago Canadensis*: ^{13}C and ^{2}H labeling studies. *Phytochemistry* 60:13-20.

[144] Stemmer, W.P.C. (1994). Rapid evolution of a protein *in vitro* by DNA shuffling. *Nature* 370: 389-391.

[145] Stoessl, A., Stothers, J.B., Ward, E.W.B. (1976). Sesquiterpenoid stress compounds of the *Solanaceae*. *Phytochemistry* 15: 855-873.

[146] Sun, C., Palmqvist, S., Olsson, H., Boren, M., Ahlandsberg, S., Jansson, C. (2003). A novel WRKY transcription factor, SUSIBA2, participates in sugar signaling in barley by binding to the sugar-responsive elements of the iso1 promoter. *Plant Cell* 15: 2076-2092.

[147] Suzuka, O. (1950). Chromosome numbers in the genus *Artemisia*. *Jap. J. Genet.* 25: 17-18.

[148] Svensson, J.T., Crosatti, C., Campoli, C., Bassi, R., Stanca, A.M., Close, T.J., Cattivelli, L. (2006). Transcriptome analysis of cold acclimation in barley albina and xantha mutants. *Plant Physiol.* 141: 257-270.

[149] Sy, L.K., Brown, G.D. (2002). The mechanism of the spontaneous autooxidation of dihydro- artemisinic acid. *Tetrahydron* 58: 897-908.

[150] Tamogami, S., Rakwal, R., Kodama, O. (1997). Phytoalexin production elicited by exogenously applied jasmonic acid in rice leaves (*Oryza sativa* L.) is under the control of cytokinins and ascorbic acid. *FEBS Lett.* 412: 61-64.

[151] Tavlik, A.F., Bishop, S.J., Ayalp, A., EL-Feraly, F.S. (1990). Effects of artemisinin, dihydroartemisinin and arteether on immune responses of normal mice. *Int. J. Immunol. Pharmacol.* 12, 385-389.

[152] Teoh, K.H., Polichuk, D.R., Reed, D.W., Nowak, G., Covello, P.S. (2006). *Artemisia annua* L. (Asteraceae) trichome- specific cDNAs reveal CYP71AV1, a cytochrome P450 with a key role in the biosynthesis of the antimalarial sesquiterpene lactone artemisinin. *FEBS Lett.* 580: 1411-1416.

[153] Terzaghi, W.B., Cashmore, A.R. (1995). Light-regulated transcription. *Annu. Rev. Plant Physiol. Plant Mol. Biol.* 46: 445-474.

[154] Torrell, M., Vallès, J. (2001). Genome size in 21 *Artemisia* L. species (Asteraceae, Anthemideae): Systematic, evolutionary, and ecological implications. *Genome* 44: 231–238.

[155] Towler, M.J., Weathers, P.J. (2007). Evidence of artemisinin production from IPP stemming from both the mevalonate and the nonmevalonate pathways. *Plant Cell Rep.* 26:2129-2136.

[156] Trigg, P.I. (1990). Qinghaosu (Artemisinin) as an antimalarial drug. *Econ. Med. Plant Res.* 3: 20–55.

[157] Ueda, J., Kato, J. (1980). Isolation and identification of a senescence-promoting substance from wormwood (*Artemisia absinthium* L.). *Plant Physiol.* 66: 246-249.

[158] Valles, J. (1987). Aportacion al conocimiento citotaxonomico de ocho taxones Ibericos del genero *Artemisia* L. (Asteraceae, Anthemideae). *Anales Jard Bot Madrid* 44: 79-96.

[159] Vallès, J., Torrell, M., Garnatje, T., Garcia-Jacas, N., Vilatersana, R., Susanna, A. (2003). The genus *Artemisia* and its allies: phylogeny of the subtribe Artemisiinae (Asteraceae, Anthemideae) based on nucleotide sequences of nuclear ribosomal DNA internal transcribed spacers (ITS). *Plant Biol.* (Stuttgart), 5: 274-284.

[160] Volkova, S.A., Boyko, E.V. (1986). Chromosome numbers in some species of *Artemisia* from the southern part of the Soviet Far East. *Bot. Zurn.* 71: 1693.

[161] Vogeli, U., Chappell, J. (1998). Induction of sesquiterpene cyclase and suppression of squalene synthase activities in plant cell cultures treated with fungal elicitor. *Plant Physiol.* 88: 1291-1296.

[162] Wallaart, T.E., van Uden, W., Lubberink, H.G.M., Woerdenbag HJ, Pras N, Quax WJ (1999) Isolation and identification of dihydroartemisinic acid from *Artemisia annua* and its role in the biosynthesis of artemisinin. *J. Nat. Prod.* 1999, 62: 430-433.

[163] Wallaart, T.E., Pras, N., Quax, W.J. (1999). Isolation and identification of dihydroartemisinic acid hydroperoxide from *Artemisia annua*: a novel biosynthetic precursor of artemisinin. *J. Nat. Prod.* 62: 1160-1162.

[164] Wallaart, T.E., Pras, N., Beekman, A.C., Quax, W.J. (2000). Seasonal variation of artemisinin and its biosynthetic precursors in plants of *Artemisia annua* of different geographical origin: proof for the existence of chemotypes. *Plant Med.* 66: 57-62.

[165] Wallaart, T.E., Bouwmeester, H.J., Hille, J., Poppinga, L., Maijers, N.C.A. (2001). Amorpha-4, 11-diene synthase: cloning and functional expression of a key enzyme in the biosynthetic pathway of the novel antimalarial drug artemisinin. *Planta* 212: 460-465.

[166] Wang, C.W. (1961). The forests of China, with a survey of grassland and desert vegetations. Harvard University Maria Moors Cabot Foundation No 5, pp. 171–187. Harvard University Cambridge, MA.

[167] Wang, E.M., Wang, R., DeParasis, J., Loughrin, J.II., Gan, S.S., Wagner, G.J. (2001) Suppression of a P450 hydroxylase gene in plant trichome glands enhances natural-product-based aphid resistance. *Nat. Biotechnol.* 19: 371-374.

[168] Wang, H., Ye, H.C., Li, G.F., Liu, B.Y., Chong, K. (2000). Effects of fungal elicitors on cell growth and artemisinin accumulation in hairy root cultures of *Artemisia annua*. *Acta Bot. Sin.* 42: 905-909.

[169] Wang, H., Ge, L., Ye, H.C., Chong, K., Liu, B.Y., Li, G.F. (2004). Studies on the effects of *fpf1* gene on *Artemisia annua* flowering time and on the linkage between flowering and artemisinin biosynthesis. *Plant Med.* 70: 347-352.

[170] Wang, H., Liu, Y., Chung, K., Liu, B.Y., Ye, H.C., Li, Z.Q., Yan, F., Li, G.F. (2007). Earlier flowering induced by over-expression of CO gene does not accompany increase of artemisinin biosynthesis in *Artemisia annua*. *Plant Biol.* (Stuttgart) 9: 442-446.

[171] Wang, J.W., Zhang, Z., Tan, R.X. (2001). Stimulation of artemisinin production in *Artemisia annua* hairy roots by the elicitor from the endophytic *Colletotrichum* sp. *Biotechnol Lett.* 23: 857-860.

[172] Wang, J.W., Xia, Z.H., Tan, R.X. (2002). Elicitation on artemisinin biosynthesis in *Artemisia annua* hairy roots by the oligosaccharide extract from the endophytic *Colletotrichum* sp. B501. *Acta Bot. Sin.* 44: 1233-1238.

[173] Wang, J.W., Wu, J.Y. (2004). Involvement of nitric oxide in elicitor-induced defense responses and secondary metabolism of *Taxus chinensis* cells. *Nitric Oxide* 11: 298-306.

[174] Wang, Y., Weathers, P.J. (2007). Sugars proportionately affect artemisinin production. *Plant Cell Rep* 26: 1073-1081.

[175] Wang, Y., Xia, Z.Q., Zhou, F.Y., Wu, Y.L., Huang, J.J., Wang, Z.Z. (1988). Studies on the biosynthesis of arteannuin III. Arteannuin acid as a key intermediate in the biosynthesis of arteannuin and arteannuin B. *Acta Chim. Sin.* 46: 386-387.

[176] Wang, Y., Xia, Z.Q., Zhou, F.Y., Wu, Y.L., Huang, J.J., Wang, Z.Z. (1993). Studies on the biosynthesis of arteannuin IV. The biosynthesis of arteannuin and arteanniun B by the leaf homogenate of *Artemisia annua* L. *Chin J. Chem.* 11: 457-463.

[177] Walker, T.S., Bais, H.P., Vivanco, J.M. (2002). Jasmonic acid induced hypercin production in *Hypericum perforatum* L. (St. John wort). *Phytochemistry* 60: 289-293.

[178] Watson, L.E., Bates, P.L., Evans, T.M., Unwin, M.M., Estes, J.R. (2002). Molecular phylogeny of Subtribe Artemisiinae (Asteraceae), including *Artemisia* and its allied and segregate genera. *BMC Evol. Biol.* 2: 17-28.

[179] Weathers, P.J., Cheetham, R.D., Follansbee, E., Theoharides, K. (1994). Artemisinin production by transformed roots of *Artemisia annua*. *Biotechnol. Lett.* 16: 1281–1286.

[180] Weathers, P.J., Bunk, G., McCoy, M.C. (2005). The effect of phytohormones on growth and artemisinin production in *Artemisia annua* hairy roots. *In Vitro Cell Dev. Biol. Plant* 41: 47-53.

[181] Weathers, P.J., Elkholy, S., Wobbe, K.K. (2006). Artemisinin: the biosynthetic pathway and its regulation in *Artemisia annua*, a terpenoids-rich species. *In Vitro Cell Dev. Biol. Plant* 42: 309-317.

[182] WHO (2001) Antimalarial drug combination therapy: report of a WHO technical consultation. WHO/CDS/ RBM/2001/35, reiterated in 2003.

[183] WHO (2003) International pharmacopoeia, 3rd ed., Vol. 5. Geneva.

[184] WHO (2005) WHO Model List of Essential Medicines, 14th ed. (Revised March 2005). Geneva.

[185] Woerdenbag, H.J., Lugt, C.B., Pras, N. (1990). *Artemisia annua* L.: a source of novel antimalarial drugs. Pharmaceutisch Weekblad, *Sci. Ed* 12: 169-181

[186] Woerdenbag, H.J., Bos, R., Salomons, M.C., Hendrika, H., Pras, N., Malingre, T.M. (1993). Volatile constituents of *Artemisia annua* L. *Flavour Fragrance J* 8: 131-137.

[187] Woerdenbag, H.J., Lfiers, J.F.J., van Uden, W., et al. (1993). Production of the new antimalarial drug artemisinin in shoot cultures of *Artemisia annua* L. *Plant Cell Tiss Org. Cult* 32: 247-257.

[188] Woerdenbag, H.J., Pras, N., Nguyen, G.C., Bui, T.B., Bos, R., Van Uden, W., Pham, V.Y., Nguyen, V.B., Batterman, S., Lugt, C.B. (1994). Artemisin in, related sesquiterpenes, and essential oil in*Artemisia annua* during a vegetation period in Vietnam. *Plant Med.* 60, 272-275.

[189] Wu, S.Q., Schalk, M., Clark, A., Miles, R.B., Coates, R., Chapel, J. (2006) Redirection of cytosolic or plastidic isoprenoid precursors elevates terpene production in plants. *Nat Biotech.* 24: 1441-1447.

[190] Xie, Z., Zhang, Z.L., Zou, X., Huang, J., Ruas, P., Thompson, D., Shen, Q.J. (2005). Annotations and functional analyses of the rice WRKY gene superfamily reveal positive and negative regulators of abscisic acid signaling in aleurone cells. *Plant Physiol.* 137: 176-189.

[191] Xiong, L.M., Schumaker, K.S., Zhu, J.K. (2002). Cell signaling during cold, drought, and salt stress. *Plant Cell* S165-S183.

[192] Xu, M.J., Dong, J.F., Zhu, M.Y. (2004). Involvement of NO in fungal elicitor-induced activation of PAL and stimulation of taxol synthesis in *Taxus chinenesis* suspension cells. *Chinese Sci. Bull* 49: 1038-1043.

[193] Xu, M.J., Dong, J.F., Zhu, M.Y. (2005). Nitric oxide mediates the fungal elicitor-induced hypericin production of *Hypericum perforatum* cell suspension cultures through a jasmonic-acid-dependent signal pathway. *Plant Physiol.* 139: 991-998.

[194] Xu, M.J, Dong, J.F., Zhu, M.Y. (2005). Effect of nitric oxide on catharanthine production and growth of *Catharanthus roseus* suspension cells. *Biotechno. Bioeng* 89: 367-371.

[195] Xu, M.J., Dong, J.F., Zhu, M.Y. (2006). Nitric oxide mediates the fungal elicitor-induced puerarin biosynthesis in *Pueraria thomsonii* Benth. suspension cells through a salicylic acid (SA)-dependent and a jasmonic acid (JA)-dependent signal pathway. *Sci. China Ser. C* 49: 379-389.

[196] Xue, G.P. (2002). Characterisation of the DNA-binding profile of barley HvCBF1 using an enzymatic method for rapid, quatitative and high-throughput analysis of the DNA-binding activity. *Nucleic Acids Res.* 30: e77.

[197] Xue, G.P. (2003). The DNA-binding activity of an AP-2 transcriptional activator HvCBF2 involved in regulation of low-temperature responsive genes in barley is modulated by temperature. *Plant J.* 33: 373-383.

[198] Yin, L.L., Zhao, C., Huang, Y., Yang, R.Y., Zeng, Q.P. (2008). Abiotic stress-induced expression of artemisinin biosynthesis genes in *Artemisia annua* L. *Chin J. Appl. Environ. Biol.* 14 : 1-5.

[199] Yu, D., Chen, C., Chen, Z. (2001). Evidence for an important role of WRKY DNA binding proteins in the regulation of NPR1 gene expression. *Plant Cell* 13: 1527-1540.

[200] Yuan, L.Z., Rouviere, P.E., Larossa, R.A., Suh, W. (2006). Chromosomal promoter replacement of the isoprenoid pathway for enhancing carotenoids production in *E.coli. Metab. Eng.* 8: 79-90.

[201] Zhang, Y.S., Ye, H.C., Liu, B.Y., Wang, H., Li, G.F. (2005). Exogenous GA$_3$ and flowering induce the conversion of artemisinic acid to artemisinin in *Artemisia annua* plants. *Russ J. Plant Physiol.* 52: 58-62.

[202] Zhang, Z.L., Xie, Z., Zou, X., Casaretto, J., Ho, T.H., Shen, Q.J. (2004). A rice WRKY gene encodes a transcriptional repressor of the gibberellin signaling pathway in aleurone cells. *Plant Physiol.* 134: 1500-1513.

[203] Zhao, J., Sakai, K. (2001). Multiple signaling pathways mediate fungal elicitor induced *h*-thujaplicin accumulation in *Cupressus lusitanica* cell cultures. *J. Exp. Bot.* 54: 647-656.

[204] Zeng, Q.P., Qiu, F., Yuan, L. (2008a). Production of artemisinin by genetically modified microbes. *Biotechnol. Lett.* 30: 581-592.

[205] Zeng, Q.P., Zhao, C., Yin, L.L., Yang, R.Yi., Zeng, X.M., Feng, L.L., Yang, X.Q. (2008b). Artemisinin biosynthetic cDNA and novel EST cloning and their quantification on low temperature-induced overexpression. *Sci. China Ser. C* 51: 232-244.

[206] Zheng, G. Q. (1994) Cytotoxic terpenoids and flavonoids from *Artemisia annua. Plant Med* 60, 54-57.

[207] Zheng, L.P., Guo, Y.T., Wang, J.W., Tan, R.X. (2007). Nitric oxide potentates oligosaccharide- induced artemisinin production in *Artemisia annua* hairy roots. *J Integr Plant Biol* 50: 49–55.

In: Biotechnology: Research, Technology and Applications   ISBN 978-1-60456-901-8
Editor: Felix W. Richter   © 2008 Nova Science Publishers, Inc.

Chapter 8

Attitude Toward Bioethics and Acceptance of Biotechnology

*Yutaka Tanaka**

Osaka Gakuin University, Osaka, Japan

Abstract

Bioethics and public acceptance of biotechnology are important social issues in this century. It appears that people's attitude toward bioethics is closely related to the acceptance of biotechnology. However, few psychological studies have examined this relationship or attitude structure of bioethics in the present circumstances. The following three psychological studies investigated these matters.

In study 1, attitudes toward bioethics and the acceptance of various biotechnologies were investigated with psychological scales. The participants were 231 Japanese undergraduate students. A cognitive map of biotechnology was constructed from these attitude scores. The cognitive map showed that people oppose to biotechnologies that are perceived as unethical, but approve of biotechnologies that are perceived as ethical. This study also indicated that people's attitude toward biotechnology, for example, toward gene recombination technology, differs considerably according to whether the technology is applied to human beings, animals or plants.

In study 2, based on Tanaka's study (2004), a structural equation model was set up, in which the common factor of "bioethics," affects three factors, namely, "human dignity," "ethically right or wrong for researching and developing biotechnologies," and "nature and the natural order." The participants were 154 Japanese undergraduate students who answered a questionnaire. Gene recombination technology toward human beings was taken up as a subject. The results confirmed the validity of the proposed model. There were also some interesting findings concerning the elements that construct each of the three factors. For example, it was suggested from the element that constructs the factor of "nature and the natural order" that people feel it is unnatural to change the human body and human nature during a short time span by using gene technology,

---

* Address correspondence to Yutaka Tanaka, Faculty of Informatics, Osaka Gakuin University, 2-36-1, Kishibe-Minami, Suita-shi, Osaka, 564-8511, JAPAN; E-mail: yutanak@utc.osaka-gu.ac.jp

because their characteristics have been formed over a very long time period since the beginning of life.

In study 3, based on Tanaka's study (2004) and the study 2 of this chapter, a hypothesis that the factor of bioethics is also important for acceptance of genetically modified foods (GM foods) was tested by using a structural equation model. The participants were 166 Japanese undergraduate students who answered a questionnaire. A causal model in which the factor of bioethics affects the factor of acceptance of GM foods was set up. The results of analysis indicated that the hypothesis was clearly supported. The result of study 3 suggests that the factor of bioethics is important for not only the acceptance of biotechnologies themselves, but also the acceptance of foods and products which are produced by use of biotechnologies.

Introduction

The development and use of biotechnology is one of the important problems in our society this century. For example, debates and disputes are going on all over the world on the pros and cons of gene-recombination, making clones and use of reproductive technology and on how to use these technologies.

As represented by the studies of Slovic (1986, 1987, 1993, 1994, 2000), many psychological studies have been conducted on people's risk perception and their attitude of acceptance of various technologies. However, few studies of the risk perception and acceptance of biotechnology exist (Siegrist, 2000). By testing his causal model, Siegrist (1999, 2000) showed that perceived risk, perceived benefit and trust in institutions determine acceptance of biotechnology. Tanaka (2002) constructed a cognitive map of biotechnology which consisted of a risk factor and a benefit factor.

The factor of bioethics is also very important for pros and cons and the public acceptance of biotechnology (Tanaka, 2004). Various definitions of bioethics have been proposed. For example, Moreno (1995) described "'Bioethics' is a popular contraction for 'biomedical ethics,' which is the study of moral values in the life sciences and in their clinical applications," and Shannon (1997) defined 'Bioethics' as follows: "bioethics examines the ethical dimension of problems at both the heart and the cutting edge of technology, medicine, and biology in their application to life." Levinson and Reiss (2003) stated that "No single definition of the word 'Bioethics' can be given. In the USA, for example, the term is still sometimes used as a synonym for 'medical ethics.' In most countries, though, it is used to encompass all the questions about ethics in biology and medicine."

Ogata (1999) mentioned that "One of the major themes of bioethics is to defend human dignity." In a discussion of human dignity and taking out a patent of human genes, Chapman (2002) stated that "The concept of the inherent dignity of the human person is the grounding for internationally recognized human rights. It is also a central concept in most Western legal systems. A number of critics have raised concerns about whether the patenting of human genes infringes respect for human dignity." "Human dignity" is thus one of the important factors of bioethics.

It has been argued that using biotechnology, such as gene-recombination technology and clone technology, is to "play God" (Steinbock, 2000). Attitudes toward whether human beings have the right to conduct gene-recombination and cloning, would be also deeply related with bioethics.

Hoban, Woodrum and Czaja (1992) conducted psychological studies on genetically modified (GM) foods and found that people tend to object to any food which is considered to be unnatural. It seems that people object not only to GM foods, but also to any biotechnology which is regarded as an unnatural technology, such as gene-recombination technology that people feel is against the natural selection process.

Sense of values, such as, "life should remain natural," and "human beings should not destroy the balance of nature," appear to be the important constituents of bioethics. Hoban et al. (1992) also showed that moral objection is the strongest predictor of opposition to genetic engineering by using multiple regression analysis.

Tanaka (2004) showed that acceptance of gene-recombination technology is explained by "perceived risk," "perceived benefit," "trust" and "sense of bioethics." Tanaka (2004) also indicated that a factor of "sense of bioethics" contains three variables, that is, "human dignity," "ethically right or wrong for researching and developing biotechnologies," and "nature and the natural order."

It appears that people's attitude toward bioethics is closely related to the acceptance of biotechnology. However, few psychological studies have examined this relationship or the attitude structure of bioethics in the present circumstances. Therefore, the following three psychological studies investigated these matters.

Study 1

The purpose of study 1 is to construct a cognitive map of biotechnology by using an attitude score toward bioethics and an attitude score toward public acceptance. Then the relationship between the attitude toward bioethics and the attitude toward public acceptance is to be examined visually based on the cognitive map.

Method

Participants

Participants were 231 undergraduate students (184 male, 47 female) who took courses in Present-day Sociology and Social Communication at a university in Osaka, Japan. The mean age was 19.07 (SD = 1.34, range of 18 to 24 years). Questionnaires were distributed and the students were asked to answer the questions during class at the university.

Subjects of Rating

The 27 subjects of bioethics which are shown in Figure1 were used as subjects of rating in this study.

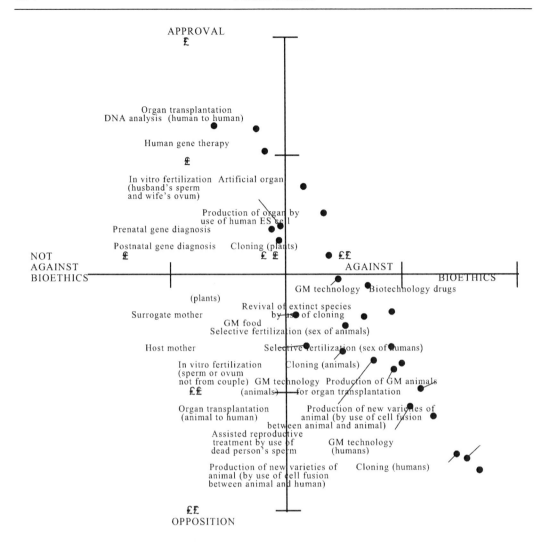

Figure 1. Cognitive map of biotechnology.

Questionnaire

Three psychological scales, that is, "human dignity," "ethically right or wrong for researching and developing biotechnologies," and "nature and the natural order," were set up to measure people's attitude toward bioethics. A scale of "pros and cons" was also set up to measure attitude toward public acceptance.

There were 4 psychological scales for each of the 27 rating subjects. Therefore, there were 108 questions (variables) in total in the questionnaire. Questions were to be answered on a 5-point bipolar scale with a midpoint that meant a middle degree or neither positive nor negative.

Results

The mean value of each of the 108 variables was calculated, and an averaged 27 (rating subjects) × 4 (scales) matrix was generated. By using mean values of the 27 rating subjects, Cronback's alpha coefficient was computed on the three scales, that is, "human dignity," "ethically right or wrong for researching and developing biotechnologies," and "nature and the natural order," showing that the coefficient was .95 and internal consistency was very high. Therefore, a mean value of these three scales was calculated for each of the 27 rating subjects, and this value was considered to be an attitude score toward bioethics.

Figure 1 is the cognitive map of biotechnology, in which the 27 rating subjects are plotted. In the map, the X axis indicates an attitude score toward bioethics and the Y axis indicates an attitude score toward public acceptance.

Although all of the 27 rating subjects are so-called "biotechnology," a glance at Figure 1 will reveal that whether people feel the technology is against bioethics or not, and whether people accept the technology into our society or not, depend on the subjects and whether the technology is applied to plants, animals or human beings. For example, the cognitive map shows that even if the technology itself is the same cloning technology, when the technology is applied to plants, people perceive this technology to be somewhat against bioethics but somewhat acceptable, on the other hand, when the technology is applied to human beings, people perceive this technology to be considerably against bioethics and by no means acceptable. Likewise the map indicates that even if the organ transplant technology itself is the same technology, when an organ is transplanted from human beings to human beings, people feel it somewhat natural and strongly agree to it, however, when the organ is transplanted from animals to human beings, people feel it unnatural and oppose it. We also see from the cognitive map on the whole that the more the biotechnology is perceived to be against bioethics, the less the technology is acceptable, and that the less the biotechnology is perceived to be against bioethics, the more the technology is acceptable.

Discussion

The results of this study supported the results of Hoban et al. (1992) and Tanaka (2004), and show that an attitude toward bioethics has a close relation to an attitude toward public acceptance. The more cloning technology and gene technology are applied to subjects close to human beings, the more they are perceived to be against bioethics, and the stronger they are opposed. For example, in the case of producing a human clone and changing human nature and the way of human reproduction in a manner which rarely or never occurs in the natural state, people appear to feel it is ethically wrong and have an aversion and show a rejection reaction to it by nature and intuitively. As for organ transplant, the organ transplant from animal to human is perceived more negatively than the organ transplant from human to human. People seem to also have a feeling of disgust and refusal by nature to unite with an animal, that is, a different species, and a part of its body.

There is a possibility that the position on the cognitive map of each biotechnology which was shown in this study, that is, the attitude toward bioethics and the attitude toward public acceptance of each biotechnology, will change with time and by culture. For example, in the case of organ transplant from human to human and in vitro fertilization (IVF), there were huge repercussions and many expressions of anxiety right after these technologies were developed and introduced into our society. However, as shown in the results of this study, these technologies are no longer considered to be against bioethics by people and are evaluated to be acceptable to our society today.

As a growing number of people use the technology and see and hear about the technology, there is a possibility that people's attitude toward the technology gradually change. However, even if the times and social conditions change significantly, it is hard to think for the time being that, for example, the innate aversion and the ethical attitude toward producing a human clone and producing a novel human being by use of gene technology, will change. Therefore, people will not always change all their attitudes toward biotechnologies into positive attitudes in the future. Furthermore, in the case that serious accidents and events occur concerning some biotechnology, there is a fair possibility that people's attitude toward the biotechnology will change into negative attitudes.

Study 2

In study 2, based on Tanaka's study (2004), the validity of the psychological model was verified by using a structural equation model, in which the common factor of "bioethics," affects three factors, namely, "human dignity," "ethically right or wrong for researching and developing biotechnologies," and "nature and the natural order." Furthermore, concrete elements of each of the three factors were investigated. Gene recombination technology applied to human beings was the content focus of the study.

Pre-Test (Open-Ended Questions)

A pre-test of open-ended questions was conducted to explore the elements of each of the three factors, "human dignity," "ethically right or wrong for researching and developing biotechnologies," and "nature and the natural order," prior to a quantitative questionnaire survey.

Method

Participants

Participants were 145 undergraduate students (61 male, 84 female) who were taking a course in Social Psychology at a university in Kyoto, Japan. The mean age was 18.45 (SD =.81, range of 18 to 24 years). Open-ended questionnaires were distributed and the students were asked to answer the questions during class.

Questionnaire

(1) Human dignity: First, participants were asked to answer the question of whether gene-recombination technology applied to human beings threatens human dignity or not on a 5-point bipolar scale with a midpoint that meant middle degree or neither positive nor negative. Then, participants were asked to write down freely the reason why they answered the way they did in their questionnaire.

(2) Ethically right or wrong for researching and developing biotechnologies: First, participants were asked to answer the question of whether or not we should engage in the field of human gene-recombination technology, on a 5-point bipolar scale with a midpoint. Then, they were asked to write down freely the reason why they answered the way they did in their questionnaire.

(3) Nature and the natural order: In the same way as (1) and (2), first, participants were asked to answer the question of whether human gene-recombination technology is against the workings of nature on a 5-point bipolar scale with a midpoint. Then, they were asked to give an account freely of the reason why they answered the way they did in their questionnaire.

Results

The results of 5-point bipolar scales were as follows: The mean value of "human dignity" on the scale (1. strongly threatening - 5. not threatening at all) was 2.13 (SD =1.00). The mean value of "ethically right or wrong for researching and developing biotechnologies" on the scale (1. quite wrong - 5. quite right) was 2.34 (SD =1.22). The mean value of "nature and the natural order" on the scale (1. completely against - 5. not against at all) was 1.95 (SD =1.06). These results indicate the majority of people consider that human gene technology threatens human dignity and human beings should not step into the field of human gene technology, and that human gene technology is against nature and the natural order. The combined results show that many people have a negative attitude toward human gene-recombination technology as a whole.

Next I summarized the answer to the open-ended questions for each of the "human dignity," "ethically right or wrong for researching and developing biotechnologies," and "nature and the natural order" categories. The results were as follows:

(3) The reason for threatening or not threatening human dignity
   (a) Because the existence of human beings itself would be changed by the technology.
   (b) Because human will and emotion would be changed by the technology.
   (c) Because if the number of genetically modified human beings increases, our society would become a bad society.
   (d) Because discrimination among human beings would increase.
   (e) Because each person's personality would be lost.

(2) The reason for ethical right or wrong in researching and developing biotechnologies

(a) Once human beings have got gene technology, the development and use of gene technology would proceed and would not be able to be stopped.

(b) Human beings should be natural and human beings should not manipulate human genes.

(c) Human beings have no right to conduct gene-recombination (Human beings are not God).

(3) The reasons for or against nature and the natural order

(a) Because human beings themselves were produced in accordance with nature and the natural order and human beings produced gene technology, gene technology itself therefore also should follow nature and the natural order.

(b) Because technology is used in gene-recombination and technology was produced by human beings, gene technology is not natural and therefore is against nature and the natural order.

(c) Because gene-recombination cannot occur in natural conditions, it is against nature and the natural order.

(d) Human beings should not change in a moment human genes which were formed naturally over a very long time.

Test of Causal Model

Based on the results of the pre-test, a structural equation model was made. The validity of the model was tested with data obtained by the questionnaire method.

Method

Participants

Participants were 154 undergraduate students (71 male, 83 female) who were taking a course in Social Psychology at an university in Kyoto, Japan. The mean age was 19.10 (*SD* = 1.06, range of 18 to 23 years). Questionnaires were distributed and the students were asked to answer the questions during class.

Causal Model and Questionnaire

In the same way as Siegrist's studies (1999, 2000) and Tanaka's study (2004), a structural equation model was used as a causal model in this study. Based on Tanaka's study (2004), the common factor of "bioethics," affects three factors, namely, "human dignity," "ethically right or wrong for researching and developing biotechnologies," and "nature and the natural order." In Tanaka's study (2004), each of the three factors, that is, "human dignity," "ethically right or wrong for researching and developing biotechnologies," and "nature and the natural order," was set up not as a factor but as a variable. However, in this study, each of the three factors was set up not as a variable but as a factor, and several concrete elements for each of the three factors were also set up in the model, in order to investigate concrete elements which consist of each of the three factors.

Based on the results of the pre-test, five elements, that is, "change of the existence of human beings itself (V1)," "change of human will and emotion (V2)," "our society becoming a good society or a bad society (V3)," "increase of discrimination in our society (V4)," and "increase or decrease of each person's personality (V5)," were set up for the factor of "human dignity." Three elements, that is, "once human beings have got gene technology, it cannot be stopped (V6)," "human beings should be natural (V7)," and "human beings have a right to conduct gene-recombination (V8)," were used for the factor of "ethically right or wrong for researching and developing biotechnologies." In the same way, four elements, that is, "human beings themselves were produced in accordance with nature and the natural order (V9)," "gene technology is against nature and the natural order (V10)," "gene-recombination's frequency of occurring in natural conditions (V11)," and "human beings should not change human genes in a moment (V12)," were set up for the factor of "nature and the natural order."

As shown in Figure 2, a psychological model with four factors and 12 variables was made. Each of these 12 variables corresponds to each of the questions in the questionnaire. Linear 5-point bipolar scales with a midpoint were used as rating scales.

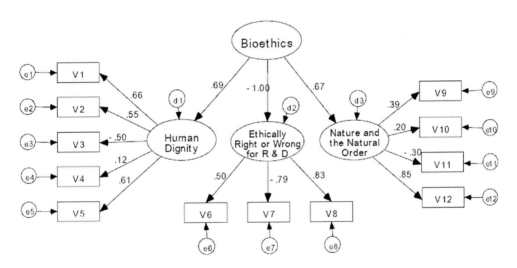

Figure 2. Initial model of attitude structure of bioethics.

Results

The program Amos was used to test the plausibility of Figure 2, that is, the initial model. The default setting, that is, the maximum likelihood (ML) method, was employed to estimate parameters in the model. As shown in Figure 2 and Table 1, the initial model has validity. The path coefficients shown in Figure 2 are the standardized coefficients.

However, in Figure 2, the value of the path coefficient from the factor of "human dignity," to the variable of "increase of discrimination in our society (V4)" is low. This shows that the variable of "increase of discrimination in our society (V4)" is not appropriate as an element of the factor of "human dignity." Therefore, the variable of "increase of discrimination in our society (V4)" was deleted. In the same way, because the value of the path coefficient from the factor of "nature and the natural order" to the variable of "gene technology is against nature and the natural order (V10)" is low in Figure 2, the variable of "gene technology is against nature and the natural order (V10)" was also deleted.

Table 1. Fit Indices of Structural Equation Model

	χ^2	df	GFI	CFI	RMSEA
Initial Model	73.07	52	.93	.94	.05
Final Model	39.34	32	.95	.98	.04

The modification indices indicated that if the correlation between the error variable of V6 (e6) and the error variable of V7 (e7) had been allowed, the goodness of fit of the model would have increased significantly. Because V6 and V7 are variables which are both included in the factor of "ethically right or wrong for researching and developing biotechnologies" and seem to have a conceptual relationship, it seemed reasonable to allow the correlation between e6 and e7. Therefore, the correlation between e6 and e7 was set up.

Using the modified model, the covariance structures analysis was conducted again. The result is shown in Table 1, and indicates that the goodness of fit of the initial model has been improved in the final model. The figure of the final model is shown in Figure 3, and the path coefficients shown in the causal model are the standardized coefficients. Table 1 and Figure 3 indicate, as hypothesized, that there is validity to the three factors, namely, "human dignity," "ethically right or wrong for researching and developing biotechnologies," and "nature and the natural order" being subordinate factors of "bioethics."

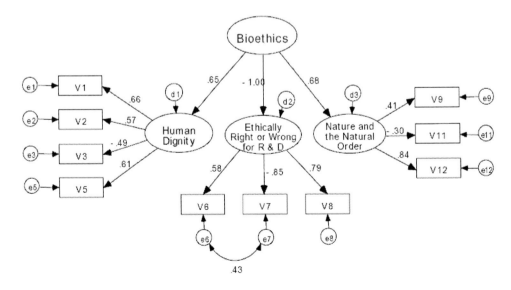

Figure 3. Final model of attitude structure of bioethics.

As specific elements of the factor of "human dignity," four elements, that is, "change of the existence of human beings itself (V1)," "change of human will and emotion (V2)," "our society becoming a good society or a bad society (V3)," and "increase or decrease of each person's personality (V5)," were shown. As specific elements of the factor of "ethically right or wrong for researching and developing biotechnologies," three elements, that is, "once human beings have got gene technology, it cannot be stopped (V6)," "human beings should be natural (V7)," and "human beings have a right to conduct gene-recombination (V8)," were indicated. As specific elements of the factor of "nature and the natural order," three elements, that is, "human beings themselves were produced in accordance with nature and the natural order (V9)," "gene-recombination's frequency of occurring in natural conditions (V11)," and "human beings should not change human genes in a moment (V12)," were shown.

Discussion

The results of study 2 confirmed again that the concept of "bioethics" consists of "human dignity," "ethically right or wrong for researching and developing biotechnologies," and "nature and the natural order." When we look at the elements which construct the factor of "human dignity," it can be observed that many of these elements are associated with the change of human nature. This finding suggests that changing human nature by gene technology, that is, by the work of human hands, has almost the same meaning as jeopardizing human dignity.

By studying the elements which construct the factor of "ethically right or wrong for researching and developing biotechnologies," we see that whether we think human beings should be natural or not is the key point when we judge the rights or wrongs of the use of gene technology. It also shows that whether we accept the fact that once human beings have got gene technology it cannot be stopped, and whether we think human beings have a right to conduct gene-recombination, are elements which affect the attitude toward "ethically right or wrong for researching and developing biotechnologies."

The most typical element in the factor of "nature and the natural order" is the element of "human beings should not change human genes in a moment." People seem to feel it is unnatural to change the human body and human nature during a short time span by using gene technology, because their characteristics have been formed over a very long time period since the beginning of life.

Participants were university students in this study. Because university students have often been used as participants in many previous psychological studies concerning risk perception and acceptance of technology (Siegrist, 2000; Tanaka, 2004), and because the validity and the generality of the results of those studies have been verified by subsequent studies (Siegrist, 2000; Tanaka, 2004), students were also used as participants in this study. However, the possibility that characteristics of university students affect the results of psychological studies cannot be entirely ruled out. Therefore, it cannot be asserted that a model which was based on and verified by student participants completely corresponds with a model which is based on and verified by randomly sampled participants. For these reasons, the results of this present study in which university students were used as participants should be seen as a pilot study.

Study 3

Based on Tanaka's study (2004) and study 2 of this chapter, the goal of study 3 is to test the hypothesis that the factor of bioethics is also important for acceptance of GM foods by using a structural equation model.

Method

Participants

Participants were 166 undergraduate students (146 male, 16 female) who were taking a course in Present-day Sociology at a university in Osaka, Japan. The mean age was 19.92 (*SD* =.95, range of 19 to 23 years). Questionnaires were distributed and the students were asked to answer the questions during class.

Causal Model and Questionnaire

Based on Tanaka's study (2004) and study 2 of this chapter, a causal model in which the factor of bioethics affects the factor of acceptance of GM foods was set up. In Tanaka's study (2004) and study 2 of this chapter, three variables or three factors, that is, "human dignity," "ethically right or wrong for researching and developing biotechnologies," and "nature and the natural order," were set up as the elements of the factor of "bioethics." However, the variable of "human dignity" is not relative to GM foods, thus "human dignity" was not used, and the other two variables, that is, "ethically right or wrong for researching and developing biotechnologies (V1)" and "nature and the natural order (V2)" were used as the elements of the factor of "bioethics." Three variables, that is, "positive or negative image of GM foods (V3)," "when buying foods, concern about GM foods (V4)," and "whether it is OK or not to eat GM foods (V5)," were set up as the elements of the factor of "acceptance."

A psychological model with two factors and 5 variables was made. Each of these 5 variables corresponds to each of the questions in the questionnaire. Questions were to be answered on a 5-point bipolar scale with a midpoint that meant a middle degree or neither positive nor negative.

Results

The program Amos was used to test the plausibility of the model. The default setting, that is, the maximum likelihood (ML) method, was employed to estimate parameters in the model. The results of analysis showed that the model has validity. However, the modification indices suggested that if the correlation between the error variable of V3 (e3) and the error variable of V4 (e4) had been allowed, the goodness of fit of the models would have been improved significantly in the model. Because V3 and V4 are both related to the acceptance of GM foods, it seemed reasonable to allow the correlation between e3 and e4. Therefore, the correlation between e3 and e4 was set up in the model as shown in Figure 4. Using this modified model, the covariance structures analysis was conducted again. The path coefficients in Figure 4 are the standardized coefficients. Figure 4 indicates that the modified model fits the data very well and clearly supports the hypothesis that the factor of bioethics is also important for acceptance of GM foods.

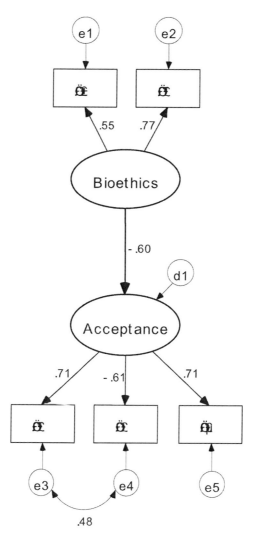

$\chi 2 = 2.674(p = .445)$ GFI $= .994$ AGFI $= .968$ RMSEA $= .000$

Figure 4. Bioethics as a determinant of acceptance of GM foods.

Discussion

The result of study 3 suggests that the factor of bioethics is important not only for the acceptance of biotechnologies themselves, such as gene recombination technology and clone technology, but also the acceptance of foods and products which are produced by use of biotechnologies. Hence it follows that researchers and enterprises that research and develop biotechnologies and the governments which support and regulate the research and development of biotechnologies need to pay adequate attention to people's attitude toward bioethics.

When people purchase and eat GM foods, whether they feel uncomfortable with GM foods as foods or not also seem to be important. Even though looks and taste appear to be the same, if people feel uncomfortable with GM foods as foods, they will hesitate to purchase and eat GM foods. Furthermore, feeling uncomfortable with GM foods seems to be related with bioethics. I would now like to go on to examine the relationship.

Conclusion

This study shows that people's attitude toward bioethics has a great effect on the acceptance of biotechnologies and foods and products which are made using biotechnologies. It also suggests that people's concept of bioethics is, in brief, that human beings, animals, and plants should be natural and human beings should not change human genes, animal genes, and plant genes in a moment.

Finally, it is necessary to consider whether the sense of bioethics which people have will be a plus or a minus for the future research and development of biotechnologies and for the prosperity of all human beings. If people and our society deny all biotechnologies, such as gene technology and clone technology, for the reason that biotechnology is against bioethics, they will simultaneously be giving up on the potentially extremely great benefits of biotechnologies, such as progress in medicine and improvement of people's health, escape from food crisis and shortage of resources, and contribution to improvement of the global environment. Therefore, taking such an attitude of rigidly and strictly observing bioethics and denying all biotechnologies will likely not bring about a good result for escaping from various crises and for the welfare of all human beings.

On the other hand, if people give up the sense of bioethics entirely and accept changes of nature and the shape of human beings, animals, and plants as people like, what kind of situation will be brought about? It is possible that a large number of human clones and animal clones will be made, and odd or super human beings and animals will also be created. Indiscreet and too frequent changing of human genes will likely cause serious social problems, such as confusion of identity of individuals and families, bad effects for human health, resulting in a minus or lethal effect on human beings and our society. Many people now hold such reservations about bioethics in common, and this appears to be serving the purpose of applying the brakes to unprincipled research and development of biotechnologies and preventing the minus effect of biotechnologies.

Among social problems concerning public acceptance of various technologies in this century, the problem of biotechnology development and use is one of the most major and controversial problems. Thus, the importance of research on people's attitude toward bioethics, which has a close relationship to the acceptance of the research and development of biotechnologies, is increasing year by year.

Acknowledgement

This study was supported by a grant from Osaka Gakuin University.

The author gratefully acknowledges Rhys Truman for his assistance and helpful comments to improve this article.

References

Chapman, A. R. (2002). Patenting human genes: ethical and policy issues. In J. Bryant., L. B. L. Velle. and J. Searle (Eds.), *Bioethics for scientists* (pp. 265-278). Hoboken: John Wiley and Sons.

Hoban, T., Woodrum, E., and Czaja, R. (1992). Public opposition to genetic engineering. *Rural Sociology, 57,* 476-493.

Levinson, R. and Reiss, M. J. (2003). Issues and scenarios. In R. Levinson and M. J. Reiss (Eds.), *Key issues in bioethics* (pp. 3-13). New York: RoutledgeFalmer.

Moreno, J. D. (1995). *Deciding together.* New York: Oxford University Press.

Ogata, K. (1999). Human dignity and bio-ethics. *Journal of Japan Association for Bioethics, 9,* 48-54. (In Japanese with English abstract).

Shannon, T. A. (1997). *An introduction to bioethics (3rd ed.).* New York: Paulist Press.

Slovic, P. (1986). Informing and educating the public about risk. *Risk Analysis, 6,* 403-415.

Slovic, P. (1987). Perception of risk. *Science, 236,* 280-285.

Slovic, P. (1993). Perceived risk,trust,and democracy. *Risk Analysis, 13,* 675-682.

Slovic, P. (1994). Perceptions of risk: paradox and callenge. In B. Brehmer, and N. E. Sahlin (Eds.), *Future risks and risk management* (pp.63-78). Boston: Kluwer Academic Publishers.

Slovic, P. (2000). The Perception of Risk. London:Earthscan.

Siegrist, M. (1999). A causal model explaining the perception and acceptance of gene technology. *Journal of Applied Social Psychology, 22,* 2093-2106.

Siegrist, M. (2000). The influence of trust and perceptions of risk and benefit on the acceptance of gene technology. *Risk Analysis, 20,* 195-203.

Steinbock, B. (2000). Cloning human beings: sorting through the ethical issues. In B. MacKinnon (Ed.), *Human Cloning* (pp. 68-84). Urbana: University of Illinois Press.

Tanaka, Y. (2002). Cognitive map of biotechnology. *Japanese Journal of Risk Analysis, 14,* 59-62. (In Japanese with English abstract.)

Tanaka, Y. (2004). Major psychological factors affecting acceptance of gene-recombination technology. *Risk Analysis, 24,* 1575-1583.

In: Biotechnology: Research, Technology and Applications   ISBN 978-1-60456-901-8
Editor: Felix W. Richter   © 2008 Nova Science Publishers, Inc.

Chapter 9

Bioprocessing Strategies for Food Technology: Immobilized Lactic Acid Bacteria and Bifidobacteria Applications

Maria Papagianni[*]

Department of Hygiene and Technology of Food of Animal Origin
School of Veterinary Medicine, Aristotle University of Thessaloniki
Thessaloniki, 54006 Greece

Abstract

Lactic acid bacteria (LAB) cultures production, as well as milk fermentations, are carried out in bioreactors operated in the batch mode using free -not immobilized- microorganisms. In recent years however, immobilized cell technologies have been applied successfully in many cases to LAB, bifidobacteria, and probiotic cultures demonstrating the importance and applications potential of this technology in the food area. As a bio-processing strategy, cell immobilization may result in comparative advantages over cell-free cultures, such as increased fermentation rates or metabolite production from lower biomass levels. Other benefits include reduced contamination possibilities, physical and chemical protection of cells and reuse of biocatalysts. Immobilized cell technologies, long-term continuous culture modes and development of controlled-release systems aim to enhance the tolerance of sensitive LAB to environmental stresses and subsequently result in new applications in the food technology, and production of high-value products with positive effect on consumers' health. This review focuses on the current status of the main bio-processing strategies in this field, their technological characteristics, applications and perspectives.

[*] Tel. +30-2310-999804, Fax +30-2310-999829, E-mail: mp2000@vet.auth.gr

1. Introduction

Lactic acid bacteria (LAB) are widely used in the production of fermented food products due to their specific metabolic activities which translate into technological, nutritional and health properties. These metabolic activities result in the production of lactic and other organic acids, which are essential for the production of fermented dairy and meat products and contribute to the development of their typical flavor. Volatile substances are also produced which are responsible for the development of flavor and aroma. Lactose fermentation improves the digestibility of fermented products, while various polysaccharides produced by many LAB contribute to desired texture development. The role of LAB cultures in fermented food preservation is well-known. Acidification by organic acids production - mainly lactate- and the production of effective antimicrobials (bacteriocins), contribute greatly to preservation by inhibiting pathogens and other contaminants.

LAB industrial fermentations are carried out for the production of pure or mixed cultures, lactic acid, polylactic acid, polysaccharides, nisin, and a plethora of fermented food products. The fermented dairy products represent a large market share of dairy products, and increased developments of products containing nutraceutical cultures are anticipated for North American, Japanese and European markets. Lactobacilli and bifidobacteria are already used in many probiotic dairy products marketed worldwide. Most fermentations are traditionally carried out in batch bioreactors using free (not immobilized) microbial cells. Recent advances however, on immobilized cell technologies applied to LAB and probiotic cultures have demonstrated the importance and interest in these new technologies. Among the reported advantages for immobilized over free suspension culture systems are high cell density and volumetric productivities, increased process stability over long fermentation periods, reuse of biocatalysts, uncoupling of biomass and metabolite production, and stimulation of secondary metabolite production. As an alternative bioprocessing strategy, cell immobilization may offer improved resistance to contamination, physical and chemical protection of cells and retention of plasmid-bearing cells. Published data for the sensitive to environmental stresses bifidobacteria have shown that immobilization and encapsulation strategies can provide the means for better incorporation and survival of their cultures into food products and subsequently in the human intestine.

2. Cell Immobilization Techniques

The literature on the use of immobilized microbial cells for the synthesis of chemicals is vast. Basically, cell immobilization involves attaching the microorganism to an insoluble matrix that may be noncellular or cellular, that is may be a synthetic polymer or biopolymer or the microorganism itself [1]. In a fermentation system, cell immobilization involves retaining microorganisms in a particular location of the system aiming at reaching high cell densities. There are six general methods for immobilization of microorganisms. These include: entrapment in a matrix, absorption on a matrix, encapsulation in a matrix, covalent binding to a matrix, chemical cross-linking of cells and flocculation. The most commonly used technique is that of entrapment. Table 1 lists the main advantages and disadvantages of various cell immobilization methods. Attachment or adsorption to various preformed carriers, entrapment in polymeric networks, membrane entrapment and encapsulation have been used for immobilization of LAB and various probiotic bacteria [2].

Table 1. Advantages and disadvantages of some cell immobilization methods

Method	Advantages	Disadvantages

Aggregation
  Covalent
  Floc formation
	Cheap, permanent
	Gentle, not denaturing
		Not applicable if cell viability required
		Impermanent /permanent if heated

Entrapment In synthetic polymers during polymerisation	Wide range of gel properties, cell growth and division possible within gel beads	Hazardous monomers, decrease enzyme activities, often expensive
In preformed synthetic polymers	Gentle, cell viability retained	Often expensive
In biopolymers	Low toxicity, gentle	Often expensive, sensitive
Alginate	Inexpensive, easy bead formation, adjustable solubility, cell viabilities retained, enzyme activities retained for extended periods	Chemical synthesis may vary
κ-carrageenan	Inexpensive, cell viabilities retained, stabilized biochemical activities	Brief contact with elevated temperatures
Chitosan	Positive charge	Inhibitory effects on certain types of cells affecting cell viability
Cellulose acetate	Cell viabilities retained, enzyme activities retained for very long periods	Cells must contact organic solvents
Collagen	Excellent for membrane manufacture	Crosslinking needed for stability
Adsorption/ colonization	Passive immobilization, very gentle and safe for cells if suitable supports are chosen for particular cell types	Cells may enter product stream
Encapsulation	Protective effects on cell viability under adverse conditions, suitable for probiotics formulations	Technological properties of cells must considered in selection of proper method

In selecting the matrix, a number of factors should be considered. These include: stability, compatibility with the microbial activity desired, and compatibility with the substrates and products of the process. It is important that enzymes not leach from the support nor should the support itself leach into the product. The carriers, when industrial applications in the food industry are the case, have to be non-toxic, readily available and of affordable cost.

Immobilization of cells, compared with freely suspended cell systems, involves a number of features which have to be judged on merit for any given application as to weather it confers advantages or disadvantages. Some of the advantages the immobilized cell processes may offer as compared with immobilized enzymes and fermentation procedures involving freely dispersed cells are the following:

- Cell separation and manipulation of cells as a discrete phase
- Higher enzyme operation stability than immobilized enzyme systems
- Endogenous cofactor regeneration
- Possible multi step enzyme reactions
- Higher volumetric productivities
- Operation of batch fermenters on a drain and fill basis
- Possibility of prolonged continuous operation (beyond the nominal washout flow rates)
- Possible higher cell density as compared to traditional fermentation
- Possible use of optimum aggregate sizes leading to maximum microbial activities
- Possible co-immobilization of different cell populations within the same reactor
- Possible effects on morphology, physiology and metabolism of cells
- Elimination of need and costs for enzyme extraction and purification
- Less plant pollution than traditional fermentation
- May reduce enzyme cost

However, the use of immobilized microorganisms may not always be a solution since:

- Side reactions may occur due to multiplicity of enzymes in microbial cells
- Specific catalytic activity may be lower than with immobilized enzymes
- Products of high molecular weight may encounter diffusion problems with certain matrices and microbial cells
- Matrices and microbial cells may impede gas transport
- Oxygen limitation may occur due to low solubility in matrices
- Products such as acids and bases may cause a detrimental pH microenvironment if diffusion in the matrix is limiting

In the final analysis, the biological approach needs to be carefully considered on a case-by-case basis. Some of the factors that determine the cost of a process with immobilized cells include the stability of the biocatalyst, the cost of immobilization, the reuse of the system, the performance of the system, the clean-up and pollution control, and finally the capital investment requirements [1].

2.1. Cell Entrapment

The most well-known and widely used technique for food applications is the entrapment of microbes inside a food-grade porous polymeric matrix. Food-grade matrices used have been Ca-alginate, sodium alginate, various modified alginates, κ-carragenan, a κ-carragenan-locust bean gum combination, gellan gum, chitosan, agarose, cellulose, cotton fabric, gluten, gelatin, a starch-gluten-milk combination. Spherical gel biocatalysts (polymer beads with the desired diameter and the immobilized active culture) are produced using mild techniques, such as extrusion or emulsification, by thermal (κ-carragenan, agarose, gellan, gelatin) or ionotropic (alginate, chitosan) gelation of droplets. These polymers are readily and widely available and accepted for use as additives in the food industry. Among various polymers, alginate is attracting increasing interest for use as immobilization matrix because of its low toxicity and biocompatibility [3, 4, 5], and ease in bead formulation by ionotropic gelation. This is a polysaccharide extracted from seaweed and is a co-polymer with alternating sequences of β-D-mannuronic and α-L-guluronic acid residues, 1,4-glycosidically linked. An important feature of alginate is its ionotropic gelation induced by bivalent or polyvalent cations, which ionically cross-link carboxylate groups in the urinate blocks of alginate at low pH, but becoming soluble at neutral or higher pHs. This behavior affords important advantages when alginate is used in certain applications, e.g. immobilization of sensitive bifidobacteria for oral administration. The alginate matrix beads are insoluble in the stomach while the immobilized cells are protected from acid shock. Moreover, its high solubility at intestinal pH conditions allows the release of viable cells into the intestinal track.

Gel entrapment is a relatively simple method that results in usually spherical beads of 0.3 to 3.0 mm diameter. Incubation of the immobilized cells in a nutritive medium results in the formation of a high cell density region that extends from the bead surface to a radial position where cell growth becomes limiting due to lack of substrate and other unfavorable conditions for growth, e.g. pH, accumulation of inhibitory products, and others [6, 7]. In the case of LAB that are sensitive to inhibitions by product formation, factors such as product concentration and pH profiles have shown to play an important role on immobilized cell growth and process productivity [6, 7, 8, 9].

Careful selection of the bead material is necessary to achieve the required mechanical stability of gel biocatalysts for long-term fermentations. An important issue however, associated with the industrialization of immobilized cells in the food industry still remains the large-scale production of polymer beads under aseptic conditions.

Cell release from the periphery of the beads into the fermentation medium is not unusual. High pressure due to cell expansion, collisions and mechanical forces due to agitation in bioreactors may result in high rates of cell release according to Sodini et al.[10], Lamboley et al. [6] and Doleyres and Lacroix [7]. Cell release may be desirable in some applications, as for example in biomass production, or in inoculation strategies. Cell release, for example, has been exploited by Duran-Paramo et al. [11] in a process, carried out in a column reactor, in which skim milk was inoculated continuously for a week for yoghurt and acid milk production. *Lactobacillus delbrueckii* spp. *bulgaricus* was immobilized in α-carrageenan beads and released from the beads, inoculating this way the milk, at a rate of almost 2×10^8 CFU/ml continuously for a week. In other applications, such as entrapped biomass production, cream fermentations, and metabolite production is not desirable. It has been shown that repeated use of the beads tends to increase the levels of free cells in the culture medium and although a number of treatments have been tested, the rate of cell release is reduced only during the early fermentation stages [12, 13].

2.1.1. Starter Cultures (Biomass)

LAB are used extensively as single or mixed cultures in the production of fermented foods, e.g. cheeses, fermented milks, yoghurts, cultured cream and butter, and sausages. The industry uses concentrated starters in order to obtain high numbers of effective cells and to ensure a particular strain balance in mixed starter cultures. LAB starters are produced traditionally in batch fermentation systems. In such systems however, end-product accumulation (e.g. lactate) may limit growth. Continuous systems may be applied to avoid growth limitation but more serious problems, associated with prolonged culture, such as contamination and plasmid stability problems may emerge [6, 14]. It has been shown that these problems can be overcome by using immobilized cell technologies in continuous systems. Entrapped cell systems may lead to high cell densities and productivities, reduced contamination risks due to the high dilution and inoculation rates provided by cell release from beads. Immobilization also improves plasmid stability [14, 15]. Diffusion limitations for substrates and inhibitory products in inoculated gel beads may be milder when these are cultivated in continuous systems. [16] have shown that immobilized cell technologies combined with continuous culture can be used to efficiently produce cells with enhanced tolerance to different environmental stresses. Also continuous immobilized cell culture systems can be used in a way as to produce cells of the exponential or the early stationary growth phase which exhibit a higher viability and metabolic activity compared to starving cells produced by conventional batch cultures.

The release of cells from the periphery of cultured gel beads is exploited in the efficient production of biomass in a liquid medium. Cell entrapment technologies are used today in the production of single or mixed strain cultures, as well as to continuously inoculate liquid ingredients in fermented food processes, e.g. fermented milk products. Based on the characteristic of cell release immobilized cell systems have been applied in many cases for starters' production.

Studies with single strain cultures and entrapment in beads with various coatings, such as alginate, modified alginate and chitosan have been reported with *Lactococcus lactis* spp. *lactis, L. lactis* spp. *cremoris, L. lactis* spp. *diacetylactis, Lb. rhamnosus, Lb. acidophilus* [5, 13, 17, 18, 19, 20]. Klinkenberg et al. [13] studied the cell release of free *L. lactis* spp. *lactis* from beads coated with chitosan and alginate in a continuous fermentation system with controlled pH and high dilution rates. The applied system permitted quantification of cell release rates without contributions from growth of free cells, as well as long-term studies of the course of the rate of cell release from beads containing immobilized bacteria. The system was used in investigations on the effects of coating on rates of cell release from alginate beads coated with chitosan, multiple coatings of chitosan and alginate, as well as alginate coatings alone. Sequential coatings with chitosan and alginate led to a significant reduction of cell release throughout fermentation while chitosan alone seemed to reduce the rate of cell release only in the early fermentation stages. Cell entrapment technologies have been successfully applied with bifidobacteria. These are fastidious and non-competitive bacteria with stringent growth requirements [21], which are now used as probiotics in a large variety of foods, pharmaceuticals, and health supplements. Bifidobacteria are very sensitive to environmental parameters, such as oxygen and pH, and require complex and expensive media for culturing and the addition of growth-promoting factors [7]. Continuous cultures of *Bifidobacterium longum* immobilized in gellan gum beads in MRS medium supplemented with whey permeate in the work of Doleyres et al. [22], showed increased biomass production (3.5 to 4.9 x 10^{12} cfu/ml for D decreasing from 2 to 0.5 h^{-1}) and maximal cell productivity of 6.9 x 10^{12} cfu/l/h (D = 2 h^{-1}), which represents the highest maximum productivity ever reported for bifidobacteria. The very high productivity reported in the work of Doleyres et al. [22], was attributed to the application of dilution rates that exceeded the maximal growth rate coefficient of cells -which would lead to washout in free-cell conditions- and the high concentration of immobilized cells (7 x 10^{10} cfu/g beads).

Cell entrapment technologies have also been successfully applied in the continuous production of mixed starters. Lamboley et al. [6] reported an effective mixed strain mesophilic lactic starter production process in supplemented whey permeate (SWP) using continuous immobilized cell cultures. Three strains of lactococci were separately immobilized in κ-carrageenan/locust bean gum (LBG) gel beads in a stirred tank bioreactor. The process showed high biological stability and the maximum cell productivity reported approximated 5.3 x 10^{12} cfu/l/h over a 50-day period. Variations in pH, dilution rate and temperature in mixed cultures led to a large range of strain ratios, while starter activity remained constant. In the work of Doleyres et al. [9] the production of a mixed lactic culture containing *L. lactis* spp. *lactis* biovar *diacetylactis* MD and *B. longum* ATCC 15707 was studied during a 17-day continuous immobilized cell culture. The two-stage fermentation system was composed of a reactor containing cells of the two strains separately immobilized in κ-carrageenan/LBG gel beads and a second reactor operated with free cells release from the first reactor. The system allowed for a continuous production of a concentrated mixed culture with a strain ratio whose composition depended on temperature and fermentation time. A stable mixed culture with a 22:1 ratio of *Lb. diacetylactis* and *B. longum* was produced at 35°C in the effluent of the second reactor, whereas the mixed culture was rapidly unbalanced in favour of *B. longum* at a higher temperature (37°C) or *Lb. diacetylactis* at a lower temperature (32°C). Cells produced by this technology also exhibited important physiological changes and increased stress tolerance.

2.1.2. Probiotics

Probiotics are now used to prepare a large variety of fermented milk products, pharmaceuticals, and nutrition supplements. Oral probiotics are living microorganisms, which when ingested at certain numbers exhibit health benefits beyond inherent basic nutrition [23]. Probiotics are mostly represented by lactic acid bacteria. Among various probiotics, bifidobacteria have been intensively investigated for their effects on diseases of the intestine [7, 24]. Potential probiotic strains should fulfill certain technological and safety criteria in order to be used in foods. A simple, large-scale production process of a viable culture concentrate, survival during preparation and storage of the carrier foods and survival in the ecosystem of the intestine are always desirable characteristics. The technological criteria are most important for large-scale applications of probiotics in food [25]. Strains with important functional health properties are often not considered for applications because of technological limitations, e.g. propagation characteristics [25].

Several approaches have been considered to improve the viability of probiotics when subjected to various inactivating factors, e.g. low pHs, oxygen, and antimicrobial compounds present in food, gastric fluids, bile acids etc. Immobilization of microorganisms by entrapment in gel beads was considered as a good method to preserve their viability for nutraceutical formulations. LAB entrapped in beads under mild conditions, are confined, being separated from their environment by a protective matrix, film or bead. Such a protective coating can extend shelf-life, preventing the exposure to oxygen during storage and improving resistance to gastric and bile acids following ingestion. Sun and Griffiths [26] immobilized LAB in gellan and xanthan gum beads, allowing protection in gastric fluid simulation medium for 30 min. Similar were the results reported by Rao et al. [27], Dinakar and Mistry [28] and Kim et al. [29] for cellulose acetate, κ-carrageenan and alginate matrices. Stadler and Vierntein [30] immobilized bacteria in hydroxypropylmethylcellulose and sodium alginate tablets. A certain resistance of tablet shape was observed after incubation for 1-2 h in artificial gastric fluid, while an important loss of viability due to mechanical compression during tablet preparation was observed.

Alginate is an immobilization matrix that offers important advantages when used in applications with sensitive probiotics (see previous section). A chitosan-alginate compact network structure has also been reported [31, 32], which makes a way to reduce diffusion and subsequently limit the access of acid to beads. Chitosan, a poly-(2-amino-2-deoxy-β-D-glucopyranose) was reported to exhibit inhibitory effects on certain types of cells, affecting their viability [33]. However, derivatization of amino groups of chitosan was found to eliminate some undesirable effects of amino groups and improve its biocompatibility [34]. Modified alginate and chitosan natural polymers derivatized by succinylation or by palmitoylation were used by Le-Tien et al. [5] in gel entrapping Lb. rhamnosus. The novelty of the approach lies in the use of functionalized immobilization matrices by succinylation (adds extra carboxylate groups, retains protons) or by palmitoylation (confers a hydrophobic character to the polymer and limits hydration and the access of gastric fluid to the bead interior). Immobilization in native alginate-based beads generated a viable cell count of 22-26% following 30 minutes incubation in simulated gastric fluid, while the level was undetectable with free cell cultures. Entrapment in succinylated alginate and chitosan beads led to 60-66% viable cell counts. Best viability (87%) was found for bacteria immobilized in N-palmitoylaminoethyl alginate, which affords a high protective effect, probably due to long alkyl pendants that improve the hydrophobicity of the beads, limiting this way hydration in acidic environments.

Cell entrapment technologies have also been successfully applied in the continuous production of mixed starters containing bifidobacteria. Because of their slow growth in milk, bifidobacteria are generally cultivated in pure cultures and eventually mixed thereafter with other LAB [7]. Stable mixed cultures however, of LAB and bifidobacteria, exhibiting increased stress tolerance, have been produced by co-cultivation of separately entrapped cells in polysaccharide gel beads and subsequent mixing of the released free cells in a second culture system [9].

2.1.3. Dairy Fermentations

Starter culture preparation is the first and most important step in the manufacture of fermented dairy products. This step is traditionally carried out in batch fermenters. The increasing demand of dairy products however has led to a requirement for large volume fermentation processes that are difficult to operate in the batch mode. Immobilized cell technologies have been used successfully in continuous systems for prefermentation of milk for yoghurt and cheese production.

Separately or co-entrapped cells of various mesophilic LAB in polymeric beads have been cultivated in stirred tank reactors in many cases. The first works on the dairy applications of immobilized cell technologies were those of Prevost and Divies [35, 36, 37]. These were on the continuous inoculation-prefermentation of milk for yoghurt production by separately entrapped cells of *Lb. delbruekii* spp. *bulgaricus* and *Streptococcus salivarious* spp. *thermophilus* in Ca-alginate beads in a stirred tank bioreactor. Cell immobilization in these studies resulted in increased biomass and lactate production and a reduction in fermentation time while the resulting yoghurt was of satisfactory quality.

The inoculation-prefermentation of milk for fresh cheese production with cultures entrapped in gel beads has been extensively studied. Prefermentation of milk using immobilized cell bioreactors has been studied by many researchers [10, 38, 39, 40, 41, 42]. An immobilized cell bioreactor can be used for a simple pre-acidification of milk (pH 6.2-5.6). Inoculation with a bulk starter is the next step. An immobilized reactor may also be used for simultaneous inoculation and acidification of milk because of the growth of microorganisms and their release in the medium. Sodini et al. [42] compared the latter with the traditional batch fermentation process and studied the acidification and coagulation kinetics, the microbiological composition and the rheological and sensory properties of the final curds of the two processes. Compared with the batch process, the continuous prefermentation of milk using an immobilized cell reactor had no significant effect on the rheological properties of the curd (susceptibility to syneresis, firmness, and modulus of elasticity) or on sensory properties of the final fresh cheese. Certain advantages of the continuous process include reduced fermentation times, reduced contamination risks with microbes and bacteriophages, and elimination of the need for preparation of bulk starter cultures, have been reported.

Cell entrapment technologies have been also applied in cream fermentations [43] and in modern manufacturing processes of kefir [44, 45].

Cell entrapment technologies have found successful applications in whey permeate fermentations for the production of LAB starters and various metabolites of LAB and yeasts. Cheese manufacturing yields large volumes of whey as a by-product. Approximately, 9 kg of whey result for every kg of cheese produced and the cost associated with disposing this large volume of whey is substantial. Moreover, the high chemical oxygen demand (COD) (50 kg O_2 / ton permeate) of whey or whey permeate makes its disposal a serious pollution problem. In many plants, whey proteins are concentrated by ultrafiltration and incorporated into cheese or dried and sold as food ingredients. The ultrafiltration step produces large quantities of the low-value whey permeate, which have limited uses. However, due to its high lactose content (~ 50 g/l) and minerals (~ 9 g/l), whey permeate has long been of industrial interest because it may be used successfully as a substrate for the cultivation of various microorganisms and the production of valuable metabolites. Examples of products resulting from whey or whey permeate include starters [22], lactic acid [46, 47, 48], acetic acid [49] propionic acid [50, 51], single cell protein [52], and the bacteriocin nisin [53]. In most cases cell entrapment applications exhibited higher lactose conversion rates and increased process stabilities.

2.1.4. Other Food Fermentations

Cell entrapment technologies have been applied in wine fermentation, in attempts to control the secondary fermentation in wine, malolactic fermentation (MLF). The LAB most commonly associated with wine belong to *Oenococcus oeni* (formely *Leuconostoc oenos*) and select *Lactobacillus* and *Pediococcus* spp. The major function of LAB is the conversion of L-malic acid to L-lactic acid during the MLF which is achieved through various metabolic pathways [54]. The reduction of wine acidity and the modification of wine flavor due to this secondary bacterial fermentation are often considered to benefit wine quality. Despite the importance of the MLF, its occurrence is both unpredictable and difficult to control or manipulate. Consequently, techniques that facilitate the efficient and complete conversion of L-malic acid to L-lactic acid in grape juice and wine have been sought. These techniques- mainly cell immobilization techniques- aim to separate the central enzyme-driven conversion from the often problematic growth of the source LAB in the wine. LAB cells immobilized alone or with yeast in various bioreactor systems have been tested by many researchers [54, 55, 56, 57, 58, 59]. Most commonly, *O. oeni* cells were entrapped in calcium alginate matrix and wine was circulated through them. Trück and Hammes [60] reported that best results were obtained with small diameter alginate beads (1.2 mm) and it was found that the pellets were able to be stored frozen in liquid nitrogen at -85°C.

Various immobilization techniques have been employed in a process developed by Kikkoman Corporation (Japan) [61] for the production of a fermented liquid seasoning, based on a hydrolyzate of soy sauce raw materials. In the conventional production of soy sauce, it has been a general practice to inoculate and cultivate LAB in the mash of raw material, or to directly add lactic acid to the mash, as a means of improving the flavor and the fermentation efficiency of the yeast. In the first case, the fermentation period required for the sufficient formation of lactic acid is at least one week, while the latter procedure, although simple in its operation, yields a product distinctly inferior to that obtained in the former case.

The process developed by Kikkoman produces efficiently, in a short period of time, a liquid seasoning of excellent flavor and taste by utilizing immobilized cells of soy sauce LAB, which greatly enhance the efficiency of the lactic acid fermentation. The liquid hydrolyzate of raw materials (pH 4.0-9.0) is contacted with immobilized cells of soy sauce LAB, at 20-35°C, for 30 minutes under anaerobic conditions. The lactic fermentation step this way is reduced significantly to 30 minutes (instead of at least one week of the conventional method) and can be transferred to the yeast fermentation step. The immobilization techniques described, include entrapment in polymer gel (alginate, κ-carageenan, polyacrylamide), physical adsorption on inorganic carriers (porous glass beads, activated carbon, porous glass, alumina, silica gel, kaoline, acid clay) activated with glutaraldehyde, or natural polymer carriers (gluten, starch, saw dusts), porous synthetic resins, and ceramics.

2.1.5. Lactic Acid Production

Lactic acid is a product that has numerous applications in the chemical, pharmaceutical, and food industries. LAB have been used widely for the production of lactic acid. Lactic acid production processes traditionally suffer from end-product inhibition. Studies on the elucidation of the inhibition mechanism have shown that this is probably related to the solubility of the undissociated lactic acid within the cytoplasmic membrane and the insolubility of dissociated lactate, which causes acidification of the cytoplasm and failure of proton motive forces. It eventually influences the membrane pH gradient and decreases the amount of energy available for cell growth [62]. Immobilization of cells has been tested in many cases as a means of high cell retention. Several materials and microorganisms have been tested, such as Ca-alginate beads [63] and *Lb. delbrueckii*, Ca-alginate beads and *Lb. reuteri* [64], sodium alginate beads and *B. longum* [48], κ-carrageenan/LBG gel beads and *Lb. helveticus* [65], in continuous systems and usually in two stage processes.

Increased productivities and long-term stability (over 100 day periods) have been obtained in the study of Norton et al. [65] during continuous fermentation of yeast extract supplemented whey permeate by *Lb. helveticus* immobilized in κ-carrageenan/LBG gel beads. Lactose conversion however was limited. Complete lactose conversion was only obtained when an additional reactor operated with free cells was placed in series with the immobilized cell reactor [66]. The overall productivity of the described two-stage continuous process reached 13.5 g/l/h in whey permeate containing 10 g/l yeast extract, with 1 g/l residual lactose at an overall dilution rate of 0.27 h⁻¹.

Senthuran et al. [67] reported the production of lactic acid by continuous culture of *Lb. casei* immobilized in poly(ethyleneimine). The system was coupled with a cell-recycle bioreactor and the authors observed that the most influencing factor for operational stability was the bead size of the immobilization matrix.

2.1.6. Bacteriocin Production

Bacteriocins are antimicrobial peptides or proteins produced by lactic acid bacteria (LAB) [68]. Their potential applications in the food and health care sectors have attracted the strong interest of academia and the industry resulting in an impressive amount of published research on bacteriocin production, purification, genetics, and applications. So far, only nisin, the most studied bacteriocin produced by some strains of *L. lactis*, is produced commercially and is an approved food additive in most major food producing countries. Immobilized cell technologies have been applied and the most important in the case of nisin are those involving continuous production in packed bed bioreactors, which will be discussed in the immobilized reactors section. An interesting example for nisin production with cell entrapment in gel beads is the study of Bertrand et al. [69] in which *L. lactis* spp. *lactis* was immobilized in κ-carrageenan/LBG beads in SWP medium. Increased nisin production was measured in the fermentation broth after 1-h cycles during repeated-cycle pH-controlled batch cultures (IC-RCB), which corresponded to a volumetric productivity of 5730 IU/ml/h. This is much higher compared to reported productivities with cell free systems.

Another well-studied bacteriocin that will likely be the next to be used by the food industry is pediocin [70, 71]. An IC-RCB system, similar to that of Bertrand et al. [69], has been employed by Naghmouchi [72] in pediocin production by *Pediococcus acidilactici*

UL5. Various aspects of pediocin production and *P. acidilactici* plasmid stability in free and immobilized continuous cultures were studied by Huang et al. [73].

2.2. Immobilized Bioreactor Systems

Various bioreactor systems operated with cells entrapped in polymeric beads have been described so far. In this section, special reactor configurations, such as membrane, packed-bed, biofilm, spiral mesh or sheet reactors and their applications will be discussed.

Membrane or packed-bed immobilized cell bioreactors have been developed and used successfully in fermentations with LAB for the production of lactic and other organic acids, polysaccharides and bacteriocins. In continuous membrane systems, cells are retained by an ultrafiltration or microfiltration membrane. Small molecules diffuse through the membrane and various inhibitory products are eliminated in the permeate. The concentrated cell fraction can be easily removed. Taniguchi et al. [74] reported a seven times higher final *B. longum* concentration in a membrane bioreactor than the obtained from free cell batch fermentations. Corre et al. [75] reported for *B. bifidum* a 15-fold improvement on free cell batch culture productivity with a membrane reactor and a whey-based medium.

2.2.1. Organic Acid Production

Increased productivities for lactic acid have been reported in many cases with membrane bioreactors. Mehaia and Cheriyan [76] used a membrane recycle bioreactor for lactic acid production from whey permeate. The cell-recycle system, together with repeated batch and continuous processes, enables the achievement of higher cell concentrations and lactic acid productivities [77]. Kwon et al. [78] reported a high-rate continuous process for the production of lactic acid from *Lb. rhamnosus* in a two-stage membrane cell-recycle bioreactor. By using a system of membrane cell-recycle bioreactors connected in series they obtained 92 g/l lactic acid corresponding to a productivity of 57 g/l/h. Oh et al. [79] reported a rate of 6.4 g/l/h for lactic acid production through a cell-recycle repeated batch fermentation. The maximum cell concentration achieved was greater that 28 g/l, which might contribute to the improvement of the productivity and reduction of nutrient supplementation. Their study also indicated that only the 26% of the yeast extract dosage, compared with the conventional batch fermentation, should be required to produce the same amount of lactic acid, a fact that might result in considerable reduction of production costs. Indicatively, other important works on the production of lactic acid using membrane bioreactors include the study of Cotton et al. [80], in which a plastic composite support biofilm rector was used in a continuous process with *Lb. casei*, the study of Jeantet et al. [81] in which nanofiltration membranes were used for cell-recycle in a continuous process with *Lb. helveticus*, and the study of Senthuran et al. [67] with *Lb. casei*.

The continuous production of lactic acid by *B. longum* was studied by Shahbazi et al. [82] in a spiral-sheet bioreactor and cheese whey as a substrate. The process was compared with fermentation of immobilized bifidobacteria in gel beads and a fluidized-bed reactor. Under controlled pH and temperature conditions, the conversion rate of lactose in the spiral-sheet reactor was 70% compared to 37% in the fluidized-bed reactor packed with gel beads.

Metabolite production by immobilized cell technologies mostly exploits the advantage of increased biomass levels. Another method to increase biomass cell concentration during fermentation is the formation of a biofilm within the fermenter. A biofilm is a natural type of cell immobilization that involves the attachment of microorganisms to fixed supports. Biofilms are used in numerous wastewater facilities in trickling filters [83] and have been also used in applications such as ethanol, acetic acid, and polysaccharides production, as well as for metal ore leaching [84]. In several fermentation systems, biofilms were found to to consistently produce increased biomass levels and increased product formation rates and final product concentrations [85,86, 87, 88]. Nice examples of applications of biofilm reactors for the production of lactic acid by *Lb. casei* are the studies of Ho et al. [86, 87, 88] who used plastic composite supports (PCS), as solid supports in the biofilm reactor. These are extruded nutrient supports containing various combinations of polypropylene and agricultural products. The polypropylene acts as a matrix to integrate the mixture of agricultural materials. The agricultural products (soybean hulls, yeast extract, bovine albumin, and soybean flour) serve to generate a porous structure on the PCS, and to provide cells with essential nutrients to sustain cell growth. Studies done by Ho et al. [87] showed that PCS releases nutrients slowly and this way lowers the nutrient requirement of the fermentation medium. Biofilm reactors with PCS have been shown to improve lactic acid production [85, 86, 87, 88].

Propionic acid production in immobilized reactors has been studied in batch and continuous cultures. Cell-recycling was applied in the studies of Colomban et al. [89] and Paik and Glatz [90] with *Propionibacterium acidipropionici*. Yang et al. [50] presented a novel recycle batch immobilized cell reactor for propionate production from whey lactose. The same group [51] also presented a novel fibrous bed bioreactor for the continuous production of propionate from whey permeate.

2.2.2. Prefermentation of Milk

A spiral mesh bioreactor construction was developed by Passos and Swaisgood [41] for use in a continuous process of inoculation and acidification of milk. *L. lactis* cells were entrapped in a calcium alginate film coating a spiral mesh and placed in column through which milk was recalculated from a reservoir. Immobilized and free cell bioreactors were compared and productivities were 1.5- to 3.5-fold increased with immobilized cell reactors than with free cells, because of higher cell densities, although specific productivities were lower for immobilized cells.

2.2.3. Extracellular Polysaccharide Production

Immobilized bioreactors have been used for the production of extracellular polysaccharides (or exopolysaccharides, EPS) from LAB. EPS production is receiving increasing attention in recent years. EPS make attractive food additives because of their contribution to texture, mouthfeel, and taste perception, and the stability of the product. Moreover, EPS may contribute to human health as prebiotics or due to their antitumor, antiulcer, immunostimulating, and cholesterol-lowering activities [91]. Bergmaier et al. [92] immobilized *Lb. rhamnosus* on solid porous supports (ImmobaSil®) and investigated the production of EPS in pH-controlled immobilized cell repeated-batch cultures. Immobilized biomass levels were very high and during repeated immobilized cell cultures, a high EPS concentration (1750 mg/l) was obtained after four cycles for a short incubation period of 7 hours. The increased biomass in the immobilized cell bioreactor, increased the maximum EPS volumetric productivity (250 mg/l/h after 7 h culture) compared with free cell batch cultures (110 mg/l/h after 18 h culture corresponding to maximum EPS concentration of 1985 mg/l). Following studies in batch culture, the same group [93] proceeded in chemostat cultures with *Lb. rhamnosus* cells immobilized on the same material. Although EPS production and volumetric productivity during continuous free-cell chemostat cultures were among the highest values reported for lactobacilli, immobilization and continuous culture resulted in low soluble EPS production. This was attributed to morphological changes, e.g. formation of large aggregates composed of biomass and non-soluble EPS, which undoubtedly changed the dynamics of the system. The study however revealed that immobilization and culture-time induced cell aggregation could be used to produce new symbiotic products with very high viable cell and EPS concentrations.

2.2.4. Bacteriocin Production

Other LAB products in which immobilized bioreactors technologies have been applied are the bacteriocins nisin and pediocin. Nisin biosynthesis was mostly studied in batch cultures using synthetic media [94, 95] and to a lesser extent in fed-batch cultures [96, 97]. Reports on continuous production are limited. Continuous culture techniques may provide certain advantages over conventional batch or fed-batch processes, such as increased productivities and reduced product inhibitions. Free-cell continuous fermentations however, are limited by cell wash-out. This drawback can be eliminated by using cell immobilization which leads to high cell concentrations. The recent study by Liu et al. [53] presented a packed-bed bioreactor which was developed for continuous nisin production by *L. lactis* on whey permeate. The fibrous matrix used for cell immobilization was made of a piece of cotton towel mounted on a stainless steel wire screen. *L. lactis* was immobilized by natural attachment to fibre surfaces and entrapment in the void volume within the spiral wound fibrous matrix. The bioreactor was operated continuously for 6 months without encountering any clogging, degeneration, or contamination problems. The cell density in the reactor reached 52 g/l with 96.4% of the cells immobilized. The study of Liu et al. [53] demonstrated the possibility of continuous production of high concentration of bacteriocins by LAB for use as food biopreservatives. An early successful attempt for pediocin production was reported by Cho et al. [98]. In that study, pediocin producer *P. acidilactici* PO2 was immobilized in a packed-bed reactor operated continuously.

Biofilm reactors with plastic composite supports have been tested by Bober and Demirci [99] for nisin production by *L. lactis* from a less expensive medium. Although the same PC support material was applied successfully in lactic acid production [85, 86, 87, 88], its use did not result to any increases in nisin production in repeated-batch fermentations.

2.2.5. Control of the Malolactic Fermentation in Wine and Cider

Immobilized bioreactor applications examples from the wine industry include bioreactor systems of various configurations, often operated continuously, comprising LAB cells immobilized alone [54, 55, 56, 57, 58, 59, 60, 100, 101, 102, 103] or co-immobilized with yeast [104, 105]. Trück and Hammes [60] employed a fluidized-bed bioreactor to carry out the MLF, with *O. oeni* cells entrapped in alginate beads. Janssen et al. [103] presented a bioreactor operated with *O. oeni* cells immobilized by absorption on oak chips. The half-life of the described fermentation system was 11 days when operated at 21°C and supplied with wine at pH 3.45 containing 13% (v/v) ethanol. Maicas [106] and Maicas et al. [107] presented a fermentation system in which *O. oeni* was immobilized by adsorption on positively charged cellulose sponges. Apart from *O. oeni*, immobilization studies have been carried out with *Lb. brevis*, *Lb. fructivorans*, and *Lb. plantarum*.

Successful laboratory trials have been carried out with co-immobilization of yeast and lactic acid bacteria in order to obtain simultaneous alcoholic and malolactic conversions in cider fermentation. *Saccharomyces cerevisiae* and *Lb. plantarum* have been co-immobilized on a sponge-like matrix [104], while *S. bayanus* and *O. oeni* have been co-immobilized in a Ca-alginate matrix in a continuous packed-bed bioreactor [106]. These studies demonstrated that the continuous immobilized fermentation system offers clear advantages in terms of volumetric productivity, stability of the process and in some cases, flavor development [104]. However, the volatile compounds profile was altered comparatively to the traditional process [106]. A lower content of higher alcohols and a higher content of diacetyl have been found and attributed to metabolism changes in immobilized cells. The organoleptic profile of the cider however, was regarded as entirely acceptable by the consumers.

2.3. Cell Encapsulation Techniques

Entrapment of cells in gel matrices and immobilized bioreactors are the most popular systems of cell immobilization mainly applied in starters and metabolites production. Encapsulation is another type of immobilization process that evolved from entrapment and is used in the probiotics industry.

Currently, probiotics have been supplemented to a large variety of products and formulations, such as capsules, tablets, creams, suspensions and powders, to create functional foods or nutraceuticals in global markets [108]. The market trend of the probiotics area is increasing and fermented and non-fermented dairy products, cereals, meat products, sous-vide products as well as prepared home meal solutions are becoming food vehicles using encapsulation technologies to protect probiotic bacteria as a means of delivering large quantities to consumers.

Viability and functional activity of probiotics are major concerns in food supplements, since a sufficient number of bacteria must resist the gastric and bile acids before reaching the colon. To provide functional properties, the minimum level of viable bacteria should be approximately 10^6 cfu/ml of product at the expiry day, while the suggested therapeutic dose is 10^8-10^9 viable cells per day [109]. This is rather difficult to achieve since studies have indicated that probiotics may not survive in sufficient numbers and retain their activity in commercial preparations or in the host gastrointestinal track [110]. Several surveys have shown large fluctuations and poor viability of probiotic bacteria, especially bifidobacteria, in yoghurt preparations which is the main probiotic carrier [111]. pH, concentration of lactic and acetic acids, hydrogen peroxide and dissolved oxygen content, have been identified to affect viability during manufacture and storage of yoghurt [112]. A number of approaches are being explored to increase viability of probiotic bacteria in commercial and experimented products, such as selection of acid and bile resistant strains, use of oxygen impermeable containers, two-stage fermentation, stress adaptation, incorporation of micronutrients and encapsulation [113]. To date, encapsulation methods have been applied widely to enhance viability of probiotic bacteria in commercial products.

In encapsulation, a continuous coating around an inner matrix is formed that is wholly contained within the capsule wall as a core of encapsulated material. Immobilization in this case refers to the trapping of material within or throughout the matrix. While in the already discussed techniques, a small percentage of immobilized material may be exposed in the surface, this is not the case for encapsulated material [114].

Encapsulation occurs naturally in LAB that produce extracellular polysaccharides. The cells are entrapped within their own secreted products that act as a protective capsule. Permeability of materials through the capsule is reduced and cells are protected from adverse environmental conditions [115]. Many LAB however that synthesize extracellular polysaccharides, make it insufficiently and they are not able to fully encapsulate themselves [116]. As an immobilization technique, encapsulation tends to stabilize cells, potentially enhancing their viability and stability in the production, storage and handling lactic cultures [29, 115]. Encapsulation has shown by several authors to enhance cell resistance to freezing and freeze-drying [7]. These important benefits of the technique -protection of cells from adverse environmental conditions and stability during handling and storage- provide solutions to the probiotics technology.

Microencapsulation is defined as a technology for packing solids, liquids or gaseous materials in miniature sealed capsules which can release their content at controlled rates under specific conditions [117]. Several studies have reported successful microencapsulation and coating of bacteria using various materials and methods (Table 2).

Table 2. Examples of successful microencapsulation and coating of bacteria using various materials and methods

Matrix	Microencapsulation technology	Microorganism	Reference
Alginate	Liquid starch core with calcium alginate membranes	*L. lactis*	Jankowski et al.,1997 [118]
Alginate	Beads/extrusion	*B. bifidum*	Khalil and Mansur, 1998 [119]
Alginate	Beads/extrusion	*B. longum*	Lee and Heo, 2000 [120]
Alginate	Beads/emulsion	*Bifidobacterium* spp.	Truelstrup Hansen et al., 2002 [121]
Alginate	Freeze-dried powder	*B. longum*	Shah and Ravula, 2000 [113]
Alginate	Freeze-dried powder	*Lb. acidophilus*	Shah and Ravula, 2000 [113]
Alginate /starch	Beads/emulsion	*B. infantis*	Godward and Kailasapathy, 2003 [122]
Alginate/ starch	Beads/extrusion	*B. lactis*	Talwalkar and and Kailasapathy, 2003 [123]

Table 2. (Continued).

Matrix	Microencapsulation technology	Microorganism	Reference
Gellan/xanthan	Beads/extrusion	B. infantis	Sun and Griffith, 2000 [26]
Starch	Freeze-dried powder	LAB	Matilla-Sandholm et al. 2002 [25]
Matrix	Microencapsulation technology	Microorganism	Reference
Starch	Spray-dried powder	*B. infantis*	Hsiao et al., 2004 [124]
Starch/gluten/milk	Dried granules	LAB	Plessas et al., 2004 [125]
Modified waxy maize starch	Spray-dried powder	*Bifidobacterium PL1*	O'Riordan et al. 2001 [126]
Milk fat/whey	Spray-dried powder	B. breve	Picot and Lacroix (2003 ; 2004) [127, 128]
Korean kimtchi	Freeze-dried granules	LAB	Jung and Choi, 2007 [129]
Oil	Artificial oil emulsions/oil bodies	*Lb. delbruecki*	Hou et al., 2003 [130]
κ-carrageenan	Beads/emulsion	*B. longum*	Adhikari et al. (2000; 2003) [131, 132]

Many of these technologies in fact rely on gel entrapment techniques. Spray drying is the most commonly used microencapsulation method in the food industry. Other methods include extrusion, emulsion and phase separation [115]. The most commonly used biopolymer for microencapsulation of bacterial cells is alginate. Artificial oil emulsions have also been tested. Hou et al. [131] developed a technique for protection of lactic acid bacteria (*Lb. delbrueckii* spp. *bulgaricus*) against simulated gastrointestinal conditions by encapsulation of bacterial cells within artificial sesame oil emulsions. Effective biocapsules were produced with increased viability under the particular conditions. It should be emphasized here that selection of materials and methods for microencapsulation of LAB should be based on the technological properties of the strains with regard to processing (e.g. heat resistance), and stress adaptation of cultures before microencapsulation. Two recent and important reviews on the technology and applications of microencapsulation in probiotics are those by Kailasapathy [115] and Doleyres and Lacroix [7].

Conclusion

Development of bioprocessing strategies, such as immobilized cell technologies and long-term continuous culture, may provide improved or new manufacturing processes with increased viability and stability of lactic acid bacteria which additionally are freed from, common in batch cultures, end-product inhibitions. Application of the reviewed research could be of great importance for the production of functional foods that contain high concentrations of viable bacteria and ingredients from LAB. Protection of cells by immobilization techniques such as microencapsulation, is an important feature, potentially useful for delivery of viable cells to the gastrointestinal track through fermented and non-fermented food products.

Lately, the advances in the field of microencapsulation have been tremendous with nutraceuticals and food ingredients. Microencasulation of live probiotic bacteria could be regarded as still developing since development of more precise machinery, capsule and delivery systems is needed. It seems however, that it is only a matter of time since probiotic therapy (which is based on the concept of healthy gut microflora) is increasing at the moment and it is expected to become very important in the future. In this respect, it is also expected that development of multiple-delivery systems will emerge as for example, co-encapsulation of probiotics, prebiotics and nutraceuticals, thus a new area of complex nutritional matrices will need to be investigated [115].

Nanotechnologies are expected to find applications in the near future in the development of designer probiotic bacterial preparations that could be delivered to certain parts of the gastrointestinal track where they interact with specific receptors. Development of such preparations (*ac de novo* vaccines), effective in modulating immune responses, will require improved techniques for delivery and sustained release and in vivo studies to confirm the efficacy of encapsulation.

References

[1]   Neidleman, SL. Industrial chemicals: Fermentation and immobilized cells. In: Moses V, Cape RE (Eds). *Biotechnology: The science and the business*. Chur, Switzerland: Harwood Academic Publishers 1994; pp. 297-310.

[2]   Lacroix, C; Grattepanche, F; Doleyres, Y; Bergmeier, D. Immobilized cell technology for the dairy industry. In: Nevidovic V, Willaert R (Eds). *Cell immobilization Biotechnology: Focus in Biotechnology*. Dortrecht, The Netherlands: Kluyver Academic Publishers, 2004; pp. 297-310.

[3]   Shapiro, L; Cohen, S. Novel alginate sponges for cell culture and transplantation. *Biomaterials* 1997, 18, 583-90.

[4]   Lee, KY; Bouhadir, KH; Mooney, DJ. Degradation behavior of covalently cross-linked of poly(aldehyde guluronate) hydrogels. *Macromolecules* 2000, 33, 97-101.

[5]   Le-Tien, C; Millette, M; Mateescu, MA; Lacroix M. Modified alginate and chitosan for lactic acid bacteria immobilization. *Biotechnol. Appl. Biochem.* 2004, 39, 347-54.

[6]   Lamboley, L; Lacroix, C; Champagne, JC. Continuous mixed strain mesophilic lactic starter production in suppmemented whey permeate medium using immobilized cell technology. *Biotechnol. Bioeng* 1997, 56, 501-16.

[7]   Doleyres, Y; Lacroix, C. Technologies with free and immobilized cells for probiotic bifidobacteria production and protection. *Int. Dairy J.* 2005, 15, 973-88.

[8]   Masson, F; Lacroix, C; Paquin C. Direct measurement of pH profiles in gel beads immobilizing Lactobacillus helveticus using a pH sensitive microelectrode. *Biotechnol. Techniques*, 1994, 8, 551-556.

[9]   Doleyres, Y; Fliss, I; Lacroix, C. Continuous production of lactic starters containing probiotics using immobilized cell technology. *Biotechnol. Prog.* 2004, 20, 145-50.

[10] Sodini, I; Boquien, CY; Corrieu, G; Lacroix C. Microbial dynamics of co- and separately entrapped mixed cultures of mesophilic lactic acid bacteria during continuous prefermentation of milk. *Enzyme Microb. Technol.* 1997, 20, 381-88.

[11] Duran-Paramo, E; Morales-Contreraw, M; Brito-Arias, MA; Robles-Martinez, F. Cell immobilization on the production and utilization of lactic acid probiotic bacteria in the dairy industry. Abstract 2-38. *27th Symposium on Biotechnology for Fuels and Chemicals.* 2005, May 1-4, Denver, Colorado, USA.

[12] Champagne, CP; Lacroix, C; Soddini-Gallot, I. Immobilized cell technology for the dairy industry. *Crit. Rev. Biotechnol.* 1994, 14, 109-134.

[13] Klinkenberg, G; Lystad, KG; Levine, DW; Dyrset, N. Cell release from alginate immobilized *Lactococcus lactis* ssp. *lactis* in chitosan and alginate coated beads. *J. Dairy Sci.* 2001, 84, 1118-1127.

[14] D'Angio, C; Beal, C; Boquien, CY; Corrieu, G. Influence of dilution rate and cell immobilization on plasmid stability during continuous cultures of recombinant strains of *Lactococcus lactis* subsp. *lactis. J. Biotechnol.* 1994, 34, 87-95.

[15] Huang, J; Lacroix, C; Daba, H; Simard, RE. Pediocin 5 production and plasmid stability during continuous free and immobilized cell cultures of *Pediococcus acidilactici* UL5. *J. Appl. Bacteriol.* 1996, 80, 635-644.

[16] Doleyres, Y; Fliss, I; Lacroix, C. Inreased stress tolerance of *Bifidocterium longum* and *Lactococcus lactis* produced during continuous mixed-strain immobilized-cell fermentation. *J. Appl. Microbiol.* 2004, 97, 527-539.

[17] Champagne, CP ; Gaudy, C ; Poncelet, D ; Neufeld, RJ. *Lactococcus lactis* release from calcium alginate beads. *Appl. Environ. Microbiol.* 1992, 58, 1429-1434.

[18] Cachon, R; Catte, M; Nomme, R; Prevost, H; Divies, C. Kinetic behavior of *Lactococcus lactis* ssp. *lactis* bv. *diacetylactis* immobilized in calcium alginate gel beads. *Process Biochem.* 1995, 30, 503-510.

[19] Gaseród, O; Smidsród, O; Skjåk-Bræk, G. Microcapsules of alginate-chitosan: A quantitative study of the interaction between alginate and chitosan. *Biomaterials* 1998, 19:1815-1825.

[20] Zhou, Y; Martins, E; Groboillot, A; Champagne, CP ; Neufeld, RJ. Spectrophotometric quantification and control of cell release with chitosan coating. *J. Appl. Microbiol.* 1998, 84, 342-348.

[21] Ibrahim, SA; Bezkorovainy, A. Growth-promoting factors for *Bifidobacterium longum*. *J. Food Sci.* 1994, 59, 189-191.

[22] Doleyres, Y; Paquin, C; LeRoy, M; Lacroix, C. *Bifidobacterium longum* ATCC 15707 cell production during free and immobilized cell cultures in MRS-whey permeate medium. *Appl. Microbiol. Biotechnol.* 2002, 60, 168-173.

[23] Guarner, F; Schaafsma, GJ. Probiotics. *Int. J. Food Microbiol.* 1998, 39, 237-238.

[24] Ouwehand, AC; Bianchi Salvadori, B; Fonden, R; Mogensen, G; Salminen, S; Sellars, R. Health effects of probiotics and culture-containing dairy products in humans. *Bull. Int. Dairy Fed* 2003, 380, 4-19.

[25] Mattila-Sandholm, T; Myllarinen, P; Crittenden, R; Mogensen, G; Fonden, R; Saarela, M. Technological challenges for future probiotic foods. *Int. Dairy J.* 2002, 12, 173-182.

[26] Sun, W; Griffiths, MW. Survival of bifidobacteria in yoghurt and simulated gastric juice following immobilization in gellan-xanthan beads. *Int. J. Food Microbiol.* 2000, 61, 17-25.

[27] Rao, AV; Shiwnarain, N; Maharaj, I. Survival of microencapsulated *Bifidobacterium pseudolongum* in simulated gastric and intestinal juices. *Can. Inst. Food Sci. Technol.* 1989, 22, 345-49.

[28] Dinakar, P; Mistry, VV. Growth and viability of *Bifidobacterium longum* in cheddar cheese. *J. Dairy Sci.* 1994, 77, 2854-2864.

[29] Kim, KI; Baek, YJ; Yoon, YH. Effects of rehydration media and immobilization in calcium-alginate on the survival of *Lactobacillus casei* and *Bifidocterium bifidum*. *Korean J. Dairy Sci.* 1996, 18, 193-198.

[30] Stadler, M; Vierntein, H. Optimization of a formulation containing lactic acid bacteria. *Int. J. Pharm.* 2003, 256, 117-122.

[31] Kokufuta E, Shimizu N, Tanaka H, Nakamura I. Use of poly-electrolyte complex stabilized calcium alginate gel for entrapment of beta-amylase. *Biotechnol. Bioeng* 1988, 32, 756-759.

[32] Kubota, N; Kikuchi Y. Macromolecular complexes of chitosan. In: Dimitriu, S (editor). *Polysaccharides: Structural diversity and functional versatility*. New York: Marcel Dekker;1998, pp. 594-628.

[33] Groboillot, AF; Champagne, CP; Darling, GD; Poncelet, D; Neufeld, RJ. Membrane formation by interfacial cross-linking of chitosan for microencapsulation of *Lactococcus lactis. Biotechnol. Bioeng* 1993, 42,1157-1163.

[34] Hirano S. Chitin biotechnology applications. *Biotechnol. Annu Rev.* 1996, 2, 237-258.

[35] Prevost, H; Divies, C. Fresh fermented cheese production with continuous prefermented milk by a mixed culture of mesophilic lactic streptococci entrapped in Ca-alginate. *Biotechnol. Lett.* 1987, 9, 789-791.

[36] Prevost, H; Divies, C. Continuous prefermentation of milk by entrapped yoghurt bacteria. I. Development of the process. *Milchwissen.* 1988, 43, 621-625.

[37] Prevost, H; Divies, C. Continuous prefermentation of milk by entrapped yoghurt bacteria. II. Data for optimization of the process. *Milchwissen.* 1988, 43, 716-719.

[38] Kim, KI; Naveh, D; Olson, NF. Continuous acidification of milk before UF by an immobilized cell bioreactor. 1. Development of the bioreactor. *Milchwissen.* 1985, 40, 605-607.

[39] Kim, KI; Naveh, D; Olson, NF. Continuous acidification of milk before UF by an immobilized cell bioreactor. 2. Factors affecting the acid production in the bioreactor. *Milchwissen.* 1985, 40, 645-649.

[40] Gobbetti, M; Rossi, J. Milk prefermentation process in fresh cheese manufacture. 2. Study of the effects of a continuous process on the quality and storage of the product. *Sci. Tec. Latt Cas* 1990, 41, 455-471.

[41] Passos, FML; Swaisgood, HE. Development of a spiral mesh bioreactor with immobilized lactococci for continuous inoculation and acification of milk. *J. Dairy Sci.* 1993, 76, 2856-5287.

[42] Sodini, I; Lagace, L; Lacroix, C; Corrieu, G. Effect of continuous prefermentation of milk with an immobilized cell bioreactor on fermentation kinetics and curd properties. *J. Dairy Sci.* 1998, 81, 631-638.

[43] Prevost, H; Divies, C. Cream fermentations by a mixed culture of *Lactococci* entrapped in two-layer calcium alginate gel beads. *Biotechnol. Lett.* 1992, 14, 583-588.

[44] Rossi, J; Gobbetti, M. Impiego di un multistarter per la producione in continuo di kefir. *Annals Microbiol.* 1991, 41, 223-226.

[45] Farnworth ER. Kefir – a complex probiotic. *FST Bulletin: Functional Foods.* 13 May 2005.

[46] Roy, D; Goulet J. Continuous production of lactic acid from whey permeate by free and calcium alginate entrapped *Lactobacillus helveticus. J. Dairy Sci.* 1987, 70, 722-756.

[47] Tejayadi, S; Cheryan, M. Lactic acid from cheese whey permeate. Productivity and economics of a continuous membrane bioreactor. *Appl. Biochem. Biotechnol.* 1995, 43, 242-248.

[48] Shahbazi, A; Mims, Li, Y; Shirley, V; Ibrahim, SA; Morris, A. Lactic acid production from cheese whey by immobilized bacteria. *Appl. Biochem. Biotechnol.* 2005, 121-124, 529-540.

[49] Yang, ST; Tang, IC; Zhu H. A novel fermentation process for calcium magnesium acetate (CMA) production from cheese whey. *Appl. Biochem. Biotechnol.* 1992, 34-35, 569-583.

[50] Yang, ST; Huang, Y; Hong, G. A novel recycle batch immobilized cell bioreactor for propionate production from whey lactose. *Biotechnol. Bioeng* 1995, 45, 379-386.

[51] Yang, ST; Huang, Y; Hong, G. Continuous propionate pproduction from whey permeate using a novel fibrous bed bioreactor. *Biotechnol. Bioeng* 1994, 43, 1124-1130.

[52] Giec, A; Kosikowski, FV. Activity of lactose-fermenting yeast in producing biomass from concentrated whey permeate. *J. Food Sci.* 1982, 47, 1992-1993.

[53] Liu, X; Chung, Y; Yang, S; Yousef, AE. Continuous nisin production in laboratory mcdia and whey permeate by immobilized *Lactococcus lactis. Process Biochem.* 2005, 40, 13-24.

[54] Matthews, A; Grimaldi, A; Walker, M; Bartowski, E; Grbin, P; Jiranek, V. Lactic acid bacteria as a potential source of enzymes for use in vinification. *Appl. Environ. Microbiol.* 2004, 70, 5715-5731.

[55] McCord, JD; Ryu DD. Development of a malolactic fermentation process using immobilized whole cells and enzymes. *Amm. J. Enol. Vitic.* 1985, 36, 214-218.

[56] Crapisi, A; Nuti, MP; Zamorani, A; Spettoli P. Improved stability of immobilized *Lactobacillus* spp. cells for the control of malolactic fermentation in wine. *Amm. J. Enol. Vitic* 1987, 38, 310-312.

[57] Crapisi, A; Spettoli, P; Nuti, MP; Zamorani A. Comparative traits of *Lactobacillus brevis, Lactobacillus fructivorans* and *Leuconostoc oenos* immobilized cells for the control of malo-lactic fermentation in wine. *J. Appl. Bacteriol.* 1987, 63, 513-521.

[58] Divies, C. On the utilization of entrapped microorganisms in the industry of fermented beverages. In: Cantarelli, C; Lanzarini G. (Eds), *Biotechnology applications in beverage production.* New York: Elsevier Science Publishers; 1989. pp. 153-167.

[59] Morenzoni R. (Editor). *Malolactic fermentation in wine*. Canada: Lallemand; 2005.

[60] Trük, HU; Hammes, WP. Die Verwndung eines Fliessbettreaktors zur Ausführung des biologischen Säureabbaus in Wein mit immobiliserten Zellen. *Chem. Kikrob. Technol. Lebensm* 1989, 12, 119-126.

[61] Kikkoman Corporation. *Process for producing liquid seasoning*. United States Patent 4587127. Publication date: 05/06/1986.

[62] Concalves, LMD; Ramos, A; Almeida, JS; Xavier, AMRB; Carrondo, MJT. Elucidation of the mechanism of lactic acid growth inhibition and production in batch cultures of *Lactobacillus rhamnosus*. *Appl. Microbiol. Biotechnol.* 1997, 48, 346-350.

[63] Göksungur, Y; Güvenc, U. Production of lactic acid from beet molasses by calcium alginate immobilized *Lactobacillus delbrueckii* IFO 3202. *J. Chem. Technol. Biotechnol.* 1999, 74, 131-136.

[64] Phetsomphou, S; Seo, CW; Shahbazi, A; Ibrahim, SA. Lactic acid production by free and immobilized *Lactobacillus reuteri* cultivated in laboratory medium supplemented with various nutrients. 17A-6, *Dairy Foods: Cheese and microbiology.* 2004 IFT Annual Meeting, July 12-16, Las Vegas, USA.

[65] Norton, S; Lacroix, C; Vuillemard; JC. Kinetic study of a continuous fermentation of whey permeate by immobilized cell technology. *Enzyme Microb. Technol.* 1994, 16, 457-66.

[66] Norton, S; Lacroix, C; Vuillemard; JC. Reduction of yeast extract supplementation in lactic acid fermentation of whey permeate by immobilized cell technology. *J. Dairy Sci.* 1994, 77, 2494-2508.

[67] Senthuran, A; Senthuran, V; Hatti-Kaul, R; Mattiasson, B. Lactic acid production by immobilized *Lactobacillus casei* in recycle batch reactor: A step towards optimization. *J. Biotechnol.* 1999, 73, 61-70.

[68] Papagianni M. Ribosomally synthesized peptides with antimicrobial properties: biosynthesis, structure, function, and applications. Biotechnol Adv 2003, 21, 465-499.

[69] Bertrand, N; Fliss, I; Lacroix, C. High nisin Z production during repeated-cycle batch cultures in supplemented whey permeate using immobilized *Lactococcus lactis* UL719. *Int. Dairy J.* 2001, 11, 953-960.

[70] Venema, K; Kok, J; Marugg, JD; Toonen, MY; Ledeboer, AM; Venema, G; Chikindas, ML. Functional analysis of the pediocin operon of *Pediococcus acidilactici* PAC1.0: PedB is the immunity protein and PedD is the precursor processing enzyme. *Mol. Microbiol.* 1995, 17, 515-522.

[71] Chikindas, ML; Venema, K; Ledeboer, AM; Venema, G; Kok, J. Expression of lactococcin A and pediocin PA-1 in heterologous hosts. *Lett. Appl. Microbiol.* 1995, 21,183-189.

[72] Nachmouchi, K. Production avec cellules immobilisées, purification et charactérisation du pouvoir immunogène de la pédiocine PA-1 de *Pediococcus acidilactici* UL5. Master Thesis, Laval University, Canada, 2003.

[73] Huang, J; Lacroix, C ; Daba, H ; Simard, RE. Pediocin production and plasmid stability during continuous free and immobilized cell cultures of *Pediococcus acidilactici* UL5. *J. Appl Bacteriol* 1996, 80, 635-644.

[74] Taniguchi, M; Kotani, N; Kobayashi, T. High concentration cultivation of *Bifidobacterium longum* in fermenter with cross-flow filtration. *Appl. Microbiol. Biotechnol.* 1987, 25, 438-441.

[75] Corre, C; Madec, M; Boyaval, P. Production of concentrated *Bifidobacterium bifidum*. *J. Chem. Technol. Biotechnol.* 1992, 53,189-194.

[76] Mehaia, MA; Cheryan, M. Lactic acid from whey permeate in a membrane recycle bioreactor. *Enzyme Microb. Technol.* 1986, 8, 289-292.

[77] Wee, YJ; Kim, JN; Ryu, HW. Biotechnological production of lactic acid and its recent applications. *Food Technol. Biotechnol.* 2006, 44, 163-172.

[78] Kwon, S; Yoo, IK; Lee, WG; Chang, HN; Chang, YK. High rate continuous production of lactic acid by *Lactobacillus rhamnosus* in a two stage membrane cell-recycle bioreactor. *Biotechnol. Bioeng* 2001, 73, 25-34.

[79] Oh, H; Wee, YJ; Yun, JS; Ryu, HW. Lactic acid production through cell-recycle repeated-batch bioreactor. *Appl. Biochem. Biotechnol.* 2003, 107, 603-613.

[80] Cotton, JC; Pometto, AL; Gvozdenovic-Jeremic; J. Continuous lactic acid fermentation using plastic composite support biofilm reactor. *Appl. Microbiol. Biotechnol.* 2001, 57, 626-630.

[81] Jeantet, R; Maubois, JL; Boyaval, P. Semicontinuous production of lactic acid in a bioreactor coupled with nanofiltration membranes. *Enzyme Microb. Technol.* 1996, 19, 614-619.

[82] Shahbazi, A; Mims, MR; Li; Y, Shirley, V; Ibrahim, SA. Lactic acid production from cheese whey by immobilized bifidobacteria. *26th Symposium on Biotechnology for Fuels and Chemicals*. 2004, 3-27, May 9-12, Chattanooga, TN, USA.

[83] Characklis, WG. Biofilm process. In: Characklis, WG; Marshall, KC (Eds). *Biofilms*. New York: Wiley, 2005; pp. 195-232.

[84] Bryers, JD. Biofilms in biotechnology. In: Characklis, WG; Marshall, KC, (Eds). *Biofilms*. New York: Wiley, 2005; pp. 733-773.

[85] Demirci, A; Pometto III, AL. Repeated-batch fermentation in biofilm reactors with plastic-composite supports for lactic acid production. *Appl. Microbiol. Biotechnol.* 1995, 43, 585-589.

[86] Ho, KLG; Pometto III, AL; Hinz, PN. Optimization of L-(+)-lactic acid production by ring and disc plastic composite supports through repeated-batch biofilm fermentation. *Appl. Environ. Microbiol.* 1997, 63, 2533-2542.

[87] Ho, KLG; Pometto III, AL; Hinz, PN; Demirci, A. Nutrient leaching and end product accumulation in plastic composite supports for L-(+)-lactic acid biofilm fermentation. *Appl. Environ. Microbiol.* 1997, 63, 2524-2532.

[88] Ho, KLG; Pometto III, AL; Hinz, PN; Demirci, A. Ingredient selection for plastic composite supports for L-(+)-lactic acid biofilm fermentation by *Lactobacillus casei* subsp. *rhamnosus. Appl. Environ. Microbiol.* 1997, 63, 2516-2523.

[89] Colomban, A; Roger, L; Boyaval, P. Production of propionic acid from whey permeate by sequential fermentation ultrafiltration, and cell recycling. *Biotechnol. Bioeng* 1993, 42, 1091-1098.

[90] Paik, HD; Glatz, BA. Propionic acid production by immobilized cells of propionate-tolerant strain of *Propionibacterium acidipropionici*. *Appl. Microbiol. Biotechnol.* 1994, 42, 22-27.

[91] Ruas-Madiedo, P; Hugenholtz, J; Zoon, P. An overview of the functionality of exopolysaccharides produced by lactic acid bacteria. *Int. Dairy J.*, 2002, 12, 163-171.

[92] Bergmaier, D; Lacroix, C ; Champagne, CP. Exopolysaccharide production during batch cultures with free and immobilized *Lactobacillus rhamnosus* RW-9595M. *J. Appl. Microbiol.* 2003, 95, 1049-1057.

[93] Bergmaier, D; Champagne, CP; Lacroix, C. Growth and exopolysaccharide production during free and immobilized cell chemostat culture of *Lactobacillus rhamnosus* RW-9595M. *J. Appl. Microbiol.* 2005, 98, 272-284.

[94] De Vuyst, L. Nutritional factors affecting nisin production by *Lactococcus lactis* subsp. *lactis* NIZO 22186 in a synthetic medium. *J. Appl. Bacteriol.* 1995, 78, 28-33.

[95] Cabo, ML; Murado, MA; Gonzalez, MP; Pastoriza, L. Effects of aeration and pH gradient on nisin production. A mathematical model. *Enzyme Microb. Technol.* 2001, 29, 264-273.

[96] Lv, W; Cong, W; Cai, Z. Effect of sucrose on nisin production in batch and fed-batch culture by *Lactococcus lactis*. *J. Chem. Technol. Biotechnol.* 2002, 80, 511-514.

[97] Papagianni, M; Avramidis, N; Filiousis, G. Investigating the relationship between the specific glucose uptake rate and nisin production in aerobic batch and fed-batch glucostat cultures of *Lactococcus lactis*. *Enzyme Microbial. Technology* 2007, 40, 1557-1563.

[98] Cho, HY; Yousef, AE; Yang, ST. Continuous production of pediocin by *Pediococcus acidilactici* pO2 in a packed bed bioreactor. *Appl. Microbiol. Biotechnol.* 1996, 45, 589-594.

[99] Bober, JA; Demirci, A. Nisin fermentation by *Lactococcus lactis* subsp. *lactis* using plastic composite supports in biofilm reactors. *Ag. Eng. Int.* 2004, Vol. VI, manuscript FP 04 001.

[100] Spettoli, P; Bottacin, A; Nuti, M.P; Zamorani, A. Immobilization of *Leuconostoc oenos* ML34 in calcium alginate gels and its application to wine technology. *Am. L Enol. Vitic* 1982, 22, 1-5.

[101] Naouri, P; Chagnaud, P; Ranaud, A; Galzy, P; Mathiu, J. A new technology for malolactic bioconversion in wine. *J. Wine Res.* 1991, 2, 5-20.

[102] Carbanes, C; Moreno, J; Mangas, JJ. Cider production with immobilized *Leuconostoc oenos*. *J. Inst. Brew*. 1998, 104, 127-130.

[103] Janssen, DE; Maddox, IS; Mawson, JA. An immobilized cell bioreactor for the malolactic fermentation of wine. *Wine Ind. J.* 2003, 161-165.

[104] Scott, JA; O'Reilly AM. Co-immobilization of selected yeast and bacteria for controlled flavour development in an alcoholic cider beverage. *Proc. Biochem.* 1996, 31, 111-117.

[105] Nedovic, VA ; Durieux, A; Van Nedervelde, L; Rosseels, P; Vandegans, J ; Plaisant, AM; Simon JP. Continuous cider fermentation with co-immobilized yeast and *Leuconostoc oenos* cells. *Enzyme Microb. Technol.* 2000, 26, 834-839.

[106] Maicas, S. The use of alternative technologies to develop malolactic fermentation in wine. *Appl. Microbiol. Biotechnol.* 2001, 56, 35-39.

[107] Maicas, S; Pardo, I; Ferrer, S. The potential of positively charged cellulose sponge for malolactic fermentation of wine, using *Oenococcus oeni*. *Enzyme Microb. Technol.* 2001, 28, 415-419.

[108] Stanton, C; Gardiner, G; Meehan, K; Collins, K; Fitzerald, P; Lynch, B; Ross, RP. Market potential of probiotics. *Am. J. Clin. Nutr.* 2001, 73, 476-483.

[109] Kurmann, JA; Rasic, JL. The health potential of products containing bifidobacteria. In: Robinson, RK (Editor). *Therapeutic properties of fermented milks*. London: Elsevier Science Publications, 1991; pp. 117-158.

[110] Hamilton-Miller, JMT; Shah, S; Winkler, JT. Public health issues arising from microbiological and labeling quality of foods and supplements containing probiotic microorganisms. *Public Health Nutr.* 1999, 2, 223-229.

[111] Shah, NP. Probiotic bacteria: Selective and enumeration and survival in dairy foods. *J. Dairy Sci.* 2000, 83, 894-907.

[112] Dave, RI; Shah, NP. Viability of yoghurt and probiotic bacteria in yoghurts made from commercial starter cultures. *Int. Dairy J.* 1997, 7, 31-41.

[113] Shah, NP; Ravula, RR. Microencapsulation of probiotic bacteria and their survival in frozen fermented dairy desserts. *Australia J. Dairy Technol.* 2000, 55,139-144.

[114] King, AH. Encapsulation of food ingredients: A review of available technology, focusing on hydrocolloids. In: Risch, S; Reineccius, GA (Eds). *Encapsulation and controlled release of food ingredients*. ACS Symposium Series 590, Washington DC: American Chemical Society, 1995; pp. 26-39.

[115] Kailasapathy, K. Microencapsulation of probiotic bacteria: Technology and potential applications. *Curr. Issues Intest. Microbiol.* 2002, 3, 39-48.

[116] Shah, NP. The exopolysaccharides production by starter cultures and their influence on textural characteristics of fermented milks. Int. Dairy Federation, 2002, 3rd June, Comwell Scanticon, Kolding, Denmark. Abstract, p5.

[117] Shahidi, F; Han, XQ. Encapsulation of food ingredients. *Crit. Rev. Food Sci. Nutr.* 1993, 33, 501-547.

[118] Jankowski, T; Zielinska, M; Wysakowska, A. Encapsulation of lactic acid bacteria with alginate/starch capsules. *Biotechnol. Tech.* 1997, 11, 31-34.

[119] Khalil, AH; Mansoor, EH. Alginate encapsulated bifidobacteria survival in mayonnaise. *J. Food Sci.* 1998, 63, 701-705.

[120] Lee, KY; Heo, TR. Survival of *Bifidobacterium longum* immobilized in calcium alginate beads in simulated gastric juices and bile salt solution. *Appl. Environ. Microbiol.* 2000, 66, 869-873.

[121] Truelstrup Hansen, L; Allan-Wojtas, PM; Jin, YL; Paulson, AT. Survival of Ca-alginate microencapsulated *Bifidobaterium* spp. in milk and simulated gastro-intestinal conditions. *Food Microbiol.* 2002, 19, 35-45.

[122] Godward, G; Kailasapathy, K. Viability and survival of free, encapsulated and co-encapsulated probiotic bacteria in ice cream. *Milchwissenschaft* 2003, 58, 161-164.

[123] Talwalkar, A; Kailasapathy, K. Effect of microencapsulation on oxygen toxicity in probiotic bacteria. *Australian J. Dairy Technol.* 2003, 58, 36-39.

[124] Hsiao, HC; Lian, WC; Choo, CC. Effect of packaging conditions and temperature on viability of microencapsulated bifidobacteria during storage. *J. Food Sci. Food Ag* *2004*, 84, 134-139.

[125] Plessas, S; Bekatorou, A; Kanellaki, M; Psarianos, C; Koutinas, A. Cells immobilized in a starch-gluten-milk matrix usable for food production. *Food Chem.* 2005, 89, 175-179.

[126] O'Reordan, K; Andrews, D; Buckle, K; Conway, P. Evaluation of microencapsulation of *Bifidobacterium* strain with starch as an approach to prolonging viability during storage. *J. Appl. Microbiol.* 2001, 91, 1059-1066.

[127] Picot, A; Lacroix, C. Production of multiphase water-insoluble microcapsules for cell microencapsulation using an emulsification/spray-drying technology. *J. Food Sci.* 2003, 68, 2693-2700.

[128] Picot, A; Lacroix, C. Encapsulation of bifidobacteria in whey protein-based microcapsules and survival in simulated gastrointestinal conditions and in yohurt. *Int. Dairy J.* 2004, 14, 505-515.

[129] Jung, YH; Choi, MK. Powder kimchi containing kimchi lactic acid bacteria. Patent No. WO 2006/123866 A1. FSTA, 2007.

[130] Hou, RCW; Lin, MY; Wang, MMC; Tzen, JTC. Increase of viability of entrapped cells of *Lactobacillus delbrueckii* spp. *bulgaricus* in artificial sesame oil emulasions. *J. Dairy Sci.* 2003, 86, 424-428.

[131] Adhikari, K; Mustapha, A; Grun, IU; Fernando, L. Viability of microencapsulated bifidobacteria in set yoghurt during refrigerated storage. *J. Dairy Sci.* 2000, 83, 1946-1951.

[132] Adhikari, K; Mustapha, A; Grun, IU. Survival and metabolic activity of microencapsulated *Bifidobacterium longum* in stirred yoghurt. *J. Food Sci.* 2003, 68, 275-280.

In: Biotechnology: Research, Technology and Applications   ISBN 978-1-60456-901-8
Editor: Felix W. Richter   © 2008 Nova Science Publishers, Inc.

Chapter 10

Recent Advances in Membrane Processes for the Preparation of Emulsions and Particles

C. Charcosset[*]

Laboratoire d'Automatique et de Génie des Procédés
UMR CNRS 5007, UCBLyon 1, ESCPE-Lyon
43 Bd du 11 Novembre 1918, 69 622 Villeurbanne Cedex, FRANCE

Abstract

Membrane emulsification has received increasing attention over the last 15 years, with applications in many fields. In the membrane emulsification process, a liquid phase is pressed through the membrane pores to form droplets at the permeate side of a membrane; the droplets are then carried away by a continuous phase flowing across the membrane surface. Compared to conventional techniques for emulsification, membrane processes offer advantages such as control of average droplet diameter by average membrane pore size, and low energy input. Under specific conditions, monodispersed emulsions and particles can be produced. The purpose of the present paper is to provide an updated review on the membrane emulsification process including: principles of membrane emulsification, experimental devices, influence of process parameters, and applications such as drug delivery systems, food emulsions, and a large range of particles.

Introduction

Membrane emulsification has received increasing attention over the last 15 years [i.e. Peng and Williams 1998, Joscelyne and Trägårdh 2000, Nakashima et al. 2000,

---

[*] Corresponding author: Catherine Charcosset, tel : +00 33 4 72 43 18 67; Fax : +00 33 4 72 43 16 99; e-mail : charcosset@lagep.univ-lyon1.fr

Vladisavljević and Schubert 2002, Altenbach-Rehm et al. 2002]. The membrane emulsification process is shown in Figure 1.

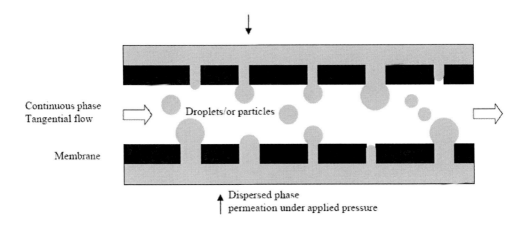

Figure 1. Schematic diagram of the membrane emulsification process.

The dispersed phase is pressed through the pores of a microporous membrane, while the continuous phase flows along the membrane surface. Droplets grow at pore openings until they detach when having reached a certain size. Surfactant molecules in the continuous phase stabilize the newly formed interface, to prevent droplet coalescence immediately after formation. The distinguishing feature is that the resulting droplet size is controlled primarily by the choice of the membrane and not by the generation of turbulent droplet break-up. The apparent shear stress is lower than in classical emulsification systems, because small droplets are directly formed by permeation of the dispersed phase through the micropores, instead of disruption of large droplets in zones of high energy density. Besides the possibility of using shear-sensitive ingredients, emulsions with narrow droplet size distributions can be produced. Furthermore, membrane emulsification processes allow the production of emulsions at lower energy input (10^4-10^6 J/m^3) compared to conventional mechanical methods (10^6-10^8 J/m^3) [Altenbach-Rehm et al. 2002].

Previous reviews on membrane emulsification [i.e. Joscelyne and Trägårdh 2000, Charcosset et al. 2004, Charcosset and Fessi 2005c, Gijsbertsen et al. 2004, Vladisavljević and Williams 2005] focused on membrane emulsification principles, influence of process parameters, comparison with other methods, and applications. Nakashima et al. [2000] provided a review recalling that membrane emulsification was introduced by these authors at the annual Meeting of the Society of Chemical Engineers, Japan, in 1988. The fundamentals of membrane emulsification were presented and the applications: food emulsions, synthesis of monodispersed microspheres and drug delivery systems were described. Gijsbertsen et al. [2004] presented a state of the art on membrane emulsification, as well as an analysis of an industrial scale production of culinary cream, for which a microsieve membrane with a low porosity was found the best suitable. Vladisavljević and Williams [2005] provided a very complete review on manufacturing emulsions and particulate products using membranes, ranging from the production of simple o/w and w/o emulsions to multiple emulsions of different types, s/o/w dispersions, coherent solids (silica particles, solid lipid microspheres, solder metal powder), and structured solids (solid lipid microcarriers, gel microbeads, polymeric microspheres, core-shell microcapsules and hollow polymeric microparticles).

The purpose of the present paper is to provide an updated review on the membrane emulsification process including the most recent findings: proposed devices, membrane emulsification modelling, and an increasing range of reported applications.

Experimental Devices

A schematic picture of a typical membrane emulsification set-up is shown in Figure 2. The system incorporates a tubular microfiltration membrane, a pump, a feed vessel, and a pressurized (N_2) oil container. The dispersed phase is pumped under gas pressure through the pores of the membrane into the continuous phase which circulates through the membrane device. The membrane should not be wetted with the dispersed phase. Therefore, at the beginning of the experiment, the membrane is wetted with the continuous phase, i.e. a hydrophilic membrane for o/w emulsions is wetted with the water phase and a hydrophobic membranes for w/o emulsions is wetted with the oil phase. At the end of the experiment, the membrane is cleaned, using an appropriate solution, until the pure water flux is restored to its original value.

The membrane emulsification process may also be carried out in batch mode without tangential flow of the continuous phase [i.e. Kukizaki and Goto 2007], or in a stirred cell configuration [i.e. Stillwell et al. 2007].

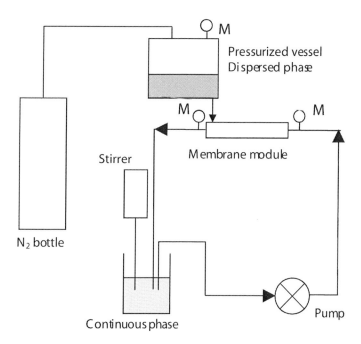

Figure 2. Typical experimental set-up for the membrane emulsification process. M: manometer.

These configurations are particularly suited for the preparation of small amounts of emulsions or microcapsules loaded with high values chemicals. A rotating membrane device was tested to increase the performances of the membrane emulsification process, especially to increase the flux through the membrane [i.e. Aryanti et al. 2006, Schadler and Windhab 2006].

Premix membrane emulsification is an other configuration of membrane emulsification. A pre-emulsion with a large droplet size is passed through the porous membrane into the continuous phase, instead of directly passing the oil or water. The droplets of the pre-emulsion are disrupted into fine droplets during their permeation through the membrane. For similar mean pore sizes, the mean droplet size resulting from premix membrane emulsification is smaller than in direct membrane emulsification, which is often an advantage [Altenbach-Rehm et al. 2002, Vladisavljević et al. 2004a]. Repeating the processes with the same membrane results in smaller mean droplet size, narrower droplet size distribution, and long-term physical stability. This technique is called repeated or multi-stage premix membrane emulsification [Vladisavljević et al. 2004b, Ribeiro et al. 2005].

Membrane Emulsification Modelling

Theoretical data for droplet formation are obtained by calculating the overall forces, assuming a rigid and spherical droplet [Schröder et al. 1998, Peng and Williams 1998, Rayner and Trägårdh 2002, De Luca et al. 2004, De Luca and Drioli 2006]. The main forces that act on the forming droplets are: (1) the interfacial tension force, F_γ, which represents the effects of dispersed phase adhesion around the edge of the pore opening; (2) the static pressure difference force, F_{SP}, due to the pressure difference between the dispersed phase and the continuous phase at the membrane surface; (3) the drag force, F_D, created by the continuous phase flowing past the droplet parallel to the membrane surface; (4) the dynamic lift force, F_L, which results from the asymmetric velocity profile of the continuous phase near the droplet; (5) the buoyancy force, F_B, due to the density difference between the continuous phase and the dispersed phase; (6) the inertial force, F_I, caused by the dispersed phase flow moving through the capillary as it inflates the droplet. The relative magnitude of these forces changes as the droplet increases in size. Peng and Williams [1998] proposed a torque balance between the drag, buoyancy, and the interfacial tension force. Schröder et al. [1998] stated that over the range of their experimental conditions, only the interfacial tension force, the dynamic effect of the pressure difference between the phases, and the drag of the continuous phase had to be taken into account. However, they did not compare experimental and calculated data. The force balance model was applied to other configurations, i.e. vibrating membranes by adding the additional forces induced by the transversal membrane movement [Kelder et al. 2007], and drop detachment from a pore by adding the hydrodynamic force due to the liquid flux outgoing from a capillary [Danov et al. 2007]. Recently, De Luca et al. [2008] compared the results obtained with two approaches (a model based on an algebraic torque balance equation (TBE) and one on a force balance (FBE)) against experimental data available in literature. The analysis showed that the FBE model yielded better results than the

TBE model under conditions of wall shear stress equal or larger than 7 Pa and membrane pore diameters below 1.5 μm.

Some numerical studies with various calculations methods have been reported on droplet formation. Abrahamse et al. [2001] simulated the process of droplet break up at a 5 μm pore in crossflow membrane emulsification using the computational fluid dynamics (CFD) softaware package CFX with the volume of fluids (VOF) method. The maximum membrane porosity was calculated to be 1.5 % to prevent coalescence of droplets growing on neighbouring pores. Quite recently, both Kobayashi et al. [2004] and Rayner et al. [2004, 2005] simulated droplet formation from straight-through microchannels using a CFD software package (CFD-ACE+) wit the so-called piecewise linear interface construction (PLIC) method and the Surface Evolver, respectively. Droplet formation was also simulated with the lattice Boltzmann method, a method suitable for modelling on the mesoscale [van de Graaf et al. 2006]. Droplet detachment in a T junction was investigated experimentally and theoretically, and both the shape of the droplet and the final droplet size were found comparable in the range of parameters investigated.

Many visualization of droplet formation using single pores, micro-engineered membranes, and microporous membranes have been performed to better understand the membrane emulsification process using high-speed camera systems or optical microscopy [i.e. Peng and Williams,1998, Abrahamse et al. 2002, Kobayashi et al. 2002, Yasuno et al. 2002, Christov et al. 2002, Scherze et al. 2005, van der Graaf et al. 2005]. Peng and Williams [1998] used a high-speed video camera to record the formation of droplets from single pores (glass capillaries of diameter ranging from 5 to 200 μm, embedded in epoxy). The droplet growth and the detachment processes were measured as a function of transmembrane pressure, membrane pore size and crossflow velocity. Abrahamse et al. [2002] visualized by microscopy the droplet formation at a micro-engineered membrane with uniform pores. Although the membrane pores were uniform in size, the obtained emulsions were polydispersed. Observations showed this to be due to steric hindrance of droplets forming simultaneously. Kobayashi et al. [2002] provided a real-time microscopic observation of emulsion droplet formation from a polycarbonate membrane. Scherze et al. [2005] developed an approach that determines the distribution of the inner aqueous phase in multiple emulsions by means of light microscopy in conjunction with image processing. Van der Graaf et al. [2005] studied droplet formation in a glass microchip with a small channel containing the dispersed phase perpendicular to a large channel with a crossflow continuous phase. In this model system, droplets are formed at a T-junction of these two rectangular channels; the droplet formation and detachment processes are studied from aside with a microscope connected to a high speed camera.

Influence of Process Parameters

Membranes Material

The most commonly used membranes for the preparation of emulsions are the Shirasu porous glass (SPG) membranes (Ise Chemical Co., Japan), because of their narrow pores size distribution and tubular shape [Nakashima et al. 1991]. The SPG membrane is synthesized from $CaO-Al_2O_3-B_2O_3-SiO_2$ type glass which is made from "Shirasu", a Japanese volcanic ash. The SPG membrane has uniform cylindrical interconnected micropores, a wide spectrum of available mean pore sizes (0.05 - 30 μm), and a high porosity (50-60 %).

The surface wettability can be changed by reaction with organic silanes, such as octadecyltrichlorosilane [Kandori et al. 1991a, Kandori et al. 1991b, Kandori et al. 1992, Fuchigami et al. 2000, Cheng et al. 2006]. The SPG membrane has been characterized by liquid permeability measurements, scanning electron microscopy, Hg porosimetry [Vladisavljević et al. 2005], high-resolution X-ray microtomography, and microscopic observation of droplet formation in real time [Vladisavljević et al. 2007].

Recently, a new type of asymmetric SPG membrane was prepared from two types of primary glass in the $NaO-CaO-Al_2O_3-B_2O_3-ZrO_2-SiO_2$ system with different growth rates of phase separation [Kukizaki and Goto 2007]. The dispersed phase flux through the asymmetric membrane increased by a factor of approximately 20 compared with that through the symmetric membrane, due to the much smaller resistance of the asymmetric membrane.

In addition to the SPG membranes, o/w emulsions were successfully prepared using silicon and silicon nitride microsieves membranes (Aquamarijn Microfiltration BV, The Netherlands) [Zhu and Barrow, 2005, Brans et al. 2006a and b, Geerken et al. 2007]. These are made by photolithographic treatment of a silicon wafer and subsequent etching, or electrochemical metal deposition on a skeleton in an electrolysis bath, respectively. These membranes have interesting properties, such as a smooth and flat surface, a very low membrane resistance and narrow pores size distribution.

Different pore geometries (circular, square, slit shaped), pore size, pore edges and membrane porosities are available. Polycarbonate track-etch membranes (Millipore, Inc.) having a very narrow pores size distribution were also tested for the preparation of particles [Tangirala et al. 2007].

Other commercial microfiltration membranes are attractive for emulsification applications because of their availability in very large surface area, and their high flux through the membrane pores: ceramic aluminium oxide ($\alpha-Al_2O_3$) membranes (Membraflow, Germany) [Schröder and Schubert 1999], α-alumina and zirconia coated membranes (SCT, France) [Joscelyne and Trägårdh 1999], and polytetrafluoroethylene (PTFE) membranes (Advantec Tokyo Ltd., Japan [Suzuki al. 1998, Kanichi et al. 2002] and Goretex Co. Ltd., Japan [Yamazaki et al. 2002]). W/o emulsions were successfully prepared using microporous polypropylene hollow fibers (Microdyn module, Wuppertal, Germany) [Sotoyama et al. 1999], macroporous silica glass membranes [Fuchigami et al., 2000], polytetrafluoroethylene (PTFE) membranes [Suzuki et al. 1998, Yamazaki et al. 2003], polyamide hollow fibers membrane [Giorno and Drioli 2003, Giorno et al. 2005], and home made silica-based monolithic membrane [Hosoya et al. 2005].

Membrane Pores Size

Several authors have shown that the average droplet diameter, $\overline{d}_d$, increases with the average membrane pore diameter, $\overline{d}_p$, by a linear relationship, for given operating conditions:

$$\overline{d}_d = c\overline{d}_p \tag{1}$$

where c is a constant. For SPG membranes, values of c range typically from 2-10. This range was explained by differences in operating conditions, and by the type of SPG membrane used [Omi 1996]. For membranes other than SPG, the values reported for c are higher, typically 3-50.

Monodispersed emulsions can be produced if the membrane pore-size distribution is sufficiently narrow. Using SPG membranes between 0.4 μm and 6.6 μm mean pore size, Vladisavljević and Schubert [2002] found that the span of the droplet size distribution was lower than that reported for ceramic membranes. Omi et al. [2000] using SPG membranes stated that fairly uniform droplets were obtained with a coefficient of variation (CV) around 10 %, due to the uniformity of the membrane pore size.

Membrane Porosity

The porosity of the membrane surface is also an important parameter for the emulsification membrane process because it determines the distance between two adjacent pores [Williams et al. 1998]. This distance is critical to ensure that two adjacent droplets do not come sufficiently close to allow contact with each other, which may lead to coalescence. Abrahamse et al. [2002] calculated the maximum membrane porosity to be 1.5 % to prevent coalescence of droplets growing on neighbouring pores of 5 μm diameter. However, a low porosity has the negative effect of a low dispersed flux.

Transmembrane Pressure

The membrane emulsification method involves using a transmembrane pressure to force the dispersed phase to permeate through the membrane into the continuous phase. The transmembrane pressure ΔP_{tm} is defined as the difference between the pressure of the dispersed phase, P_d, and the mean pressure of the continuous phase,

$$\Delta P_{tm} = P_d - \frac{(P_{c,in} + P_{c,out})}{2} \tag{2}$$

where $P_{c,in}$ and $P_{c,out}$ are the pressure of the flowing continuous phase at the inlet and at the outlet of the membrane module, respectively. The applied transmembrane pressure required to make the discontinuous phase (i.e. oil) flow can be estimated from the capillary pressure, assuming that the pores are ideal cylinders:

$$P_c = \frac{4\gamma \cos\theta}{\overline{d}_p} \tag{3}$$

where P_c is the critical pressure, γ the o/w interfacial tension, θ the contact angle of the oil droplet against the membrane surface well wetted with the continuous phase and $\overline{d}_p$ the average pore diameter. The actual transmembrane pressure required to make the discontinuous phase flow may be greater than predicted by eq. 2, due to tortuosities in the pores, irregular pore openings at the membrane surface and the significant effects of surface wettability [Williams 1998]. The dispersed phase flux J_d is related to the transmembrane pressure according to Darcy's law:

$$J_d = \frac{K \Delta P_{tm}}{\mu L} \tag{4}$$

where K is the membrane permeability, L the membrane thickness, and μ the dispersed phase flux viscosity. In cases where the membrane may be assumed to have n uniform cylindrical pores of radius r, the permeability K is given by the Hagen-Poiseuille equation:

$$K = \frac{nr^2}{8\pi} \tag{5}$$

The emulsification result is expressed in terms of the dispersed phase flux J_d, through the membrane calculated as [Schröder and Schubert 1999]:

$$J_d = \frac{M_d}{\rho_d A} \tag{6}$$

where M_d is the mass flowrate of the dispersed phase, A the membrane surface area and ρ_d the dispersed phase density. The definition of the dispersed phase flux allows the comparison of results from different types or sizes of membrane.

The dispersed phase flux is an essential parameter of the economy of the membrane emulsification process. Increasing transmembrane pressure increases the flux of dispersed phase through the membrane, according to Darcy's law. At high fluxes, the average droplet size and the size distribution tend to increase because of increased droplet coalescence at the membrane surface. Therefore, an increase in flux may be at the expense of droplet size distribution. The effect of transmembrane pressure is dependent on operating conditions, as crossflow velocity and type of surfactant [Schröder et al. 1998, Schröder and Schubert 1999, Katoh et al. 1996, Scherze et al. 1999, Yuyama et al. 2000].

Crossflow Velocity

Droplets formed at the membrane surface detach under the influence of the flowing continuous phase. The characteristic parameter of the flowing continuous phase is the cross flow velocity or the wall shear stress. It is shown that the droplet size becomes smaller as the wall shear stress increases and that the influence is greater for small wall shear stresses [Katoh et al. 1996, Williams et al. 1998, Joscelyne and Trägårdh 1999, Scherze et al. 1999, Kobayashi et al. 2002]. The effect of the wall shear stress on reducing droplet size is dependent on the membrane pore size, being more effective for smaller membrane pore sizes [Schröder and Schubert 1999].

Surfactants

The influence of the type of surfactant in the membrane emulsification process has been studied by several authors [Kandori et al. 1991, Muschiolik et al. 1997, Schröder et al. 1998, Schröder and Schubert 1999, Fuchigami et al. 2000, Yuyama et al. 2000, Kobayashi et al. 2002]. Surfactants played two main roles in the formation of an emulsion. Firstly, they lowered the interfacial tension between oil and water. This facilitated droplet distribution and in case of membranes lowers the minimum emulsification pressure. Secondly, surfactants stabilize the droplets against coalescence and/or aggregation. Schröder et al. [1998] and Schröder and Schubert [1999] showed that the type of surfactant used greatly influenced the droplet size. Droplet diameters obtained with Tween 20 were about twice the size of the droplets stabilized with SDS, in agreement with the ratio of equilibrium interfacial tensions. These authors suggested that the interfacial tension force was one of the key forces governing droplet formation during the membrane emulsification process. Van der Graaf et al. [2004] carried out droplet formation experiments with a microengineered membrane by measuring the droplet diameter and droplet formation time as a function of the surfactant concentration in the continuous phase. Their experiments confirmed that the interfacial tension influenced the process of droplet formation: higher surfactant concentrations lead to smaller droplets and shorter droplet formation times.

Viscosity

The viscosity of the dispersed phase has also an important effect on the membrane emulsification process performance. According to Darcy's law, the dispersed flux is inversely proportional to the dispersed phase viscosity. If the dispersed phase viscosity is high, then the dispersed flux will be low, and as a consequence the droplet diameter will be large compared to the mean pore diameter. Kukizaki and Goto [2006] showed that in a system composed of decane containing liquid paraffin and SDS solution containing polyethylene glycol, the resulting droplet diameter increased with increasing water-phase viscosity, while droplet diameter decreased with increasing oil-phase viscosity. Droplet diameter decreased as the ratio of oil-phase to water-phase viscosity increased. However, droplet diameter did not change in the case of a constant viscosity ratio.

Applications oF Membrane Emulsification

Emulsions

Emulsions are dispersed systems of two (or more) insoluble liquids, e.g. water and oil. Depending on which is the dispersed phase and which is the continuous phase, there are various types of emulsions: oil-in-water (o/w), water-in-oil (w/o) emulsions, and water-in-oil-in-water (w/o/w) emulsions. Emulsions play an important role in the formulation of cosmetics, pharmaceuticals, paints and foods. They are also encountered in the petroleum industry, especially during crude-oil production, as well as in some solvent extraction processes. Emulsions are usually prepared using high-pressure homogenizers, ultrasound homogenizers and rotor/stator systems, such as stirred vessels, colloid mills or toothed disc dispersing machines [Vladisavljević and Schubert 2002]. In the dispersing zone of these machines high shear stresses are applied to deform and disrupt large droplets. Therefore, high-energy inputs are required and shear-sensitive ingredients such as proteins or starches may lose functional properties. Membrane emulsification is a suitable technique for the production of single and multiple emulsions.

O/w emulsions prepared by membrane emulsification were also used recently in a multiphase reactor developed by immobilizing the lipase from *Candida rugosa* in a polymeric membrane [Giorno and Drioli 2003, Giorno et al. 2007a and b]. The configuration was a two separate phase membrane reactor constituted by an emulsion+enzyme-loaded membrane and an organic and an aqueous phase recycled along the two separate sides of the membrane. It was shown that the presence of the emulsion within the membrane improved the catalytic activity and the enantioselectivity of the immobilized enzyme as well as the transport rate of the hydrophobic reagent through the hydrophilic membrane.

Drug Delivery Systems

Various drug delivery systems were prepared by membrane emulsification. The first application proposed was an w/o/w emulsions for the treatment of liver cancer by arterial injection chemotherapy [Higashi et al. 1993, Higashi et al. 1995, Higashi et al. 1996, Higashi et al. 1999, Higashi et al. 2000]. The aqueous anticancer drug solution (epirubicin or carboplation) was mixed with the oil (iodinated poppy-seed oil or lipiodol). This mixture was then sonicated to form submicron sized w/o emulsion, and permeated through a SPG membrane into a glucose solution forming w/o/w emulsions. The efficacy of the w/o/w emulsions was proved clinically, for patients bearing hepatocellular carcinoma nodules recurrent after hepatectomy. The treatment of gastric ulcerogenicity with diclofenac sodium suspension prepared via the membrane emulsification technique was reported by Pia et al. [2007]. Uniform sized microspheres were prepared for the controlled release of anti-cancer drug anthracycline [Costa and Cardoso, 2006], carotenoid astaxanthin [Ribeiro et al. 2005], hemoglobin [Ribeiro et al. 2005], insulin [Toorisaka et al. 2003, Wang et al. 2006a, Wang et al. 2006b, Liu et al. 2006], and spironolactone for paediatric use [Limayem et al. 2006]. Drug delivery systems included microspheres of albumin [El-Mahdy et al. 1998, Muramatsy and Nakauchi 1998], calcium alginate [You et al. 2001, Liu et al. 2003], poly(D,L-lactide), copoly(lactide-glycolide) [Shiga et al. 1996], and poly(DL,lactic-co-glycolic) (PLGA) acid [Gasparini et al. 2007]. The encapsulation efficiency and/or drug release properties were measured with model drugs, i.e. recombinant human insulin (rhI) [Liu et al. 2005], blue dextran [Gasparini et al. 2007], lidocaine-hydrochloride, sodium salicylate, and 4-acetaminophen [Yun et al. 2007]. The results suggested that the release behaviour could be adjusted by changing precisely the diameters of microcapsules, prepared with the membrane emulsification technique.

Food Emulsions

Emulsions play an important role in the formulation of foods for production of o/w emulsions, i.e. dressings, artificial milks, cream liqueurs, as well as for preparation of some w/o emulsions, i.e. margarines and low fat spreads. Various preparation of food emulsions by membrane emulsification are reported [Kandori et al. 1992, Katoh et al. 1996, Muschiolik et al. 1997, Suzuki et al. 1998, Joscelyne and Trägårdh 1999, Scherze et al. 1999, Christov et al. 2002, Vladisavljević et al. 2002, Supsakulchai et al. 2002]. When preparing w/o food emulsions, Katoh et al. [1996] showed that the dispersed phase flux was increased 100 times using a hydrophilic membrane pre-treated by immersion in the oil phase. They prepared a low fat spread with a fat content of 25 % (v/v), and concluded that the membrane emulsification process was suitable for preparation of large scale w/o food emulsions. Using SPG membranes, Scherze et al. [1999] and Muschiolik et al. [1997] prepared o/w emulsions with liquid butter fat or sunflower oil as the dispersed phase and a continuous phase containing milk proteins. The emulsions so obtained were characterised by particle size distribution, creaming behaviour and protein adsorption at the dispersed phase. The advantage of membrane emulsification was pointed out to be the low shear forces on the physicochemical and molecular properties of the proteins.

Microspheres

Polymeric microspheres with a narrow size distribution are needed in industrial applications, such as chromatographic packing materials, dry and liquid toners for electrophotography and drug delivery devices. Omi et al. [2000] synthesized various microspheres with diameters from 2 to 100 μm in aqueous solution, by combining the membrane emulsification process and subsequent polymerization. Monomer droplets containing solvents, crosslinkers and an initiator, as well as a water-insoluble oligomer, are formed by permeation through the membrane pores, suspended in the aqueous solution of stabilizing agents, and transferred to a reactor for polymerization. These preparations include polystyrene [Omi et al. 1994, Omi 1996, Yuyama et al. 2000a, Ma and Omi 2002, Ma and al.

2003], polymethacrylate [Tawonsree et al. 2000], poly(styrene-*co*-methyl methacrylate) [Omi et al. 1997, Nuisin et al. 2000, Kiatkamjornwong et al. 2000], poly(methyl methacrylate) [Omi and al. 1995a and b, Yuyama et al. 2000, Omi and al. 2001], poly(styrene-*co*-2-hydroxyethyl methacrylate) [Ma and al. 2002], poly(styrene-*co*-*N*-dimethylaminoethyl methacrylate) [Ma and Omi 2002], poly(glycidyl methacrylate) microspheres [Wang et al. 2006], and titanium dioxide-polystyrene particles [Supsakulchai et al. 2003]. Microspheres were also prepared by the same authors, by combining the membrane emulsification process and subsequent solvent evaporation, to prepare poly(styrene)-poly(methyl methacrylate) [Ma et al. 1999a and b, Ma et al. 2001a and b], poly(lactide) [Ma et al. 1999c], polyurethane urea [Yuyama et al. 2000b], magnetite [Omi et al. 2001], and titanium dioxide microcapsules [Supsakulchai et al. 2002]. The last technique reported is the droplet swelling method, to prepare hydrophilic poly(methylmethacrylate) [Omi et al. 1995], poly(methylmethacrylate-*co*-2-hydroxyethyl methacrylate) [Ma et al. 2002], poly(styrene-*co*-divinylbenzene) [Omi et al. 1997], polystyrene-polyimide [Omi et al. 1999], and p(HEMA-*co*-EDMA) microspheres [Qu et al. 2002].

Preparations from other authors include poly(acrylamide-*co*-acrylic acid) [Nagashima et al. 1998a and b, Nagashima et al. 1999], poly(*N*-isopropylacrylamide-*co*-acrylic acid) [Makino et al. 2000], polystyrene [Hatate et al. 1995, Dowding et al. 2001], poly(D,L-lactide) and copoly(lactide-glycolide) [Shiga et al. 1996], agarose [Zhou et al. 2007], and chitosan microspheres [Park et al. 2004].

For toner applications, poly(styrene-butylacrylate) and poly(butadiene-styrene) particles containing carbon black were prepared [Ha et al. 1998, Ha et al. 1999], and poly(styrene-*co*-divinylbenzene) microspheres containing a colorant and a charge control agent [Hatate et al. 1997, Yoshizawa et al. 1996]. For size-exclusion chromatography applications, Yoshizawa et al. [2004] prepared polyethylene glycol monomethacrylate microspheres by the combination of the SPG membrane emulsification technique and the emulsion swelling method.

Other Particles

A large range of particles has been prepared with the membrane emulsification technique. Uniformly sized solid particles and microcapsules were prepared using a high melting point lipid as the encapsulating material, followed by cooling of the preparation at room temperature [Charcosset et al. 2005, El-Harati et al. 2006, Kukizaki and Goto 2007]. Polymeric nanoparticles were obtained successfully by the nanoprecipitation or the interfacial polymerization methods [Charcosset and Fessi 2005a and b]. Magnetic polymer particles at nano- and micro-scales were prepared by encapsulating magnetic components with dissolved or *in situ* formed polymers [Yuan and Williams 2007]. Other particles included silica hydrogel particles [Kandori et al. 1991], microcapsules of *Lactobacillus casei* [Song et al. 2003], SiO_2 nanoparticles [Yanagishita et al. 2004], core-shell enzyme-loaded microcapsules with environment-responsive membranes [Akamatsu and Yamaguchi 2007], monodisperse thermo-sensitive poly(*N*-isopropylacrylamide) hollow microcapsules [Cheng et al. 2007], pH-sensitive particles [Sheibat-Othman et al. 2007], and ferrofluid droplets [Bădescu and Bădescu 2007].

Industrial Applications

Membrane systems are particularly suitable for large scale production because they are easy to scale-up, by adding more membranes to a module. Williams et al. [1998] have shown that pilot-scale membrane emulsification can be operated successfully in both batch and semi-continuous mode. However, to date, there is only one documented product produced using membrane emulsification ("Yes light", a very low fat spread, Moringa Milk Industry, Japan) [Nakashima et al. 2000].

Some authors have compared the emulsification result using membranes and stirring methods [Okochi and Nakano 1997, Dowding et al. 2001, Supsakulchai et al. 2002]. Okochi and Nakano [1997] provided a comparative study of properties of w/o/w emulsions prepared by membrane emulsification and a two-stage stirring method. The w/o/w emulsion prepared by membrane emulsification was found to be more homogeneous and provided slower release of low molecular weight drugs. Dowding et al. [2001] found that the droplet size distribution obtained by membrane emulsification was narrower than that obtained by stirring.

The main disadvantage of the membrane emulsification process may be associated a low dispersed phase flux through the membrane. Using SPG membranes, oil fluxes for preparation of o/w emulsions range typically from 5 to 15 $dm^3.m^{-2}.h^{-1}$ with a 0.2 μm membrane and 10 to 45 $dm^3.m^{-2}.h^{-1}$ with a 0.5 μm membrane [Scherze et al. 1999]. Higher fluxes are obtained with ceramic membranes, i.e. 50-250 $dm^3.m^{-2}.h^{-1}$ using a 0.2 μm α-alumina membrane [Katoh et al. 1996]. For preparation of w/o emulsions, water fluxes of 200 $dm^3.m^{-2}.h^{-1}$ are achieved using a 0.5 μm oil-pretreated hydrophilic SPG membrane and 2 300 $dm^3.m^{-2}.h^{-1}$ using a 1 μm membrane [Katoh et al. 1996]. The flux through the membrane is determined by the applied transmembrane pressure, the permeability of the membrane and the number of active pores (Darcy 's law). The SPG membranes have a low permeability, because they are quite thick (0.45-0.75 mm) and are homogeneous in structure. The number of active pores is reported to be very low, i. e. 0.3-0.5 % [Yasuno et al. 2002] and 2 % [Vladisavljević and Schubert 2002]. For ceramic membranes, the permeability is expected to be higher, because the toplayer with the smallest pores is very thin; this layer being mechanically supported by layers with larger pore diameters [Gijsbersten-Abrahamse et al. 2002].

The dispersed phase flux can be increased using repeated premix membrane emulsification [Park et al. 2001, Altenbach-Rehm et al. 2002, Suzuki et al. 1998]. In this process, the preliminary emulsified o/w or w/o emulsion is dispersed into the continuous phase via the membrane. Oil fluxes as high as 20 000 $dm^3.m^{-2}.h^{-1}$ are obtained by single premix emulsification using a 1 μm hydrophilic PTFE membrane [Altenbach-Rehm et al. 2002].

Conclusion

Membrane emulsification was introduced over fifteen years ago, as a new emulsification technique based on microporous membranes. Many studies have been carried out in this area, especially on special devices, influence of process parameters, and possible applications. However, very few industrial applications have been reported to date. The main limitation of membrane emulsification may be the low flux associated with monodispersed emulsions. It appears more and more clearly that special membranes have to be developed to fulfil properties required for industrial applications: high fluxes, availability of large membrane area, and potentially of scaling-up. Membrane emulsification should then appear as a very interesting industrial technique.

References

Abrahamse AJ, van der Padt A, Boom RM and Heij WBC, Process fundamentals of membrane emulsification: Simulation with CFD. *AIChE J.* 47:1285-1291 (2001).

Abrahamse AJ, van Lierop R, van der Sman RGM, van der Padt A and Boom RM, Analysis of droplet formation and interactions during cross-flow membrane emulsification. *J. Membr. Sci.* 204:125-137 (2002).

Altenbach-Rehm J, Suzuki K and Schubert H, Production of O/W-emulsions with narrow droplet size distribution by repeated premix membrane emulsification. 3ième Congrès Mondial de l'Emulsion, 24-27 September 2002, Lyon, France.

Aryanti N, Williams RA, Hou R and Vladisavljevic GT, Performance of rotating membrane emulsification for o/w production. *Desalination* 200:572-574 (2006).

Asano Y and Sotoyama K, Viscosity change in oil/water food emulsions prepared using a membrane emulsification system. *Food Chem.* 66:327-331 (1999).

Bădescu V and Bădescu R, Microscopic observation of ferrofluid droplet formation from a cylindrical channels membrane. *J. Optoelectronics Advanced Mater.* 9:949-951 (2007).

Brans G, Kromkamp J, Pek N, Gielen J, Heck J, van Rijn CJM, van der Sman RGM, Schröen CGPH and Boom RM, Evaluation of microsieve membrane design. *J. Membrane Sci.* 278:344-348 (2006).

Brans G, van der Sman RGM, Schröen CGPH, van der Padt A and Boom RM, Optimization of the membrane and pore design for micro-machined membranes, *J. Membrane Sci.* 278:239-250 (2006).

Charcosset C and Fessi H, A new process for drug loaded nanocapsules preparation using a membrane contactor. *Drug Dev. Ind. Pharm.* 31:987-992 (2005b).

Charcosset C and Fessi H, Membrane emulsification and microchannel emulsification processes. *Reviews Chem. Eng.* 21:1-32 (2005c).

Charcosset C and Fessi H, Preparation of nanoparticles with a membrane contactor. *J Membrane Sci.* 266:115-120 (2005a).

Charcosset C, El-Harati AA and Fessi, Preparation of solid lipid nanoparticles for controlled drug delivery using a membrane contactor. *J. Control Release* 108:112-120 (2005).

Charcosset C, Limayem I, Fessi H, The membrane emulsification process - A review. *J. Chem. Biochem. Technol.* 79:209-218 (2004).

Cheng CJ, Chu LY and Xie R, Preparation of highly monodisperse W/O emulsions with hydrophobically modified SPG membranes. *J. Colloid Interface Sci.* 300:375-382 (2006).

Cheng CJ, Chu LY, Ren PW, Zhang J and Hu L, Preparation of monodisperse thermo-sensitive poly(N-isopropylacrylamide) hollow microcapsules. *J. Colloid Interface Sci.* 313:383-388 (2007).

Christov NC, Ganchev DN, Vassileva ND, Denkov ND, Danov KD and Kralchevsky PA, Capillary mechanisms in membrane emulsification: oil-in-water emulsions stabilized by Tween 20 and milk proteins. *Colloids Surf. A* 209:83-104 (2002).

Costa MS and Cardoso MM, Effect of uniform sized polymeric microspheres prepared by membrane emulsification technique on controlled release of anthracycline anti-cancer drugs. *Desalination* 200:498-500 (2006).

Danov KD, Danova DK and Kralchevsky PA, Hydrodynamic forces acting on a microscopic emulsion drop growing at a capillary tip in relation to the process of membrane emulsification. *J. Colloid Interface Sci.* In press (2007)

De Luca G and Drioli E, Force balance conditions for droplet formation in cross-flow membrane emulsifications, *J. Colloid Interface Sci.* 294:436-448 (2006).

De Luca G, Di Maio FP, Di Renzo A, Droplet detachment in cross-flow membrane emulsification: comparison among torque- and force-based models, *Chem. Eng. Processing* (2007) in press

De Luca G, Sindona A, Giorno L and Drioli E., Quantitative analysis of coupling effects in cross-flow membrane emulsification, *J. Membrane Sci.* 229:199-209 (2004).

Dowding PJ, Goodwin JW and Vincent B, Production of porous suspension polymer beads with a narrow size distribution using a cross-flow membrane and a continuous tubular reactor. *Colloids Surf. A* 180:301-309 (2001).

El-Harati AA, Charcosset C and Fessi H, Influence of the formulation for solid lipid nanoparticles prepared with a membrane contactor. *Pharm. Dev. Technol.* 11:153-157 (2006).

El-Mahdy M, Ibrahim ES, Safwat S, El-Sayed A, Ohshima H, Makino K, Muramatsu N and Kondo T, Effects of preparation conditions on the monodispersity of albumin microspheres. *J. Microencapsulation* 15:661-673 (1998).

Fuchigami T, Toki M and Nakanishi K, Membrane emulsification using sol-gel derived macroporous silica glass. *J. Sol-Gel Sci. Technol.* 19:337-341 (2000).

Geerken MJ, Lammertink RGH and Wessling M, Tailoring surface properties for controlling droplet formation at microsieve membranes. *Colloids Surfaces A* 292:224-235 (2007).

Gijsbersten-Abrahamse AJ, van der Padt A and Boom RM, Membrane emulsification: the influence of pore geometry and wall contact angle on process performance. 3ième Congrès Mondial de l'Emulsion, 24-27 September 2002, Lyon, France.

Gijsbersten-Abrahamse AJ, van der Padt A and Boom RM, Status of cross-flow membrane emulsification and outlook for industrial application. *J. Membrane Sci.* 230:149-159 (2004).

Giorno L, Li N and Drioli E, Preparation of oil-in-water emulsions using polyamide 10 kDa hollow fiber membrane. *J. Membrane Sci.* 217:173-180 (2003).

Giorno L, Mazzei R, Oriolo M, De Luca G, Davoli M and Drioli E, Effects of organic solvents on ultrafiltration polyamide membranes for the preparation of oil-in-water emulsions. *J. Colloid Interface Sci.* 287:612-623 (2005).

Giorno L, Piacentini E, Mazzei R and Drioli E, Membrane emulsification as a novel method to distribute phase-transfer biocatalysts at the oil/water interface in bioorganic reactions. *J. Membrane Sci.* In press (2007a).

Giorno L, Piacentini E, Mazzei R and Drioli E, Use of membrane emulsification to distribute lipase at oil/water interface in heterogeneous biotransformations. *Membrane News* 72:25-26 (2007b).

Giorno L., Li N., Drioli E., Use of stable emulsion to improve stability, activity, and enantioselectivity of lipase immobilized in a membrane reactor, *Biotech. Bioeng.* 84: 677-685 (2003).

Ha YK, Lee HJ and Kim JH, Large and monodispersed microspheres with high butadiene rubber content via membrane emulsification. *Colloids Surf. A* 145:281-284 (1998).

Ha YK, Song HS, Lee HJ and Kim JH, Preparation of core particles for toner application by membrane emulsification. *Colloids Surf. A* 162:289-293 (1999).

Hatate Y, Ohta H, Uemura Y, Ijichi K and Yoshizawa H, Preparation of monodispersed polymeric microspheres for toner particles by the Shirasu porous glass membrane emulsification technique. *J. Appl. Polym. Sci.* 64:1107-1113 (1997).

Hatate Y, Uemura Y, Ijichi K, Kato Y, Hano T, Baba Y and Kawano Y, Preparation of GPC packed polymer beads by a SPG membrane emulsifier. *J. Chem. Eng. Japan* 28:656-659 (1995).

Higashi S and Setoguchi T, Hepatic arterial injection chemotherapy for hepatocellular carcinoma with epirubicin aqueous solution as numerous vesicles in iodinated poppy-seed oil microdroplets: clinical application of water-in-oil-in-water emulsion prepared using a membrane emulsification technique. *Adv. Drug Deliv. Rev.* 45: 57-64 (2000).

Higashi S, Maeda Y, Kai M, Kitamura T, Tsubouchi H, Tamura S and Setoguchi T, A case of hepatocellular carcinoma effectively treated with epirubicin aqueous vesicles in monodispersed iodized poppy-seed oil microdroplets. *Hepato-Gastroenterology* 43:1427-1430 (1996).

Higashi S, Shimizu M and Setoguchi T, Preparation of new lipiodol-emulsion containing water soluble anticancer agent by membrane emulsification technique. *Drug Deliv. Systems* 8:59-61 (1993).

Higashi S, Shimizu M, Nakashima T, Iwata K, Uchiyama F, Tateno S, Tamura S and Setogushi T, Arterial-injection chemotherapy for hepatocellular carcinoma using monodispersed poppy-seed oil microdroplets containing fine aqueous vesicles of epirubicin, Initial medical application of a membrane-emulsification technique. *Cancer* 75:1245-1254 (1995).

Higashi S, Tabata N, Kondo KH, Maeda Y, Shimizu M, Nakashima T and Setoguchi T, Size of lipid microdroplets effects results of hepatic arterial chemotherapy with an anticancer agent in water-in-oil-in-water emulsion to hepatocellular carcinoma. *J. Pharmacol. Exp. Ther.* 289:816-819 (1999).

Hosoya K, Bendo M, Tanaka N, Watabe Y, Ikegami T, Minakuchi H and Nakanishi K, An application of silica-based monolithic membrane emulsification technique for easy and efficient preparation of uniformly sized polymer particles. *Macromol. Mater. Eng.* 290:753-758 (2005).

Joscelyne SM and Trägårdh G, Food emulsions using membrane emulsification: conditions for producing small droplets. *J. Food Eng.* 39:59-64 (1999).

Joscelyne SM and Trägårdh G, Membrane emulsification-a literature review. *J. Membr. Sci.* 169:107-117 (2000).

Kandori K, Kishi K and Ishikawa T, Formation mechanisms of monodispersed W/O emulsions by SPG filter emulsification method. *Colloids Surf.* 61:269-27(1991a).

Kandori K, Kishi K and Ishikawa T, Preparation of monodispersed W/O emulsions by Shirasu-porous-glass filter emulsification technique. *Colloids Surf.* 55:73-78 (1991b).

Kandori K, Kishi K and Ishikawa T, Preparation of uniform silica hydrogel particles by SPG filter emulsification method. *Colloids Surf.* 62: 259-262 (1992).

Kanichi S, Yuko O and Yoshio H, Properties of solid fat O/W emulsions prepared by membrane emulsification method combined with pre-emulsification. 3ième Congrès Mondial de l'Emulsion, 24-27 September 2002, Lyon, France.

Katoh R, Asano Y, Furuya A, Sotoyama K and Tomita M, Preparation of food emulsions using a membrane emulsification system. *J. Membr Sci.* 113:131-135 (1996).

Kelder JDH, Janssen JJM and Boom RM, Membrane emulsification with vibrating membranes: A numerical study. *J. Membr. Sci.* 304:50-59 (2007).

Kiatkamjornwong S, Nuisin R, Ma GH and Omi S, Synthesis of styrenic toner particles by SPG emulsification technique. *Chinese J. Polym Sci.* 18:309-322 (2000).

Kobayashi I, Mukataka S and Nakajima M, CFD simulation and analysis of emulsion droplet formation from straight-through microchannels. *Langmuir* 20:9868-9877 (2004).

Kobayashi I, Yasuno M, Iwamoto S, Shono A, Satoh K and Nakajima M, Microscopic observation of emulsion droplet formation from a polycarbonate membrane. *Colloids Surf. A* 207:185-196 (2002).

Kukizaki M and Goto M, Effects of interfacial tension and viscosities of oil and water phases on monodispersed droplet formation using a Shirasu-porous-glass (SPG) membrane. *Membrane* 31:215-220 (2006).

Kukizaki M and Goto M, Preparation and characterization of a new asymmetric type of Shirasu porous glass (SPG) membrane used for membrane emulsification. *J. Membrane Sci.* 299:190-199 (2007).

Kukizaki M and Goto M, Preparation and evaluation of uniformly sized solid lipid microcapsules using membrane emulsification. *Colloids Surf. A* 293:87-94 (2007).

Limayem I, Charcosset C, Sfar S and Fessi H, Preparation and characterization of spironolactone-loaded nanocapsules for paediatric use. *Int. J. Pharm.* 325:124-131 (2006).

Liu R, Huang SS, Wan YH, Ma GH and Su ZG, Preparation of insulin-loaded PLA/PLGA microcapsules by a novel membrane emulsification method and its release in vitro. *Colloids Surfaces B* 51:30-38 (2006).

Liu R, Ma GH, Wan YH and Su ZG, Influence of process parameters on the size distribution of PLA microcapsules prepared by combining membrane emulsification technique and double emulsion-solvent evaporation method. *Colloids Surf. B* 45:144-153 (2005).

Liu XD, Bao DC, Xue WM, Xiong Y, Yu WT, Yu XJ, Ma XJ and Yuan Q, Preparation of uniform calcium alginate gel beads by membrane emulsification coupled with internal gelation. *J. Appl. Polym. Sci.* 87:848-852 (2003).

Ma GH and Omi S, Mechanism of formation of monodisperse polystyrene hollow particles prepared by membrane emulsification technique. Effect of hexadecane amount on the formation of hollow particles. *Macromol. Symp.* 179:223-240 (2002).

Ma GH, Chen AY, Su ZG and Omi S, Preparation of uniform hollow polystyrene particles with large voids by a glass-membrane emulsification technique and a subsequent suspension polymerization. *J. Appl. Polym. Sci.* 87: 244-251 (2003).

Ma GH, Nagai M and Omi S, Effect of lauryl alcohol on morphology of uniform polystyrene-poly(methyl methacrylate) composite microspheres prepared by porous glass membrane emulsification technique. *J. Colloid Interface Sci.* 219:110-128 (1999a).

Ma GH, Nagai M and Omi S, Preparation of uniform poly(lactide) microspheres by employing the Shirasu Porous Glass (SPG) technique. *Colloids Surf. A* 153:383-394 (1999c).

Ma GH, Nagai M and Omi S, Study on morphology control of uniform composite microspheres prepared by SPG (Shirasu Porous Glass) membrane emulsification technique. *Current Topics in Colloid and Interface Science* 4:15-33 (2001a).

Ma GH, Nagai M and Omi S, Study on preparation and morphology of uniform artifical polystyrene-poly(methyl methacrylate) composite microspheres by employing the SPG (Shirasu Porous glass) membrane emulsification technique. *J. Colloid Interface Sci.* 214:264-282 (1999b).

Ma GH, Nagai M and Omi S, Study on preparation of monodispersed poly(styrene-co-N-dimethylaminoethyl methacrylate) composite microspheres by SPG (Shirasu Porous Glass) emulsification technique. *J. Appl. Polym Sci.* 79:2408-2424 (2001b).

Ma GH, Omi S, Dimonie VL, Sudol ED and El-Aasser MS, Study of the preparation and mechanism of formation of hollow monodisperse polystyrene microspheres by SPG (Shirasu Porous Glass) emulsification technique. *J. Appl. Polym. Sci.* 85:1530-1543 (2002).

Makino K, Agata H and Ohshima H, Dependence of temperature-sensitivity of poly(N-isopropylacrylamide-co-acryl acid) hydrogel microspheres upon their sizes. *J. Colloid Interface Sci.* 230:128-134 (2000).

Mine Y, Shimizu M and Nakashima T, Preparation and stabilization of simple and multiple emulsions using a microporous glass membrane. *Colloids Surf. B* 6:261-268 (1996).

Muramatsu N and Nakauchi K, A novel method to prepare monodisperse microparticles. *J. Microencapsulation* 15:715-723 (1998).

Muschiolik G, Dräger S, Scherke I, Rawel HM and Stang M, Protein-stabilized emulsions prepared by the micro-porous glass method, in *Food Colloids: Proteins, lipids and polysaccharides*, ed by Dickinson. Royal Society of Chemistry, Cambridge, pp 393-400 (1997).

Nagashima S, Ando S, Makino K, Tsukamato T and Ohshima H, Size dependence of polymer composition in the surface layer of poly(acrylamide-co-acrylic acid) hydrogel microspheres. *J. Colloid Interface Sci.* 197:377-382 (1998).

Nagashima S, Ando S, Tsukamoto T, Ohshima H and Makino K, Preparation of monodisperse poly(acrylamide-co-acrylic acid) hydrogel microspheres by a membrane emulsification technique and their size-dependent surface properties. *Colloids Surf. B* 11:47-56 (1998).

Nagashima S, Koide M, Ando S, Makino K, Tsukamoto T and Ohshima H, Surface properties of monodisperse poly(acrylamide-co-acrylic acid) hydrogel microspheres prepared by a membrane emulsification technique. *Colloids Surf A* 153:221-227 (1999).

Nakashima T, Shimizu M and Kukizaki M, Membrane emulsification by microporous glass. *Key Eng Mat* 61-62:513-516 (1991).

Nakashima T, Shimizu M and Kukizaki M, Particle control of emulsion by membrane emulsification and its application. *Adv. Drug Deliv. Rev.* 45:47-56 (2000).

Nuisin R, Ma GH, Omi S and Kiatkamjornwong S, Dependence of morphological changes of polymer particles on hydrophobic/hydrophilic additives. *J. Appl. Polym. Sci.* 77:1013-1028 (2000).

Okochi H and Nakano M, Comparative study of two preparation methods of w/o/w emulsions: stirring and membrane emulsification. *Chem. Pharm. Bull.* 45:1323-1326 (1997).

Omi S, Kaneko K, Nakayama A, Katami K, Taguchi T, Iso M, Nagai M and Ma GH, Application of porous microspheres prepared by SPG (Shirasu Porous Glass) emulsification as immobilizing carriers of glucoamylase (GluA). *J. Appl. Polym. Sci.* 65:2655-2664 (1997).

Omi S, Kanetaka A, Shimamori Y, Supsakulchai A, Nagai M and Ma GH, Magnetite (Fe_3O_4) microcapsules prepared using a glass membrane and solvent removal. *J. Microencaspulation* 18:749-765 (2001).

Omi S, Katami K, Taguchi T, Kaneko K and Iso M, Synthesis and applications of porous SPG (Shirasu Porous Glass) microspheres. *Macromol. Symp.* 92:309-320 (1995a).

Omi S, Katami K, Taguchi T, Kaneko K and Iso M, Synthesis of uniform PMMA microspheres employing modified SPG (Shirasu Porous Glass) emulsification technique. *J. Appl. Polym. Sci.* 57:1013-1024 (1995b).

Omi S, Katami K, Yamamoto A and Iso M, Synthesis of polymeric microspheres employing SPG emulsification technique. *J Appl Polym Sci* 51:1-11 (1994).

Omi S, Ma GH and Nagai M, Membrane emulsification -a versatile tool for the synthesis of polymeric microspheres. *Macromol Symp* 151:319-330 (2000).

Omi S, Matsuda A, Imamura K, Nagai M and Ma GH, Synthesis of monodisperse polymeric microspheres including polyimide prepolymer by using SPG emulsification technique. *Colloids Surf. A* 153:373-381 (1999).

Omi S, Preparation of monodisperse microspheres using the Shirasu porous glass emulsification technique. *Colloids Surf. A* 109:97-107 (1996).

Omi S, Senba T, Nagai M and Ma GH, Morphology development of 10-μm scale polymer particles prepared by SPG emulsification and suspension polymerization. *J. Appl. Polym. Sci* 79:2200-2220 (2001).

Omi S, Taguchi T, Nagai M and Ma GH, Synthesis of 100 μm uniform porous spheres by SPG emulsification with subsequent swelling of the droplets. *J. Appl. Polym. Sci.* 63:931-942 (1997).

Park SB, Jeon YJ, Haam S, Park HY and Kim WS, Preparation of chitosan microspheres using membrane emulsification and its size modelling. *J. Microencapsulation* 21:539-552 (2004).

Park SH, Yamaguchi T and Nakao SI, Transport mechanism of deformable droplets in microfiltration of emulsions. *Chem. Eng. Sci.* 56:3539-3548 (2001).

Peng SJ and Williams RA, Controlled production of emulsions using a crossflow membrane. *Chem. Eng. Res. Des.* 76:894-901 (1998).

Peng SJ and Williams RA, Controlled production of emulsions using a crossflow membrane. *Part Part Syst Charact* 15:21-25 (1998).

Qu H, Gong F, Ma G and Su Z, Preparation and characterization of large porous poly(HEMA-co-EDMA) microspheres with narrow size distribution by modified membrane emulsification method. *Applied Polymer Sci.* 105:1632-1641 (2007).

Rayner M and Trägårdh G, Membrane emulsification modelling: how can we get from characterisation to design? *Desalination* 145:165-172 (2002).

Rayner M, Trägårdh G, Trägårdh C and Dejmek P, Using the Surface Evolver to model droplet formation processes in membrane emulsification. *J. Colloid Interface Sci.* 279:175-185 (2004).

Rayner M, Trägårdh G and Trägårdh C, The impact of mass transfer and interfacial expansion rate on droplet size in membrane emulsification processes, *Colloid Surfaces A* 266:1-17 (2005).

Ribeiro AJ, Silva C, Ferreira D and Veiga F, Chitosan-reinforced alginate microspheres obtained through the emulsification/internal gelation technique. *Eur. J. Pharm. Sci.* 25:31-40 (2005).

Ribeiro HS, Rico LG, Badolato GG and Schubert H, Production of O/W emulsions containing astaxanthin by repeated premix membrane emulsification. *J. Food Sci.* 70: E117-E123 (2005).

Schadler V and Windhab EJ, Continuous membrane emulsification by using a membrane system with controlled pore distance. *Desalination* 189:130-135 (2006).

Scherze I, Knöfel R and Muschiolik G, Automated image analysis as a control tool for multiple emulsions, *Food Hydrocolloids* 19:617-624 (2005).

Scherze I, Marzilger K and Muschiolik G, Emulsification using micro porous glass (MPG): surface behaviour of milk proteins. *Colloids Surf. B* 12:213-221 (1999).

Schröder V and Schubert H, Influence of emulsifier and pore size on membrane emulsification. *Spec Publ -R Soc Chem* 227:70-80 (1999).

Schröder V and Schubert H, Production of emulsions using microporous, ceramic membranes. *Colloids Surf. A* 152:103-109 (1999).

Schröder V, Behrend O and Schubert H, Effect of dynamic interfacial tension on the emulsification process using microporous, ceramic membranes. *J. Colloid Interface Sci.* 202:334-340 (1998).

Sheibat-Othman N, Burne T, Charcosset C and Fessi H, Preparation of pH-sensitive particles by membrane emulsification. *Colloids Surf. A* (2007) in press.

Shiga K, Muramatsu N and Kondo T, Preparation of poly(D,L-lactide) and copoly(lactide-glycolide) microspheres of uniform size. *J. Pharm. Pharmacol.* 48:891-895 (1996).

Song SH, Cho YH and Park J, Microencapsulation of Lactobacillus casei YIT 9018 using a microporous glass membrane emulsification system. *J. Food Sci.* 68:195-200 (2003).

Sotoyama K, Asano Y, Ihara K, Takahashi K and Doi K, Water/Oil emulsions prepared by the membrane emulsification method and their stability. *J. Food Sci.* 64:211-215 (1999).

Stillwell MT, Holdich RG, Kosvintsev SR, Gasparini G and Cumming IW, Stirred cell membrane emulsification and factors influencing dispersion drop size and uniformity. *Ind. Eng. Chem. Res.* 46:965-972 (2007).

Supsakulchai A, Ma GH, Nagai M and Omi S, Preparation of uniform titanium dioxide (TiO₂) polystyrene-based composite particles using the glass membrane emulsification process with a subsequent suspension polymerization. *J. Microencapsulation* 20:1-18 (2003).

Supsakulchai A, Ma GH, Nagai M and Omi S, Uniform titanium dioxide (TiO₂) microcapsules prepared by glass membrane emulsification with subsequent solvent evaporation. *J. Microencapsulation* 19:425-449 (2002).

Supsakulchai A, Ma GH, Nagai M and Omi S, Uniform titanium dioxide (TiO₂) microcapsules prepared by glass membrane emulsification with subsequent solvent evaporation. *J. Microencapsulation* 19:425-449 (2002).

Suzuki K, Fujiki I and Hagura Y, Preparation of corn oil/water and water/corn oil emulsions using PTFE membranes. *Food Sci. Technol.* 4:164-167 (1998).

Tangirala R, Revanur R, Russell TP and Emrick T, Sizing nanoparticle-covered droplets by extrusion through track-etch membranes. *Langmuir* 23:965-969 (2007).

Tawonsree S, Omi S and Kiatkamjornwong S, Control of various morphological changes of poly(meth)acrylate microspheres and their swelling degrees by SPG emulsification. *J. Polym. Sci.* 38:4038-4056 (2000).

Toorisaka E, Ono H, Arimori K, Kamiya N and Goto M, Hypoglycemic effect of surfactant-coated insulin solubilized in a novel solid-in-oil-in-water (S/O/W) emulsion. *Int. J. Pharm.* 252:271-274 (2003).

van der Graaf S, Nisisako T, Schröen CGPH, van der Sman RGM and Boom RM, Lattice Boltzmann simulations of droplet formation in a T-shaped microchannel. *Langmuir* 22:4144-4152 (2006).

van der Graaf S, Schröen CGPH, van der Sman RGM and Boom RM, Influence of dynamic interfacial tension on droplet formation during membrane emulsification, *J. Colloid Interface Sci.* 277:456-463 (2004).

Van der Graaf S, Steegmans MLJ, van de Sman RGM, Schröen CGPH and Boom RM, Droplet formation in a T-shaped microchannel junction: A model system for membrane emulsification. *Colloids Surf. A* 266:106-116 (2005).

Vladisavljević G and Schubert H, Production of emulsions with a narrow droplet size distribution using a crossflow Shirasu porous glass (SPG) membrane. 3ième Congrès Mondial de l'Emulsion, 24-27 September 2002, Lyon, France.

Vladisavljević GT, Lambrich U, Nakajima M and Schubert H, Production of O/W emulsions using SPG membranes, ceramic α-aluminium oxide membranes, microfluidizer and a silicon microchannel plate- a comparative study. *Colloids Surf. A* 232:199-207 (2004a).

Vladisavljević GT, Shimizu M and Nakashima T, Preparation of monodisperse multiple emulsions at high production rates by multi-stage premix membrane emulsification. *J. Membrane Sci.* 244: 97-106 (2004b).

Vladisavljević GT and Schubert H, Preparation and analysis of oil-in-water emulsions with a narrow droplet size distribution using Shirasu-porous-glass (SPG) membranes. *Desalination* 144:167-172 (2002).

Vladisavljević GT and Williams RA, Recent developments in manufacturing emulsions and particulate products using membranes. *Adv. Colloid Interface Sci.* 113:1-20 (2005).

Vladisavljevic GT, Kobayashi I, Nakajima M, Williams RA, Shimizu M and Nakashima T, Shirasu porous glass membrane emulsification: Characterization of membrane structure by high-resolution X-ray microtomography and microscopic observation of droplet formation in real time. *J. Membrane Sci.* 302:243-253 (2007).

Vladisavljevic GT, Shimizu M and Nakashima T, Permeability of hydrophilic and hydrophobic Shirasu-porous-glass (SPG) membranes to pure liquids and its microstructure. *J. Membrane Sci.* 250:69-77 (2005).

Vladisavljević GT, Shimizu M and Nakashima T, Production of multiple emulsions for drug delivery systems by repeated SPG membrane homogenization: Influence of mean pore size, interfacial tension and continuous phase viscosity. *J. Membrane Sci.* 284:373-383 (2006).

Vladisavljević GT, Tesch S and Schubert H, Preparation of water-in-oil emulsions using microporous polypropylene hollow fibers: influence of some operating parameters on droplet size distribution. *Chem. Eng. Process* 41:231-238 (2002).

Wang LY, Gu YH, Su ZG and Ma GH, Preparation and improvement of release behavior of chitosan microspheres containing insulin. *Int. J. Pharm.* 311 (2006a) 187-195

Wang LY, Gu YH, Zhou QZ, Ma GH, Wan YH and Su ZG, Preparation and characterization of uniform-sized chitosan microspheres containing insulin by membrane emulsification and a two-step solidification process. *Colloids Surfaces B* 50:126-135 (2006b).

Wang R, Zhang Y, Ma G and Su Z, Preparation of uniform poly(glycidyl methacrylate) porous microspheres by membrane emulsification – Polymerization Technology. *J. Applied Polym. Sci.* 102:5018-5027 (2006).

Williams RA, Peng SJ, Wheeler DA, Morley NC, Taylor D, Whalley M and Houldsworth DW, Controlled production of emulsions using a crossflow membrane, Part II: Industrial scale manufacture. *Chem. Eng. Res. Des.* 76:902-910 (1998).

Yamazaki N, Yuyama H, Nagai M, Ma GH and Omi S, A comparison of membrane emulsification obtained using SPG (Shirasu Porous Glass) and PTFE [poly(tetrafluoroethylene)] membranes. *J. Dispersion Sci. Technol.* 23:279-292 (2002).

Yamazaki N., Naganuma K., Nagai M., Ma G.-H., Omi S., Preparation of w/o (water-in-oil) emulsions using a PTFE (polytetrafluoroethylene) membrane- A new emulsification device, *J. Dispersion Sci. Technol.* 24:249-257 (2003).

Yanagishita T, Tomabechi Y, Nishio K and Masuda H, Preparation of monodisperse SiO_2 nanoparticles by membrane emulsification using ideally ordered anodic porous alumina. *Langmuir* 20:554-555 (2004).

Yasuno M, Nakajima M, Iwamoto S, Maruyama T, Sugiura S, Kobayashi I, Shono A and Satoh K, Visualization and characterization of SPG membrane emulsification. *J. Membr Sci* 210:29-37 (2002).

Yoshizawa H, Maruta M, Ikeda S, Hatate Y and Kitamura Y, Preparation and pore-size control of hydrophilic monodispersed polymer microspheres for size-exclusive separation of biomolecules by the SPG membrane emulsification technique. *Colloid Polym. Sci.* 282:965-971 (2004).

Yoshizawa H, Ohta H, Maruta M, Uemura Y, Ijichi K and Hatate Y, Novel procedure for monodispersed polymeric microspheres with high electrifying additive content by particle-shrinking method via SPG membrane emulsification. *J. Chem. Eng. Jpn* 29:1027-1029 (1996).

You JO, Park SB, Park HY, Haam S, Chung CH and Kim WS, Preparation of regular sized Ca-alginate microspheres using membrane emulsification method. *J. Microencapuslation* 18: 521-532 (2001).

Yuan Q and Williams RA, Large scale manufacture of magnetic polymer particles using membranes and microfluidic devices. *China Particuology* 5:26-42 (2007).

Yun TH, Kim KS, Cho SH and Youm KH, Preparation of polycaprolactone microcapsules by membrane emulsification method and its drug release properties. *Membrane J.* 17:67-69 (2007).

Yuyama H, Hashimoto T, Ma G-H, Nagai M and Omi S, Mechanism of suspension polymerization of uniform monomer droplets prepared by glass membrane (Shirasu Porous Glass) emulsification technique. *J. Appl. Polym. Sci.* 78:1025-1043 (2000a).

Yuyama H, Watanabe T, Ma G-H, Nagai M and Omi S, Preparation and analysis of uniform emulsion droplets using SPG membrane emulsification technique. *Colloids Surf. A* 168:159-174 (2000b).

Yuyama H, Yamamoto K, Shirafuji K, Nagai M, Ma G-H and Omi S, Preparation of polyurethaneura (PUU) uniform spheres by SPG membrane emulsification technique. *J Appl Polym Sci* 77:2237-2245 (2000c).

Zhou QZ, Wang LY, Ma GH and Su ZG, Preparation of uniform-sized agarose beads by microporous membrane emulsification technique. *J. Colloid Interface Sci.* 311:118-127 (2007).

Zhu J and Barrow D, Analysis of droplet size during crossflow membrane emulsification using stationary and vibrating micromachined silicon nitride membranes. *J. Membrane Sci.* 261:136-144 (2005).

In: Biotechnology: Research, Technology and Applications   ISBN 978-1-60456-901-8
Editor: Felix W. Richter   © 2008 Nova Science Publishers, Inc.

Chapter 11

Fast and Robust Electrical Detection of Biomaterials

B. S. Kang[1], H. T. Wang[1], F. Ren[1] and S. J. Pearton[2]
[1]Department of Chemical Engineering
University of Florida, Gainesville, FL 32611
[2]Department of Materials Science and Engineering
University of Florida, Gainesville, FL 32611

Abstract

Chemical sensors can be used to analyze a wide variety of environmental and biological gases and liquids for properties of interest. For many applications, sensors need to be sensitive and able to detect a target analyte selectively. Sensors are used to detect analytes in a wide range of samples including environmental samples and biological samples. Several different methods including gas chromatography (GC), chemiluminescence, selected ion flow tube (SIFT), and mass spectroscopy (MS) have been used to measure different biomarkers. These methods show variable results in terms of sensitivity and they cannot satisfy all the requirements for a handheld biosensor.

The desired sensors should be small in size, inexpensive and capable of real time detection without consumable carrier gases. One promising new sensing technology utilizes AlGaN/GaN high electron mobility transistors (HEMTs) [1-16]. HEMT structures have been developed for use in microwave power amplifiers as well as gas and liquid sensors due to their high two dimensional electron gas (2DEG) mobility and saturation velocity. The conducting 2DEG channel of GaN/AlGaN HEMTs is very close to the surface and extremely sensitive. HEMT sensors can be used for detecting a variety of analytes including, for example, gases, ions, pH values, proteins, and DNA. In this chapter we review recent progress on functionalizing the surface of HEMTs for specific detection of glucose, kidney marker injury molecules, prostate cancer and other common substances of interest in the biomedical field.

Introduction

Chemical sensors can be used to analyze a wide variety of environmental and bodily gases and fluids for properties of interest [17-21]. For example, exhaled breath condensate (EBC) is widely known to be one of the most important bodily fluids that can be safely collected. In particular, the breath from deep within the lungs (alveolar gas) is in equilibrium with the blood, and therefore the concentrations of molecules present in the breath is highly correlated with those found in the blood at any given time. Analysis of molecules in exhaled breath condensate is a promising method that can provide information on the metabolic state of the human body, including certain signs of cancer, respiratory disease and liver and kidney function. Several different analysis methods including gas chromatography (GC), chemiluminescence, selected ion flow tube (SIFT), and mass spectroscopy (MS) have been used to measure different exhaled biomarkers, including hydrogen peroxide, nitrogen oxide, aldehydes, and ammonia, for example. However, these methods show greatly variable results in terms of sensitivity.

Another example of sensing application using body fluid is detecting breast cancer with saliva. The mortality rate in breast cancer patients can be reduced by increasing the frequency of screening. The overwhelming majority of patients are screened for breast cancer by mammography. This procedure involves a high cost to the patient. Moreover, the use of invasive radiation limits the frequency of screening. Recent evidence suggests that salivary testing for markers of breast cancer may be used in conjunction with mammography. Saliva based diagnostics for the protein c-erbB-2, a prognostic breast marker assayed in tissue biopsies of women diagnosed with malignant tumors, has shown tremendous potential. Soluble fragments of the c-erbB-2 oncogene and the cancer antigen 15-3 were found to be significantly higher in the saliva of women who had breast cancer than in those patients with benign tumors. Another recent study concluded that epidermal growth factor (EGF) is a promising marker in saliva for breast cancer detection.

Pilot studies indicate that the saliva test is both sensitive and reliable and is potentially useful in initial detection and follow-up screening for breast cancer. However, currently saliva samples are typically obtained from a patient in a dentist's office, then sent to a testing lab; it typically takes a few days to get the test results. To fully realize the potentials of sensors for environmental, health related, chemical and biomedical applications, technologies are needed that will enable easy, sensitive, and specific detection of chemical or biomolecules at home or elsewhere. In many applications, it would be desirable if the testing device allowed for concomitant wireless data transmission into preprogrammed destinations, such as transmitting breast cancer testing results to a doctor clinic. If inexpensive technologies that can detect and wirelessly transmit testing results for environmental, health related, chemical and biomedical applications. Early diagnosis of cancers or disease can significantly lower mortality and real-time wireless remote sensing for chemical and environmental applications may reduce disaster happening.

There is emerging interest in the use of III-N based wide bandgap semiconductors as sensitive chemical sensors, especially those made with piezoelectric materials [1-16]. GaN/AlGaN high electron mobility transistors (HEMTs) are very attractive for sensor applications since they form a high density electron sheet carrier concentration channel induced by piezoelectric polarization of the strained AlGaN layer and spontaneous polarization of the different ionic strength between the GaN and AlGaN layer. The conducting 2 dimensional electron gas (2DEG) channel of GaN/AlGaN based HEMTs is very close to the surface and extremely sensitive to the ambient, which enhances detection sensitivity. Gateless HEMT structures have demonstrated the ability to distinguish liquids with different polarities and to quantitatively measure pH over a broad range. The sensing mechanism for chemical adsorbates in piezoelectric materials originates from compensation of the polarization induced bound surface charge by interaction with the polar molecules in tested liquids. In gateless alumimum gallium nitride / gallium nitride (AlGaN /GaN) heterostructure transistors, the native oxide on the nitride surface is responsible for the pH sensitivity of the response of to electrolyte solutions. Bu coating a thin Sc_2O_3 layer on the gate sensing area, more sensitive and reproducible pH sensing was achieved [15]. The pH response of oxide/nitride interface has been modeled in terms of formation of hydroxyl groups that lead to a pH dependent net surface change with a resulting change in voltage drop at the semiconductor/liquid interface [2].

Biologically modified field effect transistors (bioFETs), either at conventional or nano-dimensions, have the potential to directly detect biochemical interactions in aqueous solutions for a wide variety of biosensing applications. To enhance the practicality of bioFETs, the device must be sensitive to biochemical interactions on its surface, which is functionalized to probe specific biochemical interactions and must be stable in aqueous solutions across a range of pH and salt concentrations. Typically, the gate region of the device is covered with biological probes, which are used as receptor sites for the molecules of interest. The conductance of the device is changed as reaction occurs between these probes and appropriate species in solution.

GaN-based wide energy bandgap semiconductor material systems are extremely chemically stable; this stability should minimize degradation of adsorbed cells [1-5]. The bond between Ga and N is ionic and proteins can easily attach to the GaN surface. This is one of the key factors for making a sensitive biosensor with a useful lifetime. HEMT sensors have been used for detecting gases, ions, pH values, proteins, and DNA temperature with good selectivity by the modification of the surface in the gate region of the HEMT. Single-nanorod, -nanowire, -nanotube, or -nanobelt based metal or semiconductor nanostructure sensors can be poorly controlled, time-consuming, and expensive. HEMT structures have been developed for use in microwave power amplifiers as well as gas and liquid sensors due to their high 2DEG mobility and saturation velocity. A 2-dimensional electron gas (2DEG) at the interface of AlGaN/GaN heterostructures is formed through the hetero-junction of AlGaN and GaN, which have different bandgaps. The 2DEG channel is connected to an Ohmic-type source and drain contacts. The source-drain current is modulated by a third contact, a Schottky-type gate, on the top of the 2DEG channel. For sensing applications, the third contact is affected by the sensing environment, i.e. the sensing targets changes the charges on the gate region and behave as a gate. When charged analytes accumulate on the gate area, these charges form a bias and alter the 2DEG resistance. This electrical detection technique is simple, fast, and convenient. The detecting signal from the gate is amplified through the drain-source current and makes this sensor to be very sensitive for sensor applications. The electric signal also can be easily quantified, recorded and transmit, unlike the fluorescence detection method needs human inspection and difficult to tO precisely quantified and transmitted.

One drawback of HEMT sensors is a lack of selectivity to different analytes due to the chemical inertness of the HEMT surface. This can be solved by surface modification with detecting receptors. Sensor devices of the present disclosure can be used with a variety of fluids having environmental and bodily origins, including saliva, urine, blood, and breath, for example. For use with exhaled breath, the device may include a AlGaN/GaN HEMT bonded on a thermal electric cooling device, which assists in condensing exhaled breath samples. The thermal electric cooling device cools the HEMT; water vapor and some volatile organic compounds condense on the HEMT surface. The ions in the condensate change the surface charge on the HEMT, thus changing the current flowing in the device at fixed applied bias voltage. An exhaled breath condensate (EBC) biosensor is small so that it can be handheld, low in cost, and capable of real time detection without consumable carrier gases.

AlGaN/GaN HEMTs for Prostate Specific Antigen Detection

This section discusses AlGaN/GaN HEMTs as health related sensors for Prostate Specific Antigen (PSA) detection [7]. The PSA was specifically recognized through PSA antibody, anchored to the gate area in the form of carboxylate succinimdyl ester. Figure 1 shows a schematic of the AlGaN/GaN HEMT PSA sensor.

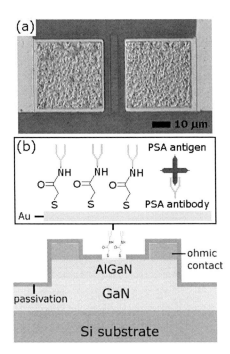

Figure 1. (a) Plan view photomicrograph of a completed device with a 5-nm Au film in the gate region. (b) Schematic of AlGaN/GaN HEMT. The Au-coated gate area was functionalized with PSA antibody on thioglycolic acid.

The HEMTs were first functionalized with thioglycolic acid through Au-S bonding to the Au surface in the gate area. The carboxylic acid functional group, COOH, of the thioglycolic acid molecules available for further chemical linking of other functional groups. For the PSA detection, the carboxylic acid functional group of the thioglycolic acid was also reacted with N, N'-dicyclohexylcarbodiimide and N-hydroxysuccinimide.

These functionalization steps resulted in the formation of succinimidyl ester groups on the gate area of AlGaN/GaN HEMT, which can react with PSA antibody. XPS and electrical measurements confirm a high surface coverage and Au-S bonding formation of the GaN surface, as shown in Figure 2.

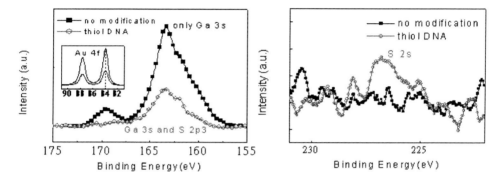

Figure 2. High resolution XPS before and after thiol-modification of Au coated GaN samples. (left) Ga 3s and S 2p3 peaks and (right) S 2s peak. Peak binding energies were referenced to the Au 4f peak at 84.0 eV (inset of the left plot).

The Ga 3S peak was at 164 eV, which was observed in the reference sample. The S 2p3 was identified at 162 eV as a shoulder of the main peak for the immobilized sample, which was consistent with peak for thiolated S-Au. The S 2s peak at 227 eV was also observed for the immobilized sample, which was shifted around 3 eV towards a lower binding energy. Figure 3 shows the real time PSA detection in PBS buffer solution through the source and drain current change with constant drain bias at 500 mV. We investigated a wide range of concentrations of PSA from to 1 µg/ml to 10 pg/ml, which is lower than the cut-off value of clinical PSA measurement requirements recently reported at 2.5 ng/ml (Healy *et al.*, 2007). No current change can be detected with both the addition of buffer solution at around 100 sec and nonspecific bovine serum albumin (BSA) around 200 sec, showing the specificity and stability of the device. By sharp contrast, the HEMT drain current showed a rapid response in less than 5 seconds when target 10 ng/ml PSA was added to the HEMT surface. The abrupt current change due to the exposure of PSA in a buffer solution could be stabilized after the PSA diffused into the buffer solution. Further real time tests to quantify the detection limit of PSA were carried out. Three different concentrations of the exposed target PSA in a buffer solution were detected, from 10 pg/ml to 1 ng/ml. In the case of 1 ng/ml, the amplitude of current change for the device exposed to PSA in a buffer solution was about 3%. The clear current decrease of 64 nA at 10 pg/ml of PSA also indicated that the detection limit could be lowered up to a few pg/ml, showing the promise of this portable electronic biological sensor for PSA screening.

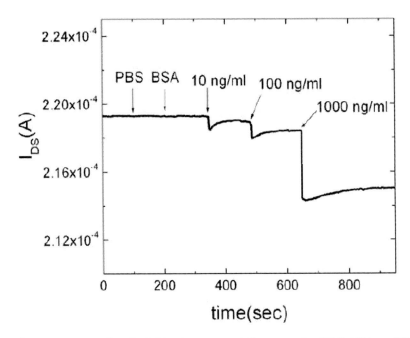

Figure 3. Drain current versus time for PSA when sequentially exposed to PBS, BSA, and PSA.

pH Sensing Using AlGaN/GaN HEMTs

Ungated AlGaN/ GaN High Electron Mobility Transistors exhibit large changes in current upon exposing the gate region to polar liquids [2,10,15]. The polar nature of the electrolyte leads to a change of surface charges, producing a change in surface potential at the semiconductor/liquid interface. The charged or polar molecules either screen or enhance the surface positive charges, which are induced by polarizations in the HEMT. These surface positive charges are produced as the counter charges of the electrons in the two-dimensional electron gas (2DEG) channel of the AlGaN/GaN HEMT, which are induced by piezoelectric and spontaneous polarization effects. Any changes in the ambient of the AlGaN/GaN HEMT affects the surface charges. These changes in the surface charge are transduced into a change in the concentration of the 2DEG.

However, ungated AlGaN/GaN HEMT based sensors have poor reproducibility, due to the poor quality of GaN native oxides. Molecular Beam Epitaxy deposited Sc_2O_3 thin films have been successfully used as the passivation layer for the HEMTs to reduce the drain current collapse and the gate oxide for AlGaN/GaN metal oxide semiconductor (MOS)HEMTs to reduce the gate leakage current and increase the gate breakdown voltage. For the pH sensing, we used these same 10 nm Sc_2O_3 films deposited on the gate region of the HEMTs.

Figure 4 shows the drain current for Sc_2O_3 capped HEMT and AlGaN native oxide HEMT at a bias of 0.25V as a function of time with the gate region exposed for 150 sec to a series of solutions whose pH was varied from 3-10. The response of the Sc_2O_3 capped HEMT is very reproducible, and the change in current was 37 μA/pH.

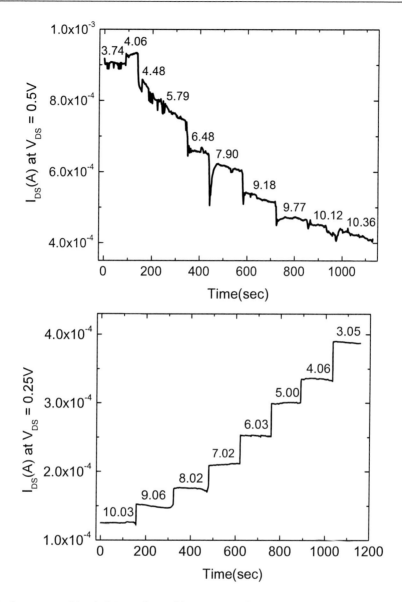

Figure 4. Drain current of (top) GaN native oxide HEMT and Sc_2O_3 capped HEMT(bottom) as a function of time from HEMTs with the gate region exposed for 150 sec to a series of solutions whose pH was varied from 3-10.

The HEMTs show stable operation with a resolution of ~0.1 pH over the entire pH range, showing the remarkable sensitivity of the HEMT to relatively small changes in concentration of the liquid. By comparison, devices with the native oxide in the gate region showed a higher sensitivity of ~70 μA/pH but a much poorer resolution of ~0.4 pH and evidence of delays in response of 10-15 secs. The GaN native oxide based HEMT may result from deep traps at the interface between the semiconductor and native oxide, whose density is much higher than at the Sc_2O_3-nitride interface. The pH range of interest for human blood is around 7-8. Figure 5 shows the drain current change in the HEMTs with Sc_2O_3 cap at a bias of 0.25V for different pH values from and to 8. Note that the resolution of the measurement is <0.1 pH.

Our results show that using a higher quality oxide encapsulation is very effective to improve the resolution of AlGaN/GaN HEMT pH sensing.

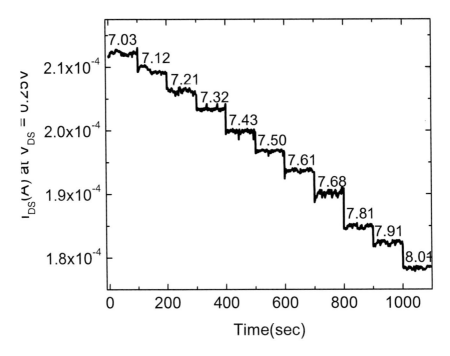

Figure 5. Change in current in gateless HEMT at fixed source-drain bias of 0.25 V with pH from 7-8.

Glucose Detection with a Novel Device Employing Integrated ZnO Nanorods on AlGaN/GaN HEMT Sensors

We recently developed a novel technique of selective-area growth of ZnO nanorods on the surfaces of other semiconductors or glass. Arrays of different patterns were easily fabricated with conventional photo-resist for masking. The resist patterned substrate was spin coated with ZnO nanocrystals used as seed materials. ZnO nanorods were grown in solution of 20 mM zinc acetate hexahydrate $(Zn(NO_3)_2 \cdot 6H_2O)$ and 20 mM hexamethylenetriamine $(C_6H_{12}N_4)$ at $95^\circ C$. The concentration of reactants, pH and temperature were carefully controlled in a flask with polypropylene autoclavable cap for 3 hour growth. Subsequently, the substrate was removed from solution, thoroughly rinsed with deioinized water to remove any residual salts and dried in air at the room temperature. After nanorod growth, negative PR was removed with standard photoresist remover in a warm bath at $60^\circ C$ for 30 minutes. Figure 6a shows SEM image for the various pattened ZnO nanorods. Figure 6b and 6c shows enlarged view and side view of patterned nanorods. As shown in Figure 6d, a photoluminescence spectrum of patterned ZnO nanorod was measured for the patterned area and substrate only area at room temperature. The patterned ZnO nanorods show a free exciton emission at 3.24 eV.

The ZnO nanorods can be integrated with AlGaN/GaN HEMT sensors. By incorporating the nano-rods on the HEMT gate sensing area, the total sensing area increases significantly as shown in Figure 7 (left). The conventional AlGaN/GaN HEMT detects the ambient changes through the "gate sensing area". This area is defined as gate length × gate width in the regular HEMT. Although, we can increase the gate width to gain higher drain current from the transistor, the sensor detection sensitivity will be the same for HEMT with both short and longer gate width. This is due to the signal and background current proportionally increasing at the same time. Iincreasing the gate length will increase the parasitic resistance of the HEMT and the drain current decreases.

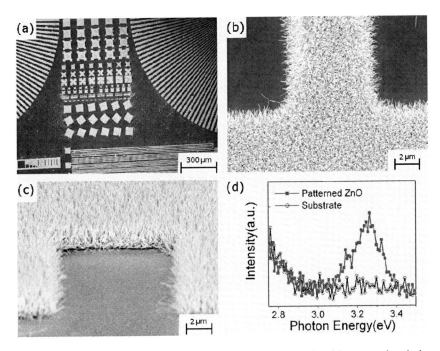

Figure 6. Field emission SEM image of patterned grown ZnO nanorods with conventional photo-resist pattern a) Low magnification and b) high magnification c) side view and d) PL spectrum of patterned grown ZnO nanorod arrays.

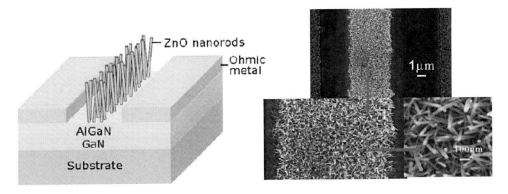

Figure 7. (right) Schematic diagram of HEMT sensor with ZnO nanorods grown on the gate sensing area. (left) SEM picture with HEMT sensor with ZnO nanorods grown on the gate sensing area. (bottom) Closer view of ZnO nanorods in the gate sensing area.

Thus, the detection sensitivity goes down. Therefore, the only way to increase the sensitivity with the same "gate dimension" is to grow 3D structures on the "gate sensing area" to increase the total sensing area with the area expansion to the third dimension. Figure 7(right) show SEM pictures of ZnO nanorods grown on the AlGaN/GaN HEMT gate sensing area. The ZnO nanorod matrix provides a microenvironment for immobilizing negatively charged GOx and retains its bioactivity, and passes charges produce during the GOx and glucose interaction to the AlGaN/GaN HEMT[8].

A record low detection limit, <1 nM, was achieved with this approach, as shown in Figure 8. The response is linear from 0.5 nM to 14.5 μM. With such low detection limit, it is possible to dilute <0.1 μ-liter of EBC in 100-200 μ-liter PBS and directly measure the glucose concentration to eliminate the effect of pH variation. Due to the fast response time and low volume of the EBC required for the measurement, the technology can be realized as handheld and real-time glucose sensing [17-21].

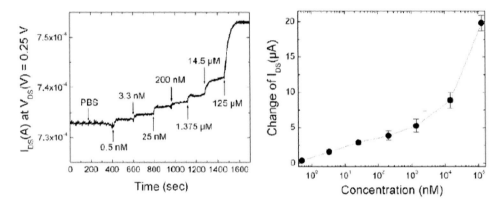

Figure 8. (right) Drain current of AlGaN/GaN HEMT sensor integrated with ZnO nanorods as a function of glucose concentrations. (left) Change of drain current as a function of glucose concentration.

Selective Detection of Hg(II) Ions from Cu(II) and Pb(II) Using AlGaN/GaN High Electron Mobility Transistors

We have also demonstrated thioglycolic acid functionalized Au-gated AlGaN/GaN HEMTs to detect mercury (II) and copper(II) ions [16]. Fast detection of less than 5 seconds was achieved for thioglycolic acid functionalized sensors and a detection limit of 10^{-7} M was achieved. The sensors had a selectivity of approximately a hundred-fold over other contaminating ions of sodium, magnesium and lead. The thioglycolic acid functionalized sensors can be easily recycled, which makes possible for realizing portable, fast response, and wireless-based heavy metal ion detectors.

The drain current of HEMT sensors decreased after exposure to Hg^{2+} ion solutions. The thioglycolic acid molecules on the Au surface align vertically with carboxylic acid functional group toward the solution. The carboxylic acid functional group of the adjacent thioglycolic acid molecules probably forms chelates (R-COO⁻(Hg^{2+})⁻OOC-R) with the Hg^{2+} ions. If the chelates are indeed forming, one would expect the charges of trapped Hg^{2+} ion in the R-COO⁻

$(Hg^{2+})^-OOC$-R will then change the polarity of the thioglycolic acid molecules. This is probably why the drain current changes in response to mercury ions. A similar type of surface functionalization was used and the detection was performed with fluorescence.

Figure 9 (left) also shows dependence of the drain current for the HEMT sensor for detecting Hg^{2+}, Cu^{2+}, and Pb^{2+} ions. A response time (less than 5 seconds) for exposure to Hg^{2+} ion solutions was obtained. The functionalized sensors could not detect Pb^{2+} ions, but could detect Cu^{2+}. However, a bare (non-thioglycolic acid functionalized) Au-gate could detect Hg^{2+} ions, but not detect the Cu^{2+} nor Pb^{2+} ions, as shown in Figure 9 (right).

With this detection difference between the bare Au-gate and the thioglycolic acid functionalized sensor, a pair of thioglycolic acid functionalized and non-functionalized sensors offers the possibility for selective detection for Hg^{2+} and Cu^{2+} ions presented in a single solution with a sensor chip containing both type of sensors. The dimension of the active area of the AlGaN/GaN HEMT sensors is less than 100 μm × 100 μm. The fabrication of both types of sensors is identical except for the thioglycolic acid functionalized sensor, which has an additional functionalization step.

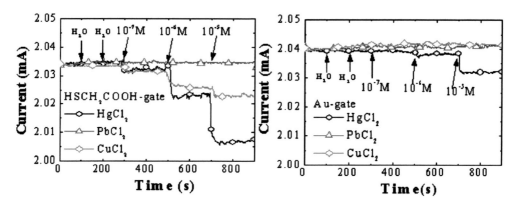

Figure 9. (left) Time dependent response of the drain current as a function of Hg^{2+}, Cu^{2+}, Pb^{2+} ion concentrations for thioglycolic acid functionalized AlGaN/GaN HEMT sensor. (right) and bare Au-gate AlGaN/GaN HEMT sensor.

This step can be accomplished with micro-inkjet system to selectively functionalize individual sensors. By integrating these two sensors on the same chip, selective detection of Hg ions over Cu and Pb ions can be achieved. The bare Au-gate and thioglycolic acid functionalized sensors also showed excellent sensing selectivity (over 100 times higher selectivity) over Na^+ and Mg^{2+} ions. Most semiconductor based chemical sensors are not reusable. However, these thioglycolic acid functionalized sensors showed good reusability.

Kidney Injury Molecules Detection

We have recently used AlGaN/GaN HEMTs for detecting an antigen involved in kidney injury [22]. This is the first clear demonstration of the potential for HEMTs in detecting antigen-antibody binding. The gate region was functionalized with a specific antibody to kidney injury molecule-1[23, 24], as shown in Figure 10.

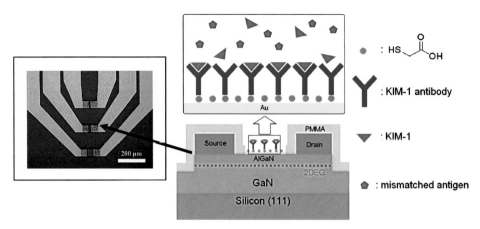

Figure 10. Schematic cross sectional view of a HEMT. The Au-coated gate area was functionalized with KIM-1 antibody on thioglycolic acid. The insert shows a plan view photomicrograph of a chip containing three devices with different dimension of gate region.

KIM-1 antigen-antibody binding was monitored through a pulsed electrical measurement. The limit of detection (LOD) was found to be 1 ng/ml with a 20×50 μm gate area, demonstrating the tremendous potential for kidney injury detection with accurate, rapid, noninvasive, and high throughput capabilities. The HEMT source-drain current showed a clear decrease as a function of KIM-1 concentration in PBS buffer. This suggests that the KIM-1 antibody modified HEMT reacts with KIM-1 and alters the surface charge on the gate region and further alters the HEMT source-drain current. Figure 11 shows the time dependent current change in KIM-1 antibody modified HEMTs upon exposure to 1 ng/ml, and then to 10 ng/ml KIM-1 in PBS buffer. The response time is around 30 seconds and the current change is proportional to KIM-1 concentration.

Finally, we could measure a clear dose-response that was statistically reliable between the current change and KIM-1 dose, as shown in Figure 11. We found similar results for the detection of biotin with immobilized streptavidin (data not shown). These two findings for detection of completely unrelated antigens (KIM-1 and biotin) suggest the potential successful use of HEMT sensors for detecting breast cancer specific biomarkers.

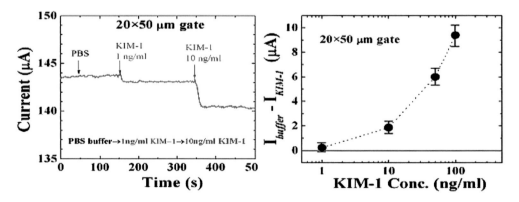

Figure 11. (right) Time dependent current signal when exposing the HEMT to 1ng/ml and 10ng/ml KIM-1 in PBS buffer. (left) Drain Current change in HEMT as a function of KIM-1 concentration.

Exhale Breath Condensate Collection

There is significant interest in developing rapid diagnosis approaches and improved sensors for determining early signs of medical problems in humans. Exhaled breath is a unique bodily fluid that can be analyzed by a wide variety of sensors and biosensors. This is a non-invasive method for taking body fluid, which will be more acceptable for patients. Typically, to collect the EBC, a cooling collector including freezing cooling tubes is used. This method is designed for off-line analysis with spectroscopic analysis. To take the advantage of the fast response (less than 1 sec) for AlGaN/GaN HEMT sensors, a real-time based EBC collector is needed. We have demonstrated the use of a thermal electric module (TEM) to cool down the AlGaN/GaN HEMT sensor and collect the EBC. TEMs utilize the effect that was discovered by Jean Peltier. The basic idea behind the Peltier effect is that whenever DC passes through the circuit of heterogeneous conductors, heat is either released or absorbed at the conductors' junctions, depending on the current polarity. The amount of heat is proportional to the current that passes through conductors, as shown in the Figure 12.

We have mounted fabricated AlGaN/GaN HEMT sensors directly on the top of a Peltier unit (TB-8-0.45-1.3 HT 232, Kryotherm), which can be cooled to precise temperatures by applying known voltages and currents to the Peltier unit. During our measurements, the hotter plate of the Peltier unit was kept at 21°C, and the colder plate was kept at 7°C by applying bias of 0.7 V and 0.2 A. The sensor takes less than 2 sec to reach thermal equilibrium with the Peltier unit. This allows the exhaled breath to immediately condense on the gate region of the HEMT sensor. It is possible to selectively collect the different chemicals present in the exhaled breath by using an array of HEMT sensors and keeping them at different temperatures with Peltier units operated at different conditions.

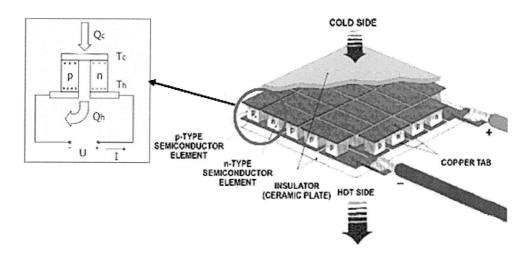

Figure 12. Schematic of a thermal electric module. The insert of the diagram shown a unit cell of the module.

For example, the less volatile chemicals will condense on the HEMT sensor set at a temperature slightly lower than room temperature. Additional sensors installed down stream of the first sensor can be set at lower temperatures to condense more volatile chemicals. Both

GaN native oxides and Sc_2O_3 capped HEMT sensors were calibrated and Sc_2O_3 capped HEMT sensors exhibited a linear change in current between pH 3-10 of 37µA/pH, compared with native oxide gated devices as shown in the inset of Figure 13. The principal component of the EBC is water vapor, which represents nearly all of the volume (>99%) of the fluid collected in the EBC. The measured current change of the exhale breath condensate can be used to determine the pH in the EBC, which is similar to the pH values of human blood.

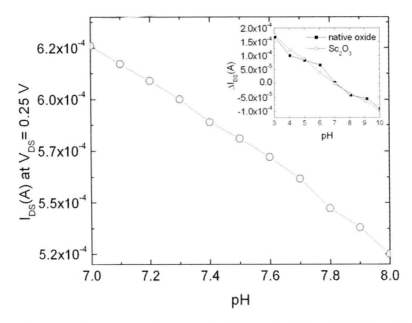

Figure 13. Drain current of Sc_2O_3 capped HEMT at fixed source-drain bias of 0.25 V as a function of pH vales. (inset) Drain current of Sc_2O_3 capped and GaN native oxide HEMT at fixed source-drain bias of 0.25 V as a function of pH value from 7 to 8.

Figure 14 shows the drain current of the Sc_2O_3 capped HEMT sensors biased at 0.5V with Sc_2O_3 as a function of time during the exposures to nitrogen, subsequently to carbon dioxide gas, then switched to a series of applications of 1 second exhaled breath pulse inputs from a human test subject. There were no changes in the drain current detected during the exposures of nitrogen or carbon dioxide to the sensor. The drain current was significantly decreased up to 340 µA upon exposure of the each exhaled breath for a time interval of 1 second. It has been reported that in aerosol or solution environments, CO_2 dissolving in water and forming H^+ and HCO_3^-, profoundly affects the pH of dilute solutions. These electrolytes are rapidly in equilibrium.

The effect of flow rate of the exhaled breath on EBC pH values was also investigated. The flow rates of exhaled breath ranged from 0.5 L/min to 3 L/min were tested, which is proportional to the intensity exhalation. Deep breath provides a higher flow rate. A similar study was conducted with pure nitrogen to eliminate the flow rates effect on the sensor sensitivity. As illustrated in Figure 15, the nitrogen did not cause any change of drain current, but the increase of exhaled breath flow rate decreased the drain current proportionally from 0.5 L/min to a saturation value of 1 L/min. For every tidal breath, the beginning portion of the exhalation is from the physiologic dead space, and the gases in this dead space do not participate in CO_2 and O_2 exchange in the lungs. Therefore, the contents in the tidal breath are diluted by the gases from the physiologic dead space. For higher flow rate exhalation, this dilution effect is less effective. Once the exhaled breath flow rate is above 1L/min, the sensor current change reaches a limit. As a result, the test subject experiences hyper ventilation and the dilution becomes insignificant.

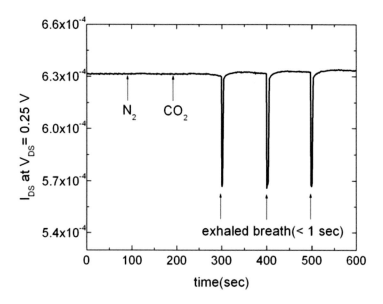

Figure 14. Changes of drain current for Sc_2O_3 capped HEMT sensor at fixed source-drain bias of 0.5 V with nitrogen or carbon dioxide gas, or a series of 1 second duration exhaled breathes at a flow rate of 1 L/min.

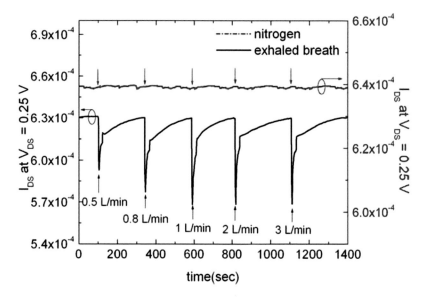

Figure 15. Changes of drain current for Sc_2O_3 capped HEMT sensor at fixed source-drain bias of 0.25 V with different flow rates of exhaled breath from tidal breath to hyper ventilation with 5 second duration for each breath.

The spike-like responses of the AlGaN/GaN HEMT sensors in Figure 14 and 15 are due to the condensed EBC gradually evaporating from the sensing area. As shown in the previous section, the volume of EBC covering the sensing area of the AlGaN/GaN HEMT is around 3 $\times$ 10^{-11} liter. Once the EBC deposits on the surface of sensing area, it can behave as an insulating layer preventing more EBC condense on the same area. On the other hand, the sensor is operated at 50 Hz and 10% duty cycle, which produces heat during the operation. The power produced by the sensor is roughly 1.4 $\times$ 10^{-5} W and the energy required to vaporize this amount of water is around and$\times$ 10^{-5} J. It only takes a few seconds for the EBC to vaporize from the sensing area. We believe that the AlGaN/GaN HEMT pH sensor is sensitive enough to detect the pH value of the EBC. An EBC collector with longer resident time of the EBC has been designed and fabricated. The current design of the EBC collector addresses the problem of spike-like output signals due to the vaporization of the EBC on the sensor. An EBC collector with a mini reservoir was designed to increase the retention time of the EBC on the sensor. The sensor array can be fabricated on a single chip to detect other biomarkers in the breath beside glucose, such as pH valued of the EBC.

Electrical Detection of Deoxyribonucleic Acid Hybridization

Biosensing of deoxyribonucleic acid (DNA) oligomer hybridization to form duplex DNA using one immobilized strand on a surface is crucial in diagnosing genetic diseases as well as sequencing the entire human genome. Several different methods have been used to detect the DNA hybridization through measuring changes in mass, optical properties or electrochemical characteristics. These approaches require pre-labeling of the DNA target, special isotopes, fluorescence tags, or re-dox indicators. These optical measurement techniques are time consuming, expensive, destructive, and may hinder portable real time detection.

To detect base pairs in DNAs rapidly with high accuracy, electronic measurements using semiconductor based field effect transistors (FETs) have been widely studied. *Baur et al.* [5,11] showed the covalent immobilization of self assembled monolayer chemicals on hydroxylated GaN and AlN surfaces and showed DNA hybridization on the functionalized GaN surface with fluorescence labeled DNA. However, obtaining direct chemical modification with hydroxyl groups on GaN surfaces with good yield and reproducibility is generally difficult due to surface roughness and unintentional contaminants. The nature of interactions between gold surfaces and DNA are well established and are used to analyze duplex DNA sequences in biotechnology.

Three synthesized label-free 15-mer oligonucleotides (Sigma Genesys) were prepared to test DNA hybridization. Oligonucleotide probes, functionalized at the 3′, ended with thiol group and connected by a tri-methylene linker, were prepared. The base sequence of this probe was 5′-AGATGATGAGAAGAA-3′-$(CH_2)_3$-thiol and the complementary target was 5′-TTCTTCTCATCATCT-3′. To test the immobilization of 3′ thiol-modified ss-DNA on the surface of the HEMT, the Au-gated area of the fabricated HEMT was exposed to buffered solutions, followed by exposure to thiol-modified DNA(~ 100 seconds) and then target DNA(~ 200 seconds). When the HEMT was dipped into thiol-modified DNA, the source-drain current of the nitride HEMT increased by around 350 µA due to the rapid adsorption of thiol functional groups onto the Au-gated area. After waiting for 100 seconds, the nitride HEMT was exposed to mismatched or matched target DNA. As shown in Figure 16, the HEMT showed a rapid reduction (~120 µA) in drain-source current and maintained fairly constant through entire testing time, suggesting that hybridization between immobilized DNA and target DNA had occurred. This explains that the current in the gate channel can be depleted by the interaction between duplex DNA molecules on the gate area. A smaller current change (~56 µA) was detected in the HEMT for mismatched target DNA because the partially modified Au-gated region was influenced by mismatched DNA molecules. This result showed that the effect of a non-specific adsorption by the target DNA was part of the overall signal [4]. The change of source-drain current due to the hybridization of mismatched or matched target DNAs was decreased to less than 30 µA (1and.8, 24 µA) after 500 seconds.

To fully immobilize the probe DNA onto the Au-gated region, the probe DNA film was exposed to a hybridization buffer containing 1 µM 3′-thiolated oligonucleotides in 0.1 M phosphate buffer, pH and(from Sigma Aldrich) for 12 hours in a test tube. After 12-hour incubation at room temperature, the samples were washed by successive immersions in 0.1 M phosphate buffer (Sigma) pH and. After probe immobilization onto the thin gold film, the samples were treated with 1 mM mercaptohexanol (MCH, Aldrich) solution in water for 2 hours. The final probe density of 2.0×10^{12} probes/cm^2 in the Au-gated region was estimated from the initial thiol DNA immobilization considering the 50% losses in the steps such as rinsing with buffer or DNA probe displacement by MCH. Hybridization solutions were prepared as 1 µM target in 1 M NaCl (Sigma) containing buffer solutions.

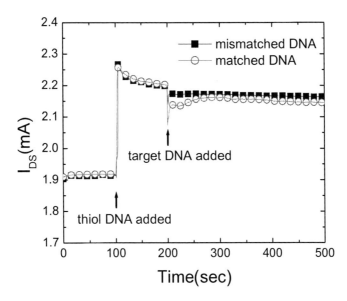

Figure 16. Change in HEMT drain-source current at $V_{DS} = 0.5$ V as a result of immobilization of single stranded DNA(15-mer 3′ thiol-modified oligonucleotide) and hybridization between immobilized DNA and matched or mismatched target DNAs.

A control experiment, using non-complementary target solution (5′-GGAGGTGACGTCCAC-3′) confirmed that nonspecific binding was completely absent. The devices with Au gate were cleaned by successive sonication in acetone and isopropanol, followed by washing thoroughly in deionized water, dried and immediately immersed in a solution of 0.1 M phosphate buffer before testing target DNAs.

A self assembled monolayer of single stranded DNA is adsorbed onto the gold gate by reaction with thiol-modified due to strong interaction between gold and the thiol-group. Figure 17 shows the time dependence of the current change in Au-gated HEMTs upon exposure to target DNA (top). The source drain current was measured at 0.5 V of drain voltage. When the HEMT dipped into the solution containing matched target DNA, after around 100 second the source-drain current abruptly decreased, as illustrated in the inset of Figure 17 (top). The drain current continuously decreased until the hybridization was completed after 20 minutes, and it was reasonable agreement with previous reports of time scales of tens of minutes. The total change of the current was around 115 µA. It was suggested that the thiol-modified probe DNA in the gate region hybridizing with target DNAs led to a double layer that alters the surface charge on the HEMT and changed the source-drain current. Totally mismatched DNA was also used as a control. It showed that nonspecific binding was absent.

The repeatability of the nitride HEMT sensor was also studied. Measurements were performed on the same DNA probe film by denaturing surface duplex by rinsing with hot water (> 60 °C) for 5 minutes. As shown in the inset of Figure 17 (top), the highly reproducible initial behavior was obtained after regenerating probe films. Through a chemical modification sequence, the Au-gated region of an AlGaN/GaN HEMT structure could be functionalized for detection of DNA hybridization. This electronic detection approach is a significant step towards integration with microfluidic channels specially designed for the detection of quick sequencing and DNA assaying by selective hybridization in DNA array chips. The device showed a great repeatability after a regenerating process [4].

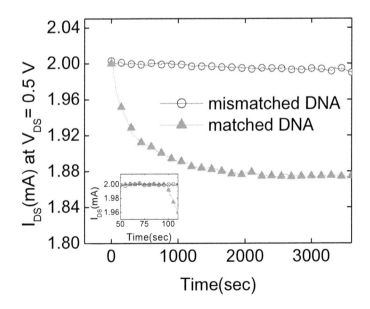

Figure 17. Change in HEMT drain-source current at V_{DS} = 0.5 V as a result of hybridization between immobilized thiol-modified DNA and matched or mismatched target DNAs.

Conclusion

The use of electrical device platforms for biological detection appears attractive from the viewpoint of response speed, sensitivity, selectivity and ability to produce arrays and integrate these with wireless communication systems that transmit the data to a central location. solid-state (semiconductor) devices for chemical, biomolecular and biomedical sensing applications applied to industries such as health care, agriculture, environmental protection, and alternative energy. The important attributes of semiconductor microelectronics, such as high performance/sensitivity, high-density integration, and mass manufacturability, will make it technologically and economically viable to widely deploy ultrasensitive, real-time, portable and inexpensive chemical and biological sensors [25-29]. To name just a few valuable applications, a nanoscale transistor with proper molecular functionalization can be used to sense protein markers for heart attacks in one's blood, breast cancer in saliva samples, prostate cancer markers, stress markers such as cortisol and the data will be transmitted wirelessly to a central location for instant diagnosis. The key aspect is that with inexpensive and simple but sensitive sensors that are networked to doctor's offices, the problem of false positive tests is largely removed, since the patient can use the sensor at home over multiple days. This provides a huge savings in both financial and emotional terms.

References

[1]   Pearton, S.J., Kang, B.S., Kim, S., Ren, F., Gila, B.P., Abernathy, C.R., Lin, J., and Chu, S.N.G.,(2004), GaN-based diodes and transistors for chemical, gas, biological and pressure sensing, *J. Phys: Condensed Matter* 16, R961-994.

[2]   Steinhoff, G., Hermann, M., Schaff, W.J., Eastmann, L.F., Stutzmann, M. and Eickhoff, M.,(2003), *p*H response of GaN surfaces and its application for *p*H-sensitive field-effect transistors, *Appl. Phys. Lett.*83, 1771-79.

[3]   Steinhoff, G., Baur, B., Wrobel, G., Ingebrandt, S., Offenhäusser, A., Dadgar, A. Krost, A., Stutzmann, M. and Eickhoff, M., (2005), Recording of cell action potentials with AlGaN/GaN field-effect transistors, *Appl. Phys. Lett.* 86, 033901.

[4]   Kang, B.S., Pearton, S.J., Chen, J.J., Ren, F.,Johnson, J.W., Therrien, R.J., Rajagopal, P., Roberts, J.C., Piner,E.L., and Linthicum, K.J.,(2006), Electrical detection of deoxyribonucleic acid hybridization with AlGaN/GaN high electron mobility transistors , *Appl. Phys. Lett.* 89,122102 .

[5]   Baur, B., Steinhoff, G., Hernando, J., Purrucker, O., Tanaka, M., Nickel, B., Stutzmann, M. and Eickhoff, M.,(2005), Chemical functionalization of GaN and AlN surfaces, *Appl. Phys. Lett.* 8and, 263901.

[6]   Kang, B.S., Wang , H.T., Tien, L.C., Ren , F.,Gila , B.P., Norton , D.P., Abernathy, C.R., Lin , J. and Pearton, S.J., (2006), *Wide Bandgap Semiconductor Nanorod and Thin Film Gas Sensors* ,Sensors 6,643.

[7]   Kang, B.S.,Wang , H.T., Lele, T.P., Tseng, Y., Ren, F., Pearton , S.J., Johnson, J., Rajagopal, P., Roberts, J.C., Piner,E.L., and Linthicum, K.J.,(2007), Prostate specific

antigen detection using AlGaN/GaN high electron mobility transistors, *Appl. Phys. Lett.* 91,222101.

[8]   Kang, B.S., Wang , H.T., Ren, F., Pearton , S.J., Morey, T.E., Dennis, D.M., Johnson, J.W., Rajagopal, P., Roberts, J.C., Piner, E.L. and Linthicum, K.L., (2007),Enzymatic glucose detection using ZnO nanorods on the gate region of AlGaN/GaN high electron mobility transistors. *Appl.Phys.Lett.*91,252103.

[9]   Kang, B.S., Heo, Y.W., Tien, L.C., Ren, F., Norton, D.P. and Pearton, S.J., (2005), UV photoresponse of single ZnO nanowires, *Appl. Phys. A* 80,497-499.

[10] Kang, B.S., Ren, F., Heo, Y.W., Tien, L.C., Norton, D.P. and Pearton, S.J., (2005), *p*H measurements with single ZnO nanorods integrated with a microchannel, *Appl. Phys. Lett.* 86, 112105.

[11] Baur, B., Howgate, J., von Ribbeck, H.G., Gawlina, Y., Bandalo, V., Steinhoff, G., Stutzmann, M., and Eickhoff, M.,(2006), Catalytic activity of enzymes immobilized on AlGaN/GaN solution gate field-effect transistors, *Appl. Phys. Lett.* 89, 183901.

[12] Kang, B.S., Kim, S., Ren, F., Gila, B.P., Abernathy, C.R., and Pearton, S.J.,(2005), AlGaN/GaN –Based Diodes and Gateless HEMTs for Gas and Chemical Sensing" *IEEE Sensors Journal* 5,677-683.

[13] Kang, B.S., Ren, F., Kang, M.C., Lofton, M., Tan, W.T., Pearton, S.J., Dabiran, A., Osinsky,A., and Chow, P.P., (2005), Electrical detection of immobilized proteins with ungated AlGaN/GaN high-electron-mobility Transistors, *Appl.Phys.Lett.*87,023508.

[14] Kang, B.S., Ren, F., Kang, M.C., Lofton, M., Tan, W.T., Pearton, S.J., Dabiran, A., Osinsky,A., and Chow, P.P.,(2005), Detection of halide ions with AlGaN/GaN high electron mobility transistors , *Appl. Phys. Lett.* 86 173502.

[15] Kang, B.S., Wang, H.T., Ren, F., Gila, B.P., Abernathy, C.R., Pearton, S.J., Johnson, J.W., Rajagopal, P., Roberts, J.C., Piner, E.L. and Linthicum, K.J.,(2007), *p*H sensor using AlGaN/GaN high electron mobility transistors with Sc$_2$O$_3$ in the gate region , *Appl. Phys. Lett.* 91, 012110.

[16] Wang, H.T., Kang, B.S., Chancellor, T., Lele, T.P. Tseng, Y. Ren, F. Pearton, S.J., Johnson, J.W., Rajagopal, P., Roberts, J.C., Piner, E.L. and Linthicum, K.J., (2007),Fast electrical detection of Hg(II) ions with AlGaN/GaN high electron mobility transistors, *Appl. Phys. Lett.* 91, 042114.

[17] Park, S., Boo, H. and Chung, T.D., (2006), Electrochemical non-enzymatic glucose sensors, *Anal. Chi. Act.*, 556, 46-54.

[18] Jung, S.K., Chae, Y.R., Yoon, J.M., Cho, B.W., Ryu, K.G., J. Microbiol. Biotechnol. , (2005), Immobilization of Glucose Oxidase on Multi-Wall Carbon Nanotubes for *Biofuel Cell Applications*, 15(2), 234-238.

[19] Burlingame, A.L., Boyd, R.K. and Gaskell, S.J.,(1996), Mass Spectrometry, *Anal. Chem.* 68, 599-652.

[20] Jackson, K.W. and Chen, G.,(1996), Atomic Absorption, Atomic Emission, and Flame Emission Spectrometry, *Anal. Chem.* 68, 231-256.

[21] Anderson, J.L., Bowden, E.F., Pickup, P.G., (1996), Dynamic Electrochemistry: Methodology and Application, *Anal. Chem.* 68, 379-444.

[22] Wang, H.T., Kang, B.S., Ren, F. Pearton, S.J., Johnson, J.W., Rajagopal, P., Roberts, J.C., Piner, E.L. and Linthicum, K.J., (2007), Electrical Detection of Kidney Injury

Molecule-1 With AlGaN/GaN High Electron Mobility Transistors , *Appl. Phys. Lett.* 91, 222101.

[23] Vaidya, V.S., Ramirez, R., Ichimura, T., Bobadilla, N.A., and Bonventre, J.V.,(2006), Urinary kidney injury molecule-1: a sensitive quantitative biomarker for early detection of kidney tubular injury, *Am. J. Physiol. Renal. Physiol.* 290, F517-529.

[24] Vaidya, V.S. and Bonventre, J.V., (2006), Mechanistic biomarkers for cytotoxic acute kidney injury, *Expert Opin. Drug Metab.* Toxicol. 2, 6977-713.

[25] Sato, M.and Webster, T. J., (2004), Nanobiotechnology: implications for the future of nanotechnology in orthopedic applications. *Expert Rev. Med. Devices* 1, (1), 105-14.

[26] Goldberg, M., Langer,R., Jia, X.,(2007), Nanostructured materials for applications in drug delivery and tissue engineering, *J. Biomater. Sci. Polymer Edn* 18(3), 241-268.

[27] Barone, P. W., Parker, R. S., Strano, M. S., (2005) In vivo fluorescence detection of glucose using a single-walled carbon nanotube optical sensor: design, fluoro-phore properties, advantages, and disadvantages, *Anal Chem,* 77, (23), 7 556-62.

[28] Stern E., Klemic J. F, Routenberg, D.A, Wyrembak P.N, (2007), Label-free immunodetection with CMOS-compatible semiconducting nanowires, *Nature* 445, 519-522.

[29] Patolsky F., Timko B.P., Yu G., Fang Y., Greytak A.B., Zheng G., Lieber C.M.(2006). Detection, Stimulation, and Inhibition of Neuronal Signals with High-Density Nanowire Transistor Arrays. *Science,* 313, 1100-1104.

In: Biotechnology: Research, Technology and Applications   ISBN 978-1-60456-901-8
Editor: Felix W. Richter   © 2008 Nova Science Publishers, Inc.

Chapter 12

The Principle of Artificial Periodic Stimulation: A Novel Idea for Bioreactor Design

Chen Hongzhang and Li Hongqiang*

National Key Laboratory of Biochemical Engineering
Institute of Process Engineering, Chinese Academy of Sciences
Beijing 100190, PR China

Abstract

In order to meet the needs of production in industry and improve the yield and productivity of the industrial strains, the metabolism of microbial cells need to be regulated and controlled. Today, the main regulations are carried out at the DNA level through genetic engineering. Many successful samples have appeared in the past 35 years, but genetic engineering does not satisfy the needs of industry without an external means of regulation. The artificial periodic stimulation theory was proposed based on the maladjustment of the classical chemical reactor theory on bioreactor design and operation. The theory emphasized that the artificial periodic stimulations enhanced the bio-reaction and mass transfer at the cell level. Toward the bio-reaction system the cell level, the periodic input regulation is a generally optimal means of control. Following this theory, the airlift loop bioreactor was built successfully and a novel solid-state fermentation bioreactor: "Gas Double-dynamic Solid-State Fermentation Bioreactor" (GDSFB) was invented. The two types of bioreactors got outstanding practice results. During the further study of GDSFB, it was validated that periodic pressure oscillation reflected as a strong normal force and weak tangential force. This was a better environment for mass and heat transfers in the bio-system. The artificial periodic stimulation can increase the respiratory intensity greatly and affect the key enzyme activity of sugar metabolism, the quantity and abundance of protein expression. The research on solid-state fermentation also discovered the unique respiratory quotient

---

* Corresponding author. Tel: +86-10-82627067, Fax: +86-10-82627071. E-mail address: hzchen@home.ipe.ac.cn
(Chen Hongzhang)

periodic phenomena not found in submerged fermentation. Many types of research are being carried out on perfecting and validating the artificial periodic stimulation theory. This theory will play a more important role in bioprocess regulation.

Keyword: artificial periodic stimulation, solid-state fermentation, bioreactor, periodic oscillation, metabolism regulation.

1. Introduction

The most ideal bioreactor should be the living body itself. For thousands of years, humans had been using the largess from nature but with many restrictions. Modern biotechnology is required to product largescale biological products in a simpler and faster manner, and it is the study object of biochemical engineering [1]. In order to satisfy this requirement, the core question to be solved was building an artificial bioreactor with perfect performance [2]. Today, the simplest artificial bioreactor was the mechanical stirring tank that had become the base of the modern fermentation industry. However, in this bioreactor, the shearing force caused by mechanical stirring can lead to damage to cells. It is also difficult to use this process on a grand scale. So building a new, more ideal type of artificial bioreactor became one of the most important study directions of modern biotechnology industry research.

2. Birth of the Novel Bioreactor Design

Charles Darwin established Evolutionism in 1869 [3]; Rudolph Clausius founded the principle of entropy increase of chemical thermodynamics in 1850. The profound conflict between Evolutionism and the principle of entropy increase puzzled scientists for a long time in history. One of the founders of quantum mechanics---Erwin Schrödinger-published a booklet named "What is Life? The physical aspects of the living cell" [4] in 1944. In the book he gave a cue first: life is an open system. Because negentropy (food) is taken in continually, bioreaction becomes a process with entropy decreasing. Until 1969, Ilya Prigogine advanced his Dissipative Structural Theory [5] and the conflict was settled in theory. An open system apart from the thermodynamics equilibrium state will form an ordered structure under nonlinear feedback, namely a self-organization phenomenon. Self-organization is a process with entropy decreasing, too. Entropy and information are the indication of matter order. Entropy and information have consanguineous relations: if information transfer stops, entropy will increase spontaneously. Dissipative Structural Theory opened a window for understanding the nature of life, for this Ilya Prigogine won the Nobel Prize in Chemistry in 1976. In the 1970s German scientists Hermann Haken and Manfred Eigen advanced synergetics [6] and the Hypercycle theory [7] separately. These theories promoted self-organization into the quantity level in concept [8]. In the 1980s scientists found the chaos characteristic of nonlinear dynamical system which made people understand the oneness between "spontaneity order" and "spontaneity disorder" in the process mechanism and the double transition relation between simpleness and complexity. These theories are called by a joint name "Complexity Science" [9]. Life process is a representative complexity science problem; it can't be linear approximated; linearly; means death. Dissipative nonlinear dynamical system has three types of math solutions: fixed point solution, periodic solution (limit cycle) and chaotic attractor. Fixed point solution is very special, periodic and chaotic solutions are the general cases. The relation of chaos and information is causality; chaos is the source of information. Because of this reason, in the past, people always said "life lies in movements", whereas today we can say "life lies in oscillation," with more accuracy.

In the different levels of a living body, from molecule and cell to apparatus, individual, colony and society, types of space-time self-organization, such as a "period" which may range from several seconds to a few days or even many years, can be observed. Such as the concentration of NADH in glycolysis surging between the range of $10^{-5} \sim 10^{-3}$ M. the Goldbeter-Lefever model had a numerical value simulation of this process [10]. Metabolism oscillation had an important effect on self-adapting and self-adjusting of a bioreaction. It is generally believed that periodic oscillation and information transfer in cells have consanguineous relations.

The essential difference between bioreaction and chemical reaction lies in that a living body can feel outside environment changes sensitively and self-adjust and self-adapt to the environment, which is the natural property of a nonlinear complex system. Many normal activities must appear in the periodic manner such as ingestion, breath, rest and exercise and so on. From the view of complex system engineering control theory, fixed point optimization control theory can show its effect on a linear system only, without effect on a nonlinear system. The control strategy of periodic import for nonlinear reaction dynamic system optimization control had received more and more regards in chemical engineering [11, 12]. For bioreaction systems at the cell level, periodic import adjusting may be a general optimization control measure [13, 14]. This periodic stimulation can be temperature, concentration, pressure, light, electricity, magnetism or sound.

From today's textbooks [15] or study papers [16] on biochemical engineering, it is not difficult to find that the study content, theory and method of a bioreactor were almost as simple as replantation of a chemical reactor's. Moreover, even most of the research was localized to dissolved oxygen (gas/liquid mass transfer) [17]. However, living body reactions perform as a nonlinear complex network system with mass, information and energy transfers at the same time, had a visible level difference with chemical reactions that perform at a molecular level. So, replanting chemical reactor theory simply was not wise. To solve the problem, a concept update was necessary, and luckily enough, the rapid development of life sciences had provided many new concepts, new methods and new theories such as the Dissipative Structural Theory [5], and chaos dynamical system theory [18], for concept updating. Bioreaction and chemical reaction (including biochemical reaction) had a natural difference at the complexity level. Bioreaction is an entropy decreasing process while a chemical reaction is an entropy increasing process, their difference is just as the difference of life and death. Bioreactor design directed by "three transfers and one reaction" theory of chemical reaction engineering must ignore the difference between life and death, so it cannot seize the key problem in any bioreactor design. Based on the above analyzes, the novel principle of artificial periodic stimulation intensification of bioreaction rate and mass transfer rate was advanced. According to the principle, new airlift double loop bioreactors [19] and large-scale solid-state fermentation (SSF) bioreactors [20, 21] have been designed and obtained very good results in industrial applications. So the principle had a general significance. This theory will play a more important role in bioprocess regulation.

3. The Principle of Artificial Periodic Stimulation A Novel Idea for Bioreactor Design

Chen and Li [22-24] summarized the application of the artificial periodic stimulation principle in SSF bioreactor design. Our SSF bioreactor research started in 1984 and persisted until today. All kinds of gas-solid reactors were used as SSF bioreactors; however, they had not been able to meet the request of long-term stability production. In 1991 according to the principle of artificial periodic stimulation, the structure of the SSF bioreactor was redesigned. Thus, the gas double-dynamic solid-state fermentation technology was born. The gas double-dynamic solid-state fermentation bioreactor (GDSFB) has undergone the development of laboratory scale (0.5 L), pilot scale (800 L) and industrial scale (50 and 100 m^3, Figure 1). Microbial pesticide (*Bacillus Thuringiensis* wettable powder and spore powder of *Beauveria bassiana*) and cellulase were the demonstration products. The demonstration factories built in 1998, 2001, 2005, and 2006 demonstrated many unsurpassable technical economic indicators to submerged fermentation.

In the traditional chemical industry, machinery agitation is generally used to enhance the mass and heat transfer in solid reactions. The aim of agitation is mixing the pellets, and accelerating the contact frequency between reaction pellets or with gas phase molecules. The agitation causes the molecular diffusion to change to convection diffusion of gas molecules between the substrate. But the shearing force caused by the excessive agitations will break the mycelium, and is disadvantageous to the microorganism's growth. So, gas phase movement is a better method than solid phase movement. In fact, in the traditional SSF equipment, heavy layer ventilation fermentation uses this method. But it often ventilates nonuniformly, with difficulty in scale-up, even with the help of agitation. The mechanical agitation of solid substrate is also easy to cause substrate aggregation and difficulty to achieve effective mixing in the single pellet scale. Therefore, the effect of changing molecular diffusion to convection diffusion is limited and could not achieve the goal of keeping a constant temperature, humidity and oxygen supply. This is the main reason for giving up the usual dynamic fermentation reactor completely.

Figure 1. The basic configuration of gas double-dynamic solid-state fermentation bioreactor.

The periodic air pressure change in the GDSFB is static state to the substrate while dynamic state to the gas phase. The mechanism is as follows: the gas molecules can enter small gaps; pressure oscillation causes the molecular diffusion to convection diffusion in the single pellet level easily. Thus, it achieves the temperature, humidity uniformity in small scale around mycelium and strengthens the rates of oxygen supply and CO_2 discharge. During the operation of discharge, the gas phase among pellets is inflated because of pressure decreasing. This results in less crowding for the solid pellets and expands the space for mycelium's reproductions, which is especially beneficial to mycelium growth. In this environment, the mycelium inside and outside the substrate is all plentiful. No matter if the microorganism is in the liquid or solid medium, sufficient moisture is necessary. The cells have the mass, energy and information exchanges with the peripheral liquid film through the cell membrane and then liquid film has the mass and energy exchanges with the solid substrate and air. The cell is a baggy living specimen with retractility. The air pressure oscillation can very quickly cause the many kinds of periodic normal forces. Because the liquid film is very thin, this normal force's transmission speed is quick, and the liquid film is more sensitive. Therefore, the artificial periodic stimulation has a more prominent and obvious effect in SSF than in submerged fermentation. The difference is not just percent but multiple.

GDSFB has some characteristics: mechanism simple, easy to seal, not easy to contaminate, only needs a low power circulating fan (1/10 of the same volume of a mechanical agitation fermentor agitation power) the asepsis air amount used is small (1/5 of the same volume of a mechanical agitation fermentor), and the total energy consumption of fermentation is low. In the GDSFB, the tray layer of substrate is generally 2-5 cm; the temperature and humidity of substrate are easy to control. The scale-up of GDSFB is easy, too. The temperature and humidity of substrate can be controlled by inside air circulation and air pressure pulse frequency. The other fermentation state parameters can be regulated by acid or base mist, and other nutriments. When the GDSFB is used in cellulase fermentation, 2-3 times the cellulase yield and 1/3 shorter fermentation time were obtained [23, 25]. When GDSFB was used in microbial pesticides fermentation production, 1/3 and 3/8 shorter fermentation times were obtained in *B. Thuringiensis* and *B. bassiana,* respectively [26]. When GDSFB was used to *B. bassiana* fermentation, living spores are generally more than 500×10^8/g dry medium.

3.1. The Macroscopic Effect of the Artificial Periodic Stimulation on Substrate

To understand thoroughly the fermentation intensification of the air pressure pulse, a video was used to observe the internal environmental variation due to the air pressure pulse.

When the normal substrate (steam-exploded wheat straw: wheat bran 4:1, the ratio [w/v] of solid substrate to inorganic salt solution, 1:3) was laid in the GDSFB, the major part of the substrate had not appeared in the macroscopic movement which one may perceive. Only a little extremely slight steam-exploded wheat straw fiber appears as small amplitude oscillation. However, intense air transport still can be observed throughout the substrate.

The volume of compressible objects will change obviously under the environment of the GDSFB.

Figure 2. The observation of medium motion state under GDSFB air charging process (0-0.2Mpa).

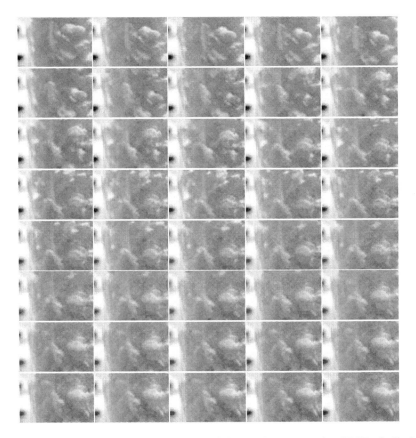

Figure 3. The observation of compressible plastic particles motion state under GDSB air discharging process (0-0.2-0 Mpa).

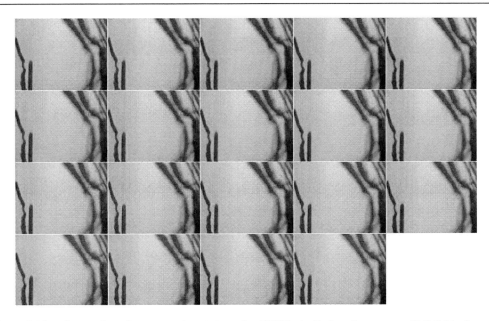

Figure 4. The observation of water motion state under GDSB air discharging process (0.2-0 Mpa).

In this part, the shape change of a bulk foam plastic was observed under the air pressure pulse. The results showed that: when air pressure increases, the volume of the foam plastic reduces; when air pressure reduces, the volume of the foam plastic is restored to the original size. The small quality and low density objects were used to test the intensity of the air pulse. These objects included ion exchange resin, pellets of foam plastic, and cotton fiber. The experimental results showed: the shear force of air pressure pulse was weak and cannot cause movement of the minimum resin pellets and thin cotton fibers; the normal force of the air pressure pulse was strong and could cause obvious compression deformation of the foam plastic pellets. The effects of air pressure pulse on liquid were tested on water. As an incompressible object, water had not experienced any change under the air pressure pulse. These phenomena can be seen in the video recorded.

These results fully display the strong normal force and weak shearing force features of air pressure oscillation.

3.2. The Effects of the Artificial Periodic Stimulation on the Quantity and Species of Protein Expression

First, *Trichoderma viride* TG-3 was cultured in static and air pressure oscillation conditions respectively. The mycelium obtained in different fermentation conditions was used for microorganism protein extraction and purification [27]. Then the mass concentration of protein, filter paper activity, endoglucanase (CMCase) and β- glucosidase were determined. From figure 5, it can be concluded that the artificial periodic stimulation has an obvious influence on protein expression quantity.

The difference of protein expression species under different fermentation conditions was compared by the SDS-PAGE analysis. The molecular weight of different protein was figured out by the electrophoresis curve of the standard protein. After 4 day-long air pressure oscillation stimulation, the microorganism cultured in the GDSFB reduced the 80400 D protein expression and increased the 28520 D protein expression.

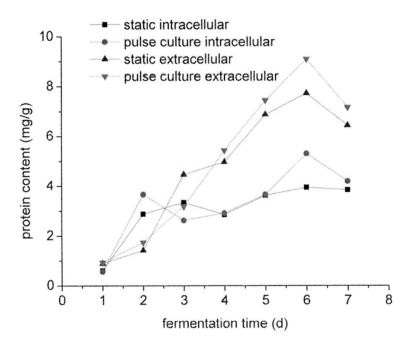

Figure 5. The effect of the artificial periodic stimulation on protein expression quantity.

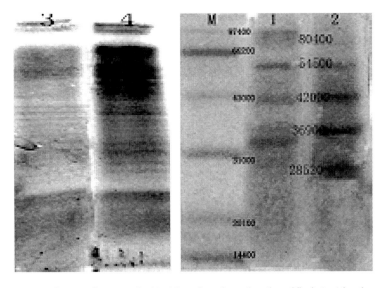

Figure 6. SDS-PAGE of extraction protein (1, 4d static cultured and purified; 2, 4d pulse cultured and purified; 3, 4d static cultured; 4, 4d pulse cultured, M, standard protein).

When the external environment changes, cells will adapt to the environment change by some means. These means include changing protein's conformation by spontaneous assembly, revolving or producing corresponding stress proteins. Heat shock proteins [28], also called stress proteins, are a group of proteins that are present in all cells in all life forms. They are induced when a cell undergoes various types of environmental stresses like heat, cold and oxygen deprivation. Under the air pressure oscillation condition, the output, enzyme activity and composition of microorganism protein have changed. Air pressure oscillation is an artificial periodic stimulation. To adapt this artificial periodic stimulation, microorganism cell protein composition changes: some protein syntheses are suppressed and others are induced. However, based on the fermentation results, the induced protein by air pressure oscillation is advantageous to microorganism growth and fermentation yield enhancement. But the detailed nature and function of these different proteins still a wait further research.

3.3. The Effect of Artificial Periodic Stimulation on Microorganism Key Enzyme Activity

The biology has the ability of responding to the outside stimulation generally. Moreover, the reaction rate of stress ability is quite quick. Take the heat shock response as an example, after being transferred to a high temperature environment, in 5 min. fungus display an obvious response. In the cell of Saccharomyces cerevisiae, the synthesis of heat shock proteins will achieve the apex in 20-30 min. The air pressure oscillation in the GDSFB is an artificial periodic stimulation. We studied the effect of short term air pressure oscillation on *T. viride* TG-3. After inoculation every 24 h, different mycelial age *T. viride* TG-3 were placed into the GDSFB for 1 h culture then used to test the effect of air pressure pulse on ATPase activity. The results (figure 7) display the decreasing trend of ATPase activity. However, the enzyme activities at the same time had no obvious differences.

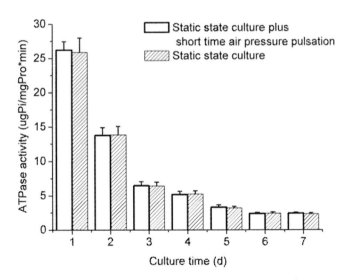

Figure 7. The effect of short term air pressure pulse on ATPase activity of T. viride TG-3, pulse parameters: pulse culture time is 1 h, pulse amplitude is 0-0.2 MPa, and pulse interval is 5 min. the data are the average values and standard deviation from three parallel measurements.

The effect of air pressure oscillation parameters such as oscillation amplitude and frequency on fermentation performance is obvious. In some ranges, increasing the oscillation amplitude and frequency is helpful in heat dissipation, and oxygen transfer then improves fermentation performance. However, this high energy consumption operation is not always better when higher. Excessive pulsation can reduce the fermentation performance instead [23, 24]. It may have to do with the air pulse influencing cell membrane structure and metabolism. The effects of different oscillation parameters on ATPase were studied. The results (figure 8) show that short term (1-2 h) air pressure pulse had few effects on ATPase. The pulse frequency also had few effects on ATPase, even a considerably high frequency [20]. Only the pulse amplitude had some effects on ATPase; high pulse amplitude reduces the ATPase activity slightly. In short, a short term pulse has limited effects on ATPase, even more violent pulse parameters than normal were adopted.

Short term air pressure oscillation has few effects on ATPase; however, air pressure oscillation culture has more notable effects on ATPase (figure 9). The effects of air pressure pulse on ATPase are complex: in the initial period of culture, air pressure pulse reduced the ATPase activity; but at the third day, ATPase activity increased and exceeded the control group. It may suggest that the air pressure pulse had disadvantageous effects on *T. viride* TG-3 at the early stage of the fermentation; however, the disadvantageous effects were adopted by *T. viride* TG-3 gradually.

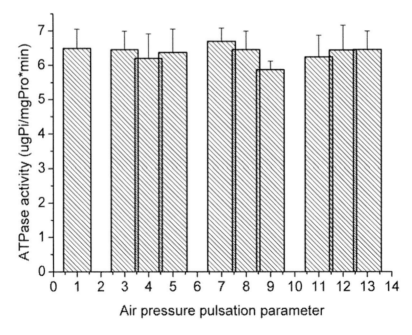

Figure 8. The effect of pulse parameters on ATPase activity of T. viride TG-3, 1 is control sample, without pulse; 3, 4, 5 with a pulse amplitude of 0-0.2 MPa, pulse interval is 5 min, pulse culture times are 1, 1.5 and 2h; 7, 8, 9 with a pulse culture time is 1h, pulse interval is 5 min, pulse amplitudes are 0-0.1, 0-0.2 and 0-0.3 MPa; 11, 12, 13 with a pulse culture time is 1 h, pulse amplitude is 0-0.2 MPa, pulse intervals are 2, 3, 5 min., all data are the average values and standard deviation from three parallel measurements.

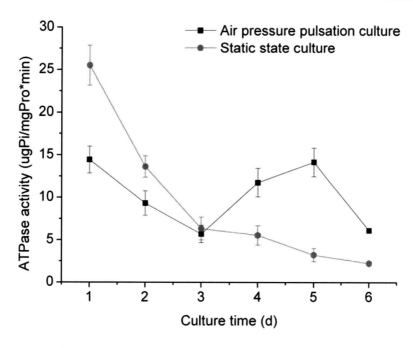

Figure 9. The effect of pulse culture on the ATPase activity of T. viride TG-3, the pulse parameters: pulse amplitude is 0-0.2 MPa, pulse interval is 20 min., all data are the average values and standard deviation from three parallel measurements.

3.4. The Effect of Artificial Periodic Stimulation on Microorganism Respiratory Intensity

We studied the effect of air pressure pulse on *Penicillium decumbens* JUA10 respiratory intensity by an exhaust analysis system coupling to the GDSFB [29]. The computer recorded the varying of CO_2 concentration synchronously. Producing CO_2 in 20 min. can be gotten by subtracting the CO_2 concentration when the low-pressure process begins from CO_2 concentration when the low-pressure process is over. The quantity of CO_2 reflects the metabolism activity of microbe in 20 min. It makes the contrast of respiration intension in static SSF and air pressure pulse SSF according to the above-mentioned idea in figure 10. Static SSF got the peak value of respiration intension in 50 periods and keeps the value a long time, but Periodic air pressure pulse SSF got the peak value in 100 periods and dropped quickly. These two kinds of curves are different completely.

The varying curves of total CO_2 metabolism in two ways were gotten by integrating CO_2 metabolism intensity curves, and the curves are shown in figure 11. Two curves are alike in the fermentation initial stage, but the metabolism activity of air pressure pulsation SSF is higher than the other fermentation way after 60 periods. The produced CO_2 (62.27) of air pressure pulsation SSF is the 3.29-folds of the static SSF (18.95). These data show artificial periodic stimulation enhanced the respiratory intensity of *P. decumbens* JUA10 greatly.

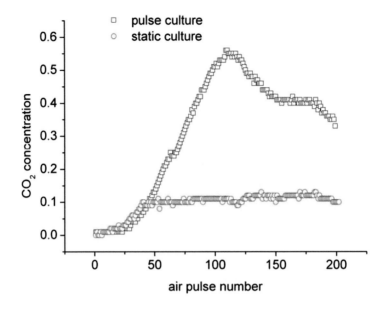

Figure 10. The respiration intensity curve of the two kinds of fermentations.

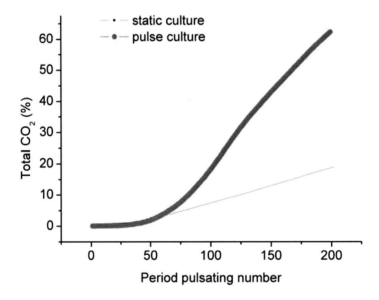

Figure 11. The total CO2 curve of the two kinds of fermentations.

3.5. The Effect of Artificial Periodic Stimulation on Respiratory Quotient

Respiratory quotient (RQ) was determined by its definition (RQ = CER/OUR) using the on-line measured OUR and CER data [29]. RQ can reflect the metabolism status of microorganism cultured. If the RQ=1 in yeast fermentation, it indicates that sugar is metabolized through aerobic metabolism pathway and the sugar is just used to produce yeast biomass and without production formation. If the RQ>1.1, it indicates that sugar is metabolized through an EMP pathway and produced ethanol. Different RQ values are obtained when the same microorganism is cultured on different substrates. In antibiotic fermentation, the RQ values are different because the biomass growth, biomass maintenance and product formation appear in a different stage [30].

The RQ oscillation phenomena were found during the study of SSF RQ. From the following two figures, the RQ curves from SSF have an essential difference from the RQ curves from submerged fermentation.

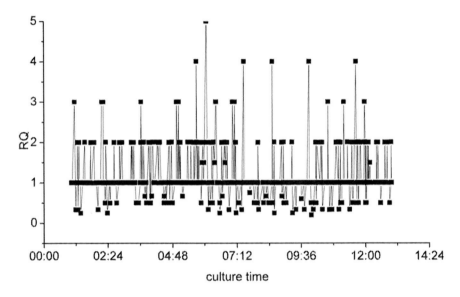

Figure 12. RQ periodic oscillation curve, 24-36 h of Penicillium decumbens JUA10 under static culture.

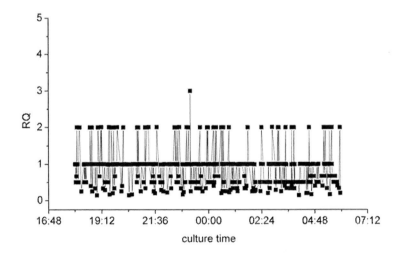

Figure 13. RQ periodic oscillation curve, 24-36 h of Penicillium decumbens JUA10 under pulse culture.

The RQ curves of submerged fermentation slow change in the scope of 0.7-1.5 [31]. The SSF especially in small-scale SSF RQ curves have bigger variation range and more complex oscillation rules. These rules are difficult to explain by the theory of submerged fermentation RQ [32]. However, these curves may contain the essential rule of microorganism RQ.

3.6. The Artificial Periodic Stimulation and Metabolic Regulation

The microorganism cell has a set of strict metabolism rules and is regulated by a complex metabolism regulation mechanism. Under the regulation of such a regulation mechanism, the microorganism can adapt the bad natural environment. The following facts embody the existence and significance of metabolism regulation: 1. when a microorganism is cultured in a synthetic medium including only sole organic compound, the synthesis speeds of macro-molecule and monomer are unisonous and a useless monomer has not been synthesized; 2. Any kind of monomer, if it can be obtained and entered from the environment, will be stopped its synthesis', moreover, the synthesis of the enzyme that synthesizes the monomer will be stopped too; 3. Microorganism only synthesizes some hydrolases whose substrates exist in the medium; 4. if there are two kinds of organic compounds in the medium, microorganism only synthesizes these enzymes related to the easily digested compound. After the easily digested compound is exhausted, the enzymes related to the hard digested compound begin to synthesize; 5. nutrient affects the growth rate of microorganism, and changes the composition of cell macro-molecules such as RNA. The regulation mechanism is absolutely necessary to the survival of the microorganism. However, it does not conform to the needs of science on industrial production strain. Scientists requesting production strain can effectively produce a specific product using specific medium. So, the artificial regulating of microorganism metabolism is needed.

The regulation of a bioprocess basically includes two aspects. One is intrinsic regulation of the strain and the other is outside environment regulation. The intrinsic regulation basically is realized through the means of genetic engineering. According to the number and complexity of gene operation, genetic engineering had developed from the second generation genetic engineering (protein engineering), the third generation genetic engineering (Metabolic engineering), and the fourth generation genetic engineering (genome engineering). Although the development is rapid, humans still have to accept these facts: the success of gene modification still depends upon luck; still could not grasp the chromosome constitution rule, even the simplest chromosome and also is very long way from the goal of artificial synthesis of life. Therefore, the regulation of microorganism metabolism has to depend on outside environment regulation to a great extent. It is to say that we must give the information to thr microorganism to make it keep the production state as long as possible. Pitifully, although the cell signal transfer is the research front and hot spot of life science, people still have not realized how to communicate with the microorganism. The have only learned several phrases possibly understood by the microorganism: temperature, pH, dissolved oxygen and nutrition concentration. The artificial periodic stimulation principle increased the possibility of having a channel used for transferring information to a microorganism and increased the means for microorganism metabolism regulation.

Conclusion

After several years development, the artificial periodic stimulation principle has obtained a series of successes in practice. The 25 m^3, 50 m^3, 75 m^3 and 100 m^3 GDSFBs built in 1998, 2001, 2005, 2006 demonstrated outstanding performance in many technical economic indicators to submerged fermentation. This principle has been primarily interpreted at the cellular and molecular level. The artificial periodic stimulation makes a better environment for mass and heat transfers on a bio-system in GDSFB. The artificial periodic stimulation can increase the respiratory intensity greatly and affect the key enzyme activity of sugar metabolism, the quantity and abundance of protein expression. The unique respiratory quotient periodic phenomena not found in submerged fermentation were discovered in SSF. These RQ curves might contain the essential rule of microorganism RQ. With this deep research, the artificial periodic stimulation principle will become more perfect and achieve more success.

References

[1]   Atkinson, B. and Mavituna, F. (1983). *Biochemical Engineering and Biotechnology Handbook:* New York: Nature Press.

[2]   Lidén, G. (2002). Understanding the bioreactor. *Bioprocess and Biosystems Engineering, 24,* 273-279.

[3]   Bowler, P. J. (1996). *Charles Darwin: The Man and His Influence:* Cambridge: Cambridge University Press.

[4]   Schrödinger, E. (1944). *What is Life?: The Physical Aspect of the Living Cell:* Cambridge: Cambridge University Press.

[5]   Prigogine, I. (1969). In M. Marois, *Theoretical physics and biology* (pp. 23-52). Amsterdam: North-Holland.

[6]   Haken, H. (1983). *Synergetics. an introduction: Springer Series in Synergetics* (3rd ed). Berlin: Springer.

[7]   Eigen, M. and Schuster, P. (1978). The Hypercycle. *Naturwissenschaften, 65,* 7-41.

[8]   Nicolis, G. and Prigogine, I. (1977). *Self-organization in nonequilibrium systems:* New York: Wiley.

[9]   Plsek, P. E. and Greenhalgh, T. (2001). Complexity science. *British Medical Journal, 323,* 625-628.

[10] Decroly, O. and Goldbeter, A. (1984). Multiple periodic regimes and final state sensitivity in a biochemical system. *Physics Letters A, 105,* 259-262.

[11] Bailey, J. E. (1974). Periodic Operation of Chemical Reactors: A Review. *Chemical Engineering Communications, 1,* 111-124.

[12] Silveston, P., Hudgins, R. R. and Renken, A. (1995). Periodic operation of catalytic reactors—introduction and overview. *Catalysis Today, 25,* 91-112.

[13] Abulesz, E. M. and Lyberatos, G. (1989). Periodic operation of a continuous culture of Baker's yeast. *Biotechnology and Bioengineering, 34,* 741-749.

[14] Constantinides, A. and Mehta, N. (1991). Periodic operation of immobilized live cell bioreactor for the production of candicidin. *Biotechnology and Bioengineering, 37,* 1010-1020.

[15] Bailey, J. E. and F., O. D. (1986). *Biochemical Engineering Fundamentals:* New York: McGraw-Hill Book Co.

[16] Menisher, T., Metghalchi, M. and Gutoff, E. B. (2000). Mixing studies in bioreactors. *Bioprocess and Biosystems Engineering, 22,* 115-120.

[17] Bhattacharya, S. K. and Dubey, A. K. (1997). Effects of dissolved oxygen and oxygen mass transfer on overexpression of target gene in recombinant E. coli. *Enzyme and Microbial Technology, 20,* 355-360.

[18] Wiggins, S. (2003). *Introduction to Applied Nonlinear Dynamical Systems and Chaos:* Springer.

[19] Xianwen, H., Zuohu, L., Jianxin, C., Jianying, h. and Naizhong, Y. (1999). A new air-lift and liquid-lift animal cell bioreactor. *Letters in Biotechnology, 10,* 100-104.

[20] Hongzhang, C., Fujian, X., Zhonghou, T. and Zuohu, L. (2002). A novel industrial-level reactor with two dynamic changes of air for solid-state fermentation. *Journal of Bioscience and Bioengineering, 93,* 211-214.

[21] Hongzhang, C. and Zuohu, L. U.S. Patent 7183074 B2, 2007.

[22] Chen, H. Z., Xu, J. and Li, Z. H. (2005). Temperature control at different bed depths in a novel solid-state fermentation system with two dynamic changes of air. *Biochemical Engineering Journal, 23,* 117-122.

[23] Fujian, X., Hongzhang, C. and Zuohu, L. (2002). Effect of periodically dynamic changes of air on cellulase production in solid-state fermentation. *Enzyme and Microbial Technology, 30,* 45-48.

[24] Tao, S., Beihui, L., Zuohu, L. and Deming, L. (1999). Effects of air pressure amplitude on cellulase productivity by Trichoderma viride SL-1 in periodic pressure solid state fermenter. *Process Biochemistry, 34,* 25-29.

[25] Haito, M., Xiaoyong, Z. and Zuohu, L. (2004). Control of gas phase for enhanced cellulase production by *Penicillium decumbens* in solid-state culture. *Process Biochemistry, 39,* 1293-1297.

[26] Xianlu, X. (2003). Studies on the technology for industrial production of beauveria. *Journal of Zhe Jiang University of Technology, 31,* 520-523.

[27] Xiaoguo, F., Hongzhang, C., Hongqiang, L. and Runyu, M. (2006). Study of microorganism protein and mechanism in solid state fermentation with periodical dynamic changes of air. *Journal of Beijing University of Chemical Technology, 33,* 42-46.

[28] Lindquist, S. and Craig, E. A. (1988). The Heat-Shock Proteins. *Annual Review of Genetics, 22,* 631-677.

[29] Hongqiang, L. and Hongzhang, C. (2005). The Periodic Change of Environment Factors in Solid State Fermentation and Effect on Microorganism Fermentation. *Chinese Journal of Biotechnology, 21,* 440-445.

[30] Juntang, Y. and Xiaoxuan, T. (1991). *Biotechnology:* Shanghai: East China University of Science and Technology Press.

[31] Zeng, A. P., Byun, T. G., Posten, C. and Deckwer, W. D. (1994). Use of respiratory quotient as a control parameter for optimum oxygen supply and scale-up of 2, 3-butanediol production under microaerobic conditions. *Biotechnology and Bioengineering, 44,* 1107-1114.

[32] Anderlei, T., Zang, W., Papaspyrou, M. and Büchs, J. (2004). Online respiration activity measurement (OTR, CTR, RQ) in shake flasks. *Biochemical Engineering Journal, 17,* 187-194.

In: Biotechnology: Research, Technology and Applications   ISBN 978-1-60456-901-8
Editor: Felix W. Richter   © 2008 Nova Science Publishers, Inc.

Chapter 13

Functional Roles of Soil Arthropods in Forest Ecosystem

B. M. Sharma
Department of Zoology, J.N.V. University, Jodhpur, India

Abstract

It is well established that soil animals are important contributors to soil health. But one must keep in mind the paucity of information regarding harsh climatic conditions. They play a key role in the decomposition process, ingesting large volumes of soil containing dead, decaying material, animal debris and associated microorganisms. Soil arthropods favour microbial activity, increase enzymatic activity, stimulate root development and maintain soil fertility via biochemical and biomechanical processes. The microbe-grazing arthropods scrape bacteria and fungi off of the root surface and complete the nutrient cycle processes. Overall, soil faunal populations influence soil biological processes, nutrient cycling, soil structure and environment regulatory functions.

Introduction

Soil is one of the most precious natural resources of planet earth. It is the main source for food and fibre, essential to the lives of the rapidly growing human population and future generations. This important resource is being over exploited for economic benefits. The need for sustainable use of soil was recognized in Agenda 21 of the United Nations Conference on Environment and Development (UNCED) at Rio de Janeiro in 1992. All the major agriculture development agencies have shown a greater concern to the indicators of soil health as a basis of sustainable management practices. Soil biological health is the ability of a soil to maintain its productive capacity over time and under agroecological change as might arise through changed economic, climatic and species choices [1, 2]. Soil biological health is used in a generic sense to describe the properties, processes and potentials of the soil system associated dead and living organic materials. The importance of soil arthropods in the design of nature has long piqued the curiosity of biologists. Soil arthropods that drive most of their sustenance from plants are usually adapted for feeding in one of two systems. One is the above ground (plant) and another is below ground (soil-litter) system. These systems consist of above-ground and below-ground faunas that interact to influence community and ecosystem level processes and properties [3]. The roles of below-ground fauna in forest soil-litter systems have been envisioned in various ways. In general, soil arthropods are viewed as regulators of the decomposition segment of forest ecosystems. These are envisioned as accelerating nutrient release from decomposing organic matter. Among other invertebrates, soil arthropods are of direct relevance and include millipedes, mites, springtails and beetles etc. Soil arthropods play a prominent role in control of population of microorganisms and humification of soil organic matter [4]. They affect plant growth both by their activity modifiers of soil infertility and water infiltration and by their effects as consumers [5]. Microbe-grazing by arthropods mineralizes nutrients to complete the nutrient recycling process and regulate decomposition process. Addition of an arthropod shredder greatly enhanced nutrient mineralization [6]. The significant effects of soil fauna on the nutrient dynamics of ecosystem have been documented [7, 8, 9, 10, 11].

Coniferous forest soils are characterized by extreme climatic conditions, recalcitrant organic matter, acid raw humus and scarcity or even absence of large detritivore animals. On the other hand, most of these forests are extensively managed for silviculture purposes. Forestry practices include harvesting, site preparation, planting prescribed burning, fertilization etc. All these management practices have the potential to affect soil decomposer organisms both directly and indirectly, and hence, through changes in soil processes, productivity and sustainable yield of forest ecosystems. The litter decomposition process is almost entirely due to microorganisms, but for these to thrive, animal activity is essential. Some microarthropods feed on decomposing litter, reducing its mass and exposing broken surfaces to increased rates of nutrient release [12]. Other feed on fungal hypae increasing nutrient cycling and soil aggregation. Field experiment showed that the rates of forces litter decomposition reduces excluding microarthropods [13, 14]. The role of soil invertebrates in forest decomposition became deemphasized because the role of soil arthropods in forest ecosystems contains various perspectives at primary to end stage. Yet such knowledge may be critical in order to explain fully the fundamental forces that shape the structure and regulate the functioning of forest ecosystems.

Faunal Distribution

Faunal populations show discontinuous distribution pattern because of their grouping behavior, limited dispersion and non-uniform resource consumption [15]. Specialization of soil organisms to particular microhabitats may be due to high species diversity in soil [16, 17, 18]. Many anthropogenic activities can create secondary discontinuity in soil. It has been shown that along heavy metal gradients not only the abundance of individuals but also the structure of the whole below-ground community is changed [19]. Species-specific distribution patterns in discontinuously contaminated soil may be due to a general tendency of organisms to disperse, adaptation to toxicants, interspecies interaction, and capability to perceive different concentrations of toxicants in their environment [19]. Dispersal of soil animals by active locomotion is inefficient for long-distance movements [20]. Soil animals may localize, colonize and utilize clean patches in discontinuously contaminated soil [21, 22]. Colonization patterns and successional development of soil animal communities have been studied in many soil types and environmental conditions [20, 23, 24, 25, 26].

Soil faunal biodiversity is crucial for pedoecosystem sustainability and minimizing risk of soil and environmental degradation. Faunal activity and diversity depend upon land use system. Some land use systems enhance biodiversity of soil fauna and others adversely affect faunal diversity [27]. Agroforestry system might be playing an important role in improving soil faunal biodiversity in tropical countries. Knowledge of soil-inhabiting arthropods is largely confined to the larger forms such as scorpions, spiders, myriapods, beetles and other soil arthropods that can be collected by hand and observed in the field [28, 29, 30]. Tenebrionid beetles are mostly back ground dwellers, detrivorous, feeding mainly on dead plant materials and highly adaptable to desert environment [31]. They consume litter and cow dung and thus increase soil fertility in ecosystem, while their larvae are mainly root feeders.

Microbial Grazing

Although the rate of microbial grazing by soil in different soil types remains largely unquantified, such grazing rates may be very high [32, 33]. More than 200,000 individual fungivorous arthropods can inhabit every square meter of conifer forest soil on average [6, 34]. The most frequently occurring fungivorous springtail, *Onychiurus*, is the prime determinant of the fungal community composition of a conifer soil in Scotland [35]. By preferentially consuming a fast growing species, it keeps fungal diversity high, when *Onychiurus* is removed, a single fungal species predominates, representing 90% of the fungal biomass. Perhaps most of the nitrogen in growing plants enters a root as the result of fauna grazing the microbes in the rhizosphere [36, 37]. Microbial grazers do not get too dense and they stimulate mycorrhizal growth and exoenzymetic activity [38].

Fauna-associated Decomposition

Soil arthropods affect decomposition process in two ways i.e., directly and indirectly. These arthropods affect directly by feeding upon organic matter and associated microflora and indirectly by channeling and mixing of the soil, improving quality of substrate for microflora, inoculation of organic debris with microbes. Forest ecosystems are vital for a better understanding of structuring forces and the functions of soil decomposer communities because their resources are manipulated under natural or semi-natural conditions. Decomposition results from an interaction between physical and biological process with regulation accomplished by soil biota [1, 39]. Input litter that must be weathered by physical processes before it becomes suitable for microbial or faunal attack (Fig. 1). Microbial immobilization and consumption of microbes by fauna may be sufficient regulatory mechanisms. Conservation of leaf litter to arthropods faeces is a form of substrate modification. Since soil arthropods consume organic debris and microflora and released inorganic substances. Nutrients accumulated by microbes may be returned via fauna to the pool of decomposing organic matter. Eventually, nutrients reach an inorganic state and may be lost from the system.

Dunger [40] is believed to produce a slow step by step humification of litter as dead plant materials, soil, faeces and microflora are ingested and reingested and showed positive feed back relationship between microflora and soil fauna. Gist and Crossley [41] estimated the consumption of litter input was 7.5 and 20% based on calcium flow and potassium flow, respectively in food web of hardwood forest floor. These nutrients flow estimates considered organic debris plus microflora as a single food source. Microflora would be more realistically considered to be an intermediary. It suggests that direct feeding is a significant role for soil arthropods.

The dynamics of microfloral and faunal interactions are probably the crux of the indirect influence of soil arthropods on the decomposition process [39]. Lumped together as indirect effects are such phenomena as conservation of litter to faeces, fragmentation of litter mixing of litter and soil, regulation of microflora and other activities. Presumably the soil fauna fragmented the litter, exposing a greater surface area to the effects of leaching by rainwater. Soil invertebrates may consume 20-100% of annual litter input and in so during produce an immense amount of excrement [42]. In forest soils, upto half of the fallen leave may be decomposed by soil mites [43]. Some scientists have been reported similar significant effects of soil arthropods [8, 11, 44]. The consensus is that indirect effects of soil arthropods on decomposition and soil fertility in general may exceed their direct effects of feeding on litter and microflora.

Soil Quality Indicator

Below-ground faunal biodiversity affect soil quality directly and indirectly both depending on their size and specific activity. It can be said that higher the activity and species diversity of soil fauna, then better the soil quality. Knoepp et al. [45] studied the biological indices of soil quality and compared with four common groups of soil biological indicators including soil microarthropods. Diverse soil fauna can reach tremendous species numbers in tropics which represents potential indicators of sustainability of rural activities [46]. Netruzhilin et al. [47] assessed agricultural impact using and morphospecies as bioindicators in the Amazonian Savanna-forest ecotone. Many other studies have also demonstrated the importance of the soil fauna as a regulator in nutrient cycling [48, 49].

Several properties or functions of soil fauna can be used to indicate soil quality such as the presence of specific organisms and their population or community analysis (functional groups and biodiversity) and biological processes including soil structure modification and decomposition rates [50]. Relative abundance of soil fauna may be linked to litter quality and nutrient cycling rates [1, 9, 11], the effects of their activities range from miner to major. Macroarthropods (millipedes, centipedes, insect larvae and others) have the ability to modify soil structure by decreasing bulk density, increasing soil pore space, mixing soil horizons and improving aggregate structure [51]. Some soil arthropods can rearrange the soil profiles, mixing soil horizons through burrowing and nest-building activities.

Shredding and Succession

The arthropod shredder is the bacteria can opener. Bacteria and fungi will eventually use all of a dead leaf, but they are far more efficient if the leaf is shredded first. Shredders (i.e., millipedes, sowbugs) crush vast quantities of plant cells from which they extract only the most readily available nutrients. The normal soil recycling chain as the shredders defecate the crushed fragments [52]. Passage through numerous shredding devices (mouthparts of smaller organisms) is required before all the resources are finally available for complete enzymatic digestion [53]. This process of continually refined shredding takes time, which accounts for the persistence of humus layers in the soil.

There may be as many as 40,000 kinds of microbes in a single teaspoon of soil [54]. These different microbes exhibit a wide variety of chemical capabilities [55, 56, 57]. This level of specialization for resources is mediated by the fauna. The vast majority of microbes is inactive at any given time, only the subset of species capable of using the actual specific chemical composition of resources currently available is metabolically active [58]. As such species grow and consequently change the chemistry of the remaining resources, the fauna consume them (presumably after most of them have reproduced), which permits other species to succeed them.

Burrowing and Transportation

Soil meso- and macro-fauna create channels, pores aggregates and mounds that profoundly influence the transport of gases and water in soil. Most soil dwellers excavate networks of tunnels immediately beneath or beside the pad, and pull down pellets of dung (Figure 2). Some beetles excise a chunk of dung and move it some distance to a dug out chamber, often within a network of tunnels [59]. This movement from pad to nest chamber may occur either by head-butting an unformed lump or by rolling moulded spherical balls over the ground to the burial site. The female lays eggs into the buried pellets and the larvae develop within the faecal food ball, eating fine and coarse particles. The adult scrubs may also feed on dung but only on the fluids and finest particulate matter and increase soil health.

Animals take mobility for granted until they become physically disabled. There may be 500 million bacteria in a teaspoon of forest soil, but each one is largely incapable of movement even though some bacteria have flagella that permit limited movement. Each bacterium needs diverse nutrients for growth and reproduction. Bacterial and fungal inocula can be carried either on the outer body surfaces of invertebrates or in their intestinal tracts [33, 60]. In general, the number of spores carried phoretically is directly proportional to the surface area of an organism. The viability of ingested inocula is generally proportional to the time it requires to pass through the length of the gut [61]. Well-fed individual invertebrates have high percentages of viable inocula in their faeces, whereas poorly fed individuals produce faeces with minimal viable inocula. These transporter arthropod species that feed largely on the soil surface, defecate deep in the soil and construct extensive tunnels in the rooting zone [62].

Predatory Nature

Forest ecosystems have much less insect pest problems than crop fields, because there are a large number of natural enemies [11]. Soil arthropods are most numerous and diverse natural enemies of forest insect pests. More than 200 families in 15 orders are entomophagous [63]. These predators may feed on several or all stages (from egg to adult) of their prey and each predator usually consumes several individual prey organisms during its life, with the predatory habitat often characterizing both immature and adult instars. Most predators are either other insects, particularly spiders, mites and beetles. They are important in regulating populations of phytophagous pests. However, some highly efficient natural enemies, especially certain predatory coccinellids, sometimes eliminate their food organisms so effectively that their own populations die out, with subsequent uncontrolled resurgence of the pest [59]. The predatory insects differ from parasite ones in that the larvae or nymphs, as the case may be, require several to many prey individuals to attain maturity. The adults generally deposit their eggs near the prey population and after hatching the active mobile immature search out and consume prey individuals. Larvae and nymphs as well as corresponding adults may be predaceous or only one stage may exhibit and habit.

Predatory insects widely recognized in pest management practices primarily are found in the order Coleoptera, Diptera, Hymenoptera and Neuroptera. More than half of all insect predators are beetles. The coccinelids (ladybettles) feed on scale insects, mealy bugs, aphids and whiteflies [64]. Smaller ladybeetles in the genus *Stethorus* are highly effective predators of spider mites. Dipteran include the Asilidae, Syrphidae and Cecidomyiidae larvae are effective predators of aphids. Hymenopterans (Formicoidea) may be highly effective predators of insect larvae, pupae and adults, but they have a negative effect if they interfere with other natural enemies or are direct pests themselves [65]. Other hymenopterans include the Vespidae and Sphecoidae, which provision their nest with insects for their progeny. Most nuropterans are predators, particularly the green and brown lacewings (Chrysopidae and Hemerobiidae), feeding on scale insects, aphids, mealy bugs and mites [66].

Interaction of Soil Fauna with Mycorrhiza

Under our feet lies a living ecosystem of biota that eat, breathe, catch prey, create and transport food and support the biosystems [67]. The most important productive component of soil is life: microorganisms such as bacteria, fungi, small arthropods or insects, and their waste products and carcasses. This underground ecosystem is vital to the health and productivity of forests. Plant roots are passive sponges because they cannot absorb the nutrients stored in fallen leaves until it has been processed by a host of other creatures. One key animal for such task is the millipede. The bright yellow and black cyanide millipede is a familiar sight on the forest floor. It is the first 'processor' of needles and leaves, shredding and breaking them down into accessible bits for others. Waste products from millipedes and others are eaten again and again by other organisms. Soil invertebrates like the millipede, springtails and tiny mites are the catalysts that control the rate of nutrient cycling throughout the forest by crushing and mixing the organic and inorganic components of the soil.

It has been shown that Collembola can affect the nutrient uptake of plants. The interaction between grazers and mycorrhiza, however, is complex and has both direct and indirect aspects. As mycorrhizas are especially important under nutrient limiting conditions more knowledge is needed of the effects of fungal feeding soil fauna on mycorrhizal functioning. The effect of soil arthropods on growth, arbuscular mycorrhizae infection, and phosphorus uptake in a common forest herb (*Geranium robertianum*) under greenhouse conditions gave an interesting insight [68]. They also showed the total and aboveground growth were greater at a particular density of Collembola than either at higher density or without Collembola. These differences were greater when the plants were grown in sand. Root mass was not affected by Collembola density. In the soil mix, root length decreased with increasing population density, but not in the sand. Total infected root length decreased linearly with increasing population density. Few significant differences in phosphorus uptake were found. The plant growth (but not phosphorus uptake) may be stimulated at low population density and inhibited at high. Mechanisms which may be responsible for this non-linear response, and the implications of the pattern of response to studies of plant competition, nutrient turnover, and revegetation need to be reinvestigated.

Since biotic component of soil includes plants, animals, fungi and bacteria and they are closely associated and combined to influence physical, chemical and biological properties of soil. This reflects that the development of one group of faunal population facilitate the development of another group by creating a conducive soil environment. Soil fauna occupy many important positions (detrivorous, omnivorous, herbivorous, predacious etc.) in the tropical levels in ecosystems [1], the present reveiw will help in developing strategies to enrich soil biodiversity and improve sustainability of pedoecosystem in forest environment. It may also help in conservation of below-ground fauna and, in turn, soil conservation on a sustainable basis.

Future Prospectus

The primary goal of the land management practices adopted is to enhance productivity of the desired forest trees and this is achieved by manipulating the habitat so as to make it favourable for the growth of those species. This frequently involves such practices as the addition of synthetic compounds, manipulation of residue and the disturbance of the soil itself. Many aspects of soil biology were put forward around the dynamics of soil communities. The interrelation between litter fauna and microorganisms are a fascinating field of study for soil zoologists and soil microbiologists.

References

[1]   B.M. Sharma, In : *Changing Faunal Ecology in the Thar Desert.* Tyagi, B.K. and Baqri, Q.H. (Eds.), Jodhpur, India Scientific Publishers, pp. 45-60, 2005.

[2]   B.M. Sharma and J.P. Gupta (Eds.) *Agroforestry for Sustained Productivity in Arid Areas.* Scientific Publishers, Jodhpur, India, 1997.

[3]   D.A. Wardle, R.D.Bardgett, J.N. Klironomos, H. Setala, W.H. van der Putten and D.H. Wall, *Science.* 304, 1629-1633, 2004.

[4]   J. Singh, In : *Advances in Management and Conservation of Soil Fauna.* Veeresh, G.K., Rajagopal, D. and Viraktamath, C.A. (Eds.), New Delhi, India, Oxford and IBH Publishing Co., pp. 127-139, 1991.

[5]   W.G. Whitford, Invertebrates : their effect on the properties and processes of desert ecosystems. In : *Ecology of Desert Environments.* Prakash, I. (Ed.), Jodhpur, India, Scientific Publishers, pp. 333-356, 2001.

[6]   A.R. Moldenke, One-hundred twenty-thousand little legs. *Wings.* 15, 11-14.

[7]   H.A. Verhoef and L. Brussaard, *Biogeochem.* 11, 175-211, 1990.

[8]   M.V. Reddy (Ed.) *Management of Tropical Agroecosystems and the Beneficial Soil Biota.* New Delhi, India, Oxford and IBH Publishing Co., p. 546, 1999.

[9]   B.M. Sharma and G. Tripathi, *Intl. Conf. MT in Trop.*, Jodhpur, India, AFRI, 2004.

[10] G. Tripathi and B.M. Sharma, *Environ. Tech.* 26, 1-11, 2005.

[11] B.M. Sharma, In : *Forest Insect Pests and Control,* Tyagi, B.K., Veer, V. and Prakash, S. (Eds.), Jodhpur, India, Scientific Publishers, pp. 111-125, 2005.

[12] J. Lussenhop, *Adv. Ecol. Res.*, 23: 1-33, 1992.

[13] T.R. Seastedt and D.A. Crossley Jr., *Soil Biol. Biochem.* 15, 159-165 (1983).

[14] J.M. Blair, D.A. Crossley Jr., and L.A. Callaham, *Biol. Fertil. Soil.* 12, 241-252, 1992.

[15] P. Kareiva, In : Community Ecology Diamond, J. and Case, T.J. (Eds.), New York, USA, Harper and Row, pp. 192-206, 1986.

[16] Anderson, J.M. In : *Progress in Soil Zoology.* Vanek, J. (Ed.), Prague, Academia Press, pp. 51-58, 1975.

[17] J.M. Anderson, *Oecologia.* 32, 341-348, 1978.

[18] W.D. Atkinson and B. Shorrocks, *J. Ani. Ecol.*, 50, 461-471, 1981.

[19] G. Bengtsson and L. Tranvik, *Water, Air, Soil Pollution.* 47, 381-417, 1989.

[20] G. Bengtsson, S. Rundgren and M. Sjogren, *Oikos.* 71, 13-23, 1994.

[21] M.S. Usher, In : *Ecological Interactions in Soil.* Fitter, A., Atkinson, D., Read, D. and Usher, M. (Eds.), Oxford, England, Blackwell Scientific Publication, pp. 243-265, 1985.

[22] H. Siepel, *Biol. Fertil. Soil.* 18, 263-278, 1994.

[23] L. Tranvik and H. Eijsackers, *Oecologia.* 80, 195-200, 1989.

[24] S. Hagvar and G. Abrahamsen, *Oikos.* 34, 245-258, 1980.

[25] B. Hudson, *J. Appl. Ecol.* 17, 255-275, 1980.

[26] B. Streit, A. Buehlmann and P. Reutimann, *Pedobiol.* 28 : 1-12, 1985.

[27] R. Lal, In : *Management of Tropical Agroecosystems and the Beneficial Soil Biota* Reddy, M.V. (Ed.), New Delhi, India, Oxford and IBH Publisingh Co., pp. 67-81, 1999.

[28] J. Singh and V.B. Lal, *A Handbook of Soil Fauna.* Technical Bulletin No. 3, Agricultural Research Station, Jaipur, India, p. 44, 2001.

[29] B.M. Sharma, S. Ram and G. Tripathi, *Natl. Symp. Biod. Manag. 21ˢᵗ Century*, Varanasi, India, Banaras Hindu University, 2003.

[30] G. Tripathi, B.M. Sharma and J. Singh, *J. Appl. Biosc.*India, 31, 68-89, 2005.

[31] H.S. Pruthi and D.R. Bhatia, *Bull. Natnl. Inst. Sci. India*, 1, 241-245, 952.

[32] D.C. Coleman and D.A. Crossley Jr., *Fundamentals of Soil Ecology.* New York, USA, Academic Press, 1996.

[33] S. Visser, In : *Ecological Interactions in the Soil* Fitter A.H. (Ed.), Oxford, England, Blackwell Scientific Publcation, pp. 297-313, 1985.

[34] H. Petersen and M. Luxton, *Oikos.* 39, 287-308, 1982.

[35] K. Newell, *Soil Biol. Biochem.* 16, 227-240, 1984.

[36] P.C. De Ruiter, J.C. Moore, K.B. Zwart, L.A. Bouwman, J. Hassink, J. Bloem, J.A. de Vos, J.C.Y. Marinissen, W.A.M. Didden, G. Lebbink and L. Brussaard, *J. Appl. Ecol.* 30, 95-106, 1993.

[37] J.C. Moore, D.E. Walter and H.W. Hunt, *Ann. Rev. Entomol.* 33, 419-439, 1988.

[38] R.D. Finlay, In : *Ecological Interactions in the Soil.* Fitter, A.H., Atkinson, D., Read, D.J. and Usher, M.B. (Eds.), Oxford, London, Blacwell Science, pp. 319-332, 1985.

[39] D.A. Crossley Jr., In : *The roles of arthropods in Forest Ecosystems.* Mattson, W.J. (Ed.), New York, USA, Springer-Verlag, pp. 49-56, 1977.

[40] W. Dunger, *J. Planz. Boden.* 82, 174-193, 1958.

[41] C.S. Gist and D.A. Crossley Jr., In : *Mineral Cycling in Southwestern Ecosystems* Howell, F.G. and Smith, M.H. (Eds.), ERDA Symp. Series., Europe, pp. 84-106, 1975.

[42] G.F. Kurcheva, *Pedology. Leningr.* 4 : 16, 1960.

[43] P. Berthet, *Mem. Inst. Roy. Sci. Nat. Belg.* 152, 1-15, 1964.

[44] G. Tripathi, B.M. Sharma and R. Deora, *J. Appl. Zool. Res.,* India, 16, 108-112, 2005.

[45] D.J. Knoepp, D.C. Coleman, D.A. Crossley Jr. and J.S. Clark, Biological indices of soil quality: an ecosystem case study of their use. *Forest Ecol. Manag.* 138, 357-368, 2000.

[46] T. Erwin, In : *Biodiversity-II.* Reaka-Kudla, M.L., Wilson, D.E. and Wilson, E.O. (Eds.). J. Henry Press, Washington, DC, pp. 27-40, 1997.

[47] Netuzhilin, P. Chacon, H. Cerda, D. Lopez-Hernandez, F. Torres, and M.G. Paoletti, In : *Management of Tropical Agroecosystems and the Beneficial Soil Biota.* Reddy, M.V. (Ed.), New Delhi, India, Oxford and IBH Publishing Co., pp. 291-352, 1999.

[48] J.M. Anderson, M.A. Leonard, P. Ineson and S.A. Huish, *Soil. Biol. Biochem.* 17, 735-737, 1985.

[49] T. Persson, *Plant Soil.* 115, 241-245, 1989.

[50] D.R. Linden, P.F. Hendrix, D.C. Coleman and P.C.J. Van Vliet, In : *Defining Soil Quality for a Sustainable Environment.* Doran, J.W., Coleman, D.C., Bezdicek, D.F. and Stewart, B.A. (Eds.), Madison, USA, Soil Sci. Soc. Publ., pp. 91-106, 1994.

[51] Abbott, In : *Animals in Primary Succession : The Role of Fauna in Reclaimed Lands* J.D. Major (Ed.), Cambridge, UK, Cambridge University Press, pp. 39-50, 1989.

[52] S.P. Hopkinand and H.J. Reed, *The biology of Millipedes.* Oxford, England, Oxford Scientific Publications, 1992.

[53] V. Dawod and E.A. FitzPatrick, *Geoderma.* 56, 173-178, 1993.

[54] J.M. Tiedje, *Am. Soc. Microbiol. News.* 60, 524-525, 1994.

[55] J.L. Fox, *Am. Soc. Microbiol. News.* 60, 533-536, 1994.

[56] D.L. Hawksworth, *Mycol. Res.* 95, 641-655, 1991.

[57] N.J. Palleroni, *Am. Soc. Microbiol. News.* 60, 537-540, 1994.

[58] P. Lavelle, *Transactions of the 15th world Cong. Soil Sci.*, Mexico, Mexi. Soci. Soil Sci., pp. 189-200, 1994.

[59] Gullan, P.J. and Cranston, P.S. (1994) (Eds.) *The Insects.* Chapman and Hall, London.

[60] J.M. Anderson, *Agric. Ecosys. Environ.* 24, 5-19, 1988.

[61] J.M. Anderson, *J. Anim. Ecol.* 44, 475-495, 1975.

[62] P. Lavelle, *Biol. Fertil. Soil*, 6, 237-251, 1988.

[63] C.P. Clausen, *Entomophagous Insects.* New York, USA, McGraw-Hill, 1940.

[64] Hodek, *Biology of Coccinellidae.* The Hague, W. Junk Press, 1973.

[65] M.J. Way and K.C. Khoo, *Annu. Rev. Entomol.* 37, 479-503, 1992.

[66] M.A. Hoy, In : *Introduction to Insect Pest Management.* Metcalf, R.L. and Luckmann, W.M. (Eds.). New York, USA, John Wiley and Sons, pp. 129-198, 1994.

[67] G. Tripathi and R. Prasad, In : *Basic Research and Applications of Mycorrhizae.* Podila, G.K. and Varma, A. (Eds.), New Delhi, India, I. K. International, pp. 297-325, 2005.

[68] K.K. Harris and R.E.J. Boerner, *Plant Soil.* 129, 203-210, 1991.

In: Biotechnology: Research, Technology and Applications   ISBN 978-1-60456-901-8
Editor: Felix W. Richter   © 2008 Nova Science Publishers, Inc.

Chapter 14

BAP-Fusion: A Versatile Molecular Probe for Biotechnology Research

*Montarop Yamabhai**

School of Biotechnology, Suranaree University of Technology
111 University Avenue
Nakhon Ratchasima, 30000 Thailand

Abstract

Bacterial alkaline phosphatase (BAP) is a useful enzyme for detection in biotechnological researches. There are vast arrays of commercial available substrates, which can be converted to soluble or precipitated products for either colorimetric or chemiluminescent detection. This research article describes the application of bacterial alkaline phosphatase fusion protein as a convenient and versatile molecular probe for direct detection of different molecular interactions. Short peptide, protein binding domain, or single chain variable fragment (scFv) of monoclonal antibody were fused to bacterial alkaline phosphatase and used as one step detection probe for the study protein-protein or antibody-antigen interactions. The BAP-fusion could be generated by cloning a gene of interest in frame of BAP gene in an *Escherichia coli* expression vector. The fusion protein contained N-terminal signal peptide for extracellular secretion and could be induced for over-expression with isopropyl-β-D- thiogalactopyranoside (IPTG), allowing simple harvesting from culture broth or periplasmic extract. The BAP-fusion that was tagged with poly-histidine could be further purified by nickel affinity chromatography. This one step detection probe generated a specific and robust signal, suitable for detection in various formats as demonstrated on microtiter plate, dot blot, or western blot. In addition, it could also be used for an estimation of binding affinity by competitive inhibition with soluble ligand.

* Tel 66 44 224152-4 Fax 66 44 224150; Email montarop@sut.ac.th

Introduction

One of the important experiments in biotechnological research is the detection of molecular interactions. These include the interactions between protein-protein and protein with other type of molecules such as DNA, lipid, carbohydrate, small chemical compounds, metals, etc. These studies are required for the understanding of biological systems, which are essential for efficient drug development, or the development of affinity reagents for various biotechnological applications. Typically, basic techniques for the detection of protein interaction with various targets are done via a secondary probe, such as specific antibody that is linked to enzyme or fluorescent probe and are used for the detection of molecular interactions in various formats, such as enzyme-link immuno assay (ELISA), western blot, dot blot, or staining of tissue or cells [1]. The limitation of these assays is the availability of specific antibody to the protein of interest.

One solution to this problem is the construction of alkaline phosphatase fusion protein, which can be used as one-step detection probe for the study of the interaction of protein of interest with various targets, as demonstrated in numerous publications. The enzyme alkaline phosphatase can be obtained from various sources such as human placenta [2, 3], calf intestine [4], or bacteria [5, 6]; however, the most common type of the enzyme is *Escherichia coli* alkaline phosphatase (BAP) [5, 7] or its derivatives [8]. A BAP-fusion protein can be obtained by simply cloning the gene of the protein of interest into an appropriate fusion vector and over express in *Escherichia coli* expression system.

Normally the protein of interest is fused to the N terminus of the enzyme, as it is far from active site and will not interfere with its catalytic activity [9]. In many cases, the fusion protein is linked to the N-terminal signal peptide for secretion, thus it can be conveniently harvested from culture broth or periplasmic extracts [10]. In addition, many BAP is linked with poly-histidine tag, allowing convenient purification with one-step immobilized metal affinity chromatography (IMAC).

In addition to basic experiment in biotechnological research, BAP-fusions are also useful for high-throughput screening of potential drug leads because of its robust and highly specific interaction [11-14] [15] [16-18]. It also has potential for the development of useful biosensor [19-23].

This research provides evidences for the versatile use of BAP-fusion protein in various biotechnological applications. The BAP-fusion vector for *E. coli* expression system, which contains signal peptide for the secretion of the fusion protein and 10xHistidine tag at the C-terminus for affinity purification by IMAC, has been constructed. The genes of interest were fused to the N-terminus of the BAP by cloning into the multiple cloning site that comprises recognition sequences for five restriction enzymes. Over expression of BAP-fusions can be induced with isopropyl-β-D-thiogalactopyranoside (IPTG).

The genes encoding short peptide, protein binding modules, and single chain variable fragment (scFv) of human antibody, have been cloned into this vector and produced as BAP-fusion proteins. Detection of molecular interactions in various formats with these constructs has been demonstrated.

Material and Method

Construction of BAP-Fusion Vector: pMY201

The BAP-fusion vector used in this study was modified from pFLAG-CTS (Sigma). The construction was done in two steps, first DNA insert containing the multiple cloning site and sequence encoding ten histidine was ligated into *Hind* III and *Sal* I sites of pFLAG-CTS. This DNA insert was generated by annealing of two oligonucleotides; pHisFLAG1up: AGC TTC GCT CGA GGA ATT CGG ATC CGG TAC CAG ATC TGT CGA CCA CCA TCA CCA TCA CCA TCA CCA TCA TCA CC; and pHisFLAG1dn : TCG AGG TGA TGA TGG TGA TGG TGA TGG TGA TGG TGG TCG ACA GAT CTG GTA CCG GAT CCG AAT TCC TCG AGC GA. The two oligonucleotides were slowly assembled in a thermo cycler machine by mixing equal molar of each strand at 95°C for 5 min, and reduce the temperature slowly to room temperature. The insert was then ligated into pFLAG-CTS that has been cut and purified by agarose gel extraction. This vector was designated as p10HisFlag. Then the gene for *E. coli* alkaline phosphatase was amplified by using specific primers containing *Kpn* I and *Sal* I site. These are APfwKpnI: CTG TGC GGT ACC ATG CCT GTT CTG GAA AAC CGGG and APrvSalI: CTGTGC GTC GAC CGG TAC TTT CAG CCC CAG AGC. The template for amplification of BAP was from a previously published AP-fusion vector; pMY101 [14, 24]. The 100 μl of PCR reaction consisted of 0.5 μM of primers, 0.2 mM dNTP, 3 units of *Pfu* DNA polymerase (Promega), and 10X the reaction buffer, provided by the manufacturer. The amplifications were done as follows: initial DNA denaturation at 95°C for 2 min.; 30 cycles of denaturation at 95°C for 45 sec., annealing at 55°C for 1 min., extension at 72°C for 3 min., and a final extension at 72°C for 10 min. PCR products were purified using PCR purification kits (Qiagen GmbH, Germany) and then cut with appropriate restriction enzymes (*Kpn*I and *Sal*I) and ligated into p10HisFlag that has been cut with corresponding enzymes. The ligation reaction was transformed into *E. coli* DH5α. The DNA sequence and the integrity of the construct were determined by automated DNA sequencing (Macrogen, Korea).

Construction of BAP-Fusion Proteins

All BAP-fusion constructs were created by PCR amplification of interested genes, using specific primers that were flanked by appropriate restriction sites, and cloned into pMY101[14] or pMY201 expression vector that has been cut with corresponding restriction enzymes. The integrity of all constructs was determined by automated DNA sequencing (Macrogen, Korea).

Expression and Purification of BAP-Fusion Proteins

Freshly transformed *E. coli* harboring BAP-fusion construct was inoculated into 5-10 ml of LB broth containing 100 μg/ml of ampicillin at 37°C for 16 hr. Then, the overnight culture was inoculated into 250-1000 ml of LB broth containing 100 μg/ml ampicillin and grown at 37°C until the optical density (O.D.) at 600 nm reached 0.9-1.2. Then, IPTG was added into the culture broth to a final concentration of 1 mM. The culture was then incubated with vigorous shaking (200 rpm) at 26°C (room temperature) for 4-6 hr. Then, the culture was collected and chilled in an ice box for 5 min and centrifuged at 8,000 rpm for 10 min at 4°C to collect cells and supernatant. To extract the periplasmic content, the cells were resuspended in 2.5 ml of cold (4°C) spheroplast buffer [100 mM Tris-HCl, pH 8.0, 0.5 mM EDTA, 0.5 mM sucrose, and 20 μg/ml phenylmethylsulfonyl fluoride (PMSF)]. After incubation for 5 min on ice, bacterial cells were collected by centrifugation at 8000g at 4°C for 10 min and re-suspended in 1-2 ml of ice-cold sterile water supplemented with 1 mM $MgCl_2$ and incubated on ice for 5 minutes with frequent shaking. The supernatant of nearly 1-2 ml was then collected by centrifugation at 8000g at 4°C for 15 min as the periplasmic fraction. To extract the cell lysate, the precipitated cells from the previous step were washed once with lysis buffer (50mM Tris-HCl + 0.5 mM EDTA), resuspended in 1-2 ml of lysis buffer, and sonicated (Ultrasonic Processor; 60 amplitude, pulser 6 sec, for 2 minutes) on ice. The cell debris was then spun down at 10,000xg and the supernatant was collected as the cell lysate.

SDS-PAGE Analysis

Denaturing sodium dodecyl sulfate-polycrylamide gel electrophoresis (SDS-PAGE) was performed according to the method of Laemmli [25], in a 12% (w/v) polyacrylamide. The protein samples were briefly heat at 100°C in a heat block (Eppendorf) for 3 mins in the loading buffer. Protein bands were visualized by staining with Coomassie brilliant blue R-250. The molecular weight markers were from Biorad.

Detection on Microtiter Plate (ELISA)

Target molecules were immobilized onto wells of a microtiter plate by incubating 1-10 μg of target molecules in 0.1M $NaHCO_3$ (pH8.5) at 4°C, overnight. The wells were then washed with TBS [25 mM Tris-HCl (pH 7.5), 145 mM NaCl, 3 mM KCl] containing 0.1% (v/v) Tween 20. To eliminate nonspecific binding, 5-10 μg of BSA in 100 μl TBS were added into each well and incubated for an additional hour. The wells were then washed and incubated with 150 μl of culture media or TBS containing appropriate BAP-fusion proteins for 1 hour at room temperature. Then, the wells were washed 5 times with TBS + 0.1% (v/v) Tween 20. The amount of bound BAP fusions was measured by adding 150 μl of p-Nitrophenyl Phosphate, pNPP (Sigma Fast) and quantifying the absorbance with a microtiter plate reader at 405 nm. To screen for peptides that can inhibit the binding of BAP fusions to target proteins, various concentration of peptides were added with BAP fusions into each well and processed as described above.

Dot Blot Analysis

Seven amino acid-long peptides were immobilized on cellulose membrane using Spot-Synthesis method [26]; whereas target proteins were dot-blotted onto Immobilon P nylon membrane strips (Millipore) with dot blot apparatus (Biorad). The membrane was first blocked with 10% BSA in TBS for 1 hour at room temperature. The membrane was then washed 3 times with TBS containing 0.1% (v/v) Tween 20 and incubated in culture media containing an appropriate BAP fusion for 1 hour at room temperature, or overnight at 4°C. The membrane was then washed 5 times with TBS + 10% (v/v) Tween 20 and incubated with Nitro Blue Tetrazolium and 5-Bromo-4-Cholor-3-Indolyl Phosphate (NBT/BCIP, Sigma Fast) for 10 minutes and washed with de-ionized water. For chemiluminescent detection of the fusion proteins, the membrane was incubated with CSPD chemiluminescent substrate (Applied Biosystem, Foster city, CA, USA) and the signal was detected with Fluor-S Multimager (BioRad, Hercules, USA)

Western Blot Analysis

Cell lysate or GST fusion proteins were electrophoresed on a 12% SDS-polyacrylamide gel and transferred onto polyvinylidene difluoride (PVDF) membrane (Immobilon-P, Millipore Corp.). The membrane was blocked in blocking solution (SuperBlock™, Pierce, Rockford, IL) at 4°C overnight, then transferred into a plastic bag containing 15 ml of culture supernatant containing appropriate BAP fusions and incubated at room temperature for 2 hours. The membrane was then washed with 1xTBS, Tween20 three times for 15 minutes each at room temperature, and incubated with CSPD chemiluminescent substrate (Applied Biosystem, Foster city, CA, USA). The signal was detected by exposure to BioMax Light-1 film (Eastman Kodax).

Result

Construction of BAP-Fusion Vector

The new BAP-fusion vector was constructed based on the fusion vector that has been previously reported (pMY101) [14]. This new vector contains five restriction sites in the multiple cloning site (MCS), allowing the generation of fusion protein to the N-terminus of bacterial (*E. coli*) alkaline phosphatase (BAP) gene. The C-terminus of the BAP gene is linked to DNA encoding ten histidine followed by a FLAG epitope (DYKDDDK). Normally, 6xhistidine is sufficient for purification with Nickel affinity chromatography; however, in some case, we found that 10xHis was a better tag for affinity purification, thus 10xHis was used instead of 6xHis. The expression of BAP fusion proteins were under the control of *tac* promoter, which is a combination of -35 region of trp promoter to end of lacI binding region. The vector contain lac I gene which encodes lacI repressor, which is over produced from lacI promoter and allow induction of the BAP-fusion construct with IPTG. The vector contains ampicillin resistance gene for cloning and maintenance of the recombinant plasmid. All of the fusion proteins are fused to the *E. coli* secretory signal peptide, ompA at the N-terminus, allowing the secretion of the fusion proteins into the periplasmic space as well as the culture media [10]. Amino acid sequence analysis of the BAP-fusion constructs can be done by using; forward sequencing primer, N-26: 5' CAT CAT AAC GGT TCT GGC AAA TAT TC 3'; and reverse sequencing primer, C-24: 5' T GTA TCA GGC TGA AAA TCT TCT CT 3'.

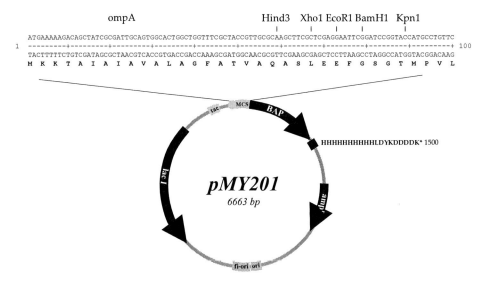

Figure 1. Map of the bacterial alkaline phosphatase fusion vector (BAP-fusion vector), pMY201. The coding region of the E. coli ompA signal/leader peptide, the multiple cloning site (MCS), and the first four amino acid of E. coli alkaline phosphatase (BAP) gene is shown. The c-terminus of BAP is linked to ten histidine tag and FLAG epitope, allowing purification with IMAC. The 6663 base-pair (bp) vector was derived from pFLAG-CTS vector. The vector also carries genes for ampicillin resistance (ampr) and the lac repressor (lacI), which is over produced from lacIp promoter and represses tac promoter. The over-expression of BAP fusion protein can be induced with IPTG.

Expression and Purification of BAP Fusion Proteins

All BAP fusion constructs used in this study could be induced for over expression by 1mM IPTG. However, we found that the proteins could also be produced without IPTG. This maybe because the *tac* promoter is relatively leaky. The levels of expression of different constructs were varied, depending on the size and the amino acid composition of the fusion protein. Figure 2 on the left panel illustrated a representative of the expression of one BAP-fusion (peptide-BAP). The cells were grown in LB media containing 100 µg/ml ampicillin until the optical density at 600 nm (OD_{600}) reached 1.2, before 1mM IPTG was added (0 hr). At this point, we could find some of the proteins accumulated in the periplasmic space. After two hours of induction by IPTG, the protein was found in culture broth, periplasm, as well as cell lysate. There was no significant different in the level of protein expression at other time points of induction. Interestingly, more protein was accumulated in the cell lysate after overnight induction. Nevertheless, it is recommended that optimization of induction condition for different protein should be performed.

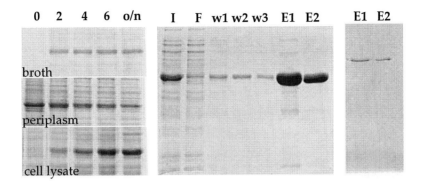

Figure 2. Expression and purification of BAP fusion proteins. An example of the expression of a BAP fusion is demonstrated in the left panel. The peptide-BAP fusion construct was generated in pMY201 and over-expressed in *E. coli* Top10. The bacteria was grown in LB medium containing 100 μg/ml ampicillin until the OD_{600nm} reach 1.2 before 1mM IPTG was added. The samples were collected at 2, 4, 6, and 20 (overnight, o/n) hours after induction. Culture broth, periplasmic extract, and cell lysate were prepared for each time point of induction, and resolved by 10% SDS-PAGE. The gel was stained with Coomassie Brilliant Blue. Equal volume of the samples was loaded onto each lane. The volume of culture broth was ten times more than the periplasm and cell lysate extracts. The mixture of periplasm and cell lysate were further purified by IMAC and resolved in 10% SDS-PAGE as shown in the middle panel. The last panel on the right illustrates the purification of scFv-BAP fusion, which was equally pure, but had much less yield. I: input, F: flow through, w1-3; first, second, and third wash, E1and2: first and second elution.

Even if all the BAP fusions used in this study were fused with the ompA secretory signal peptide, significant amounts of BAP fusion constructs were found to accumulated in the periplasmic space as well as in the cytoplasm, as shown in Figure 2, left panel. Thus, both periplasmic and cytoplasmic extracts were used for subsequent affinity purification step. Since all the BAP fusion constructs that were generated from pMY201 contains 10xHis, all of these constructs could be purified by affinity column chromatograph with Ni-NTA agarose beads. The result of the purification of the peptide-BAP construct, as seen in the left panel, was shown in Figure 2, middle panel. *E. coli* cell lysate was incubated with Ni-NTA agarose beads before washing and eluted using column chromatography. More than 90% of purified proteins can be obtained. We found that the amounts of purified BAP-fusions from one-liter culture media were varied greatly, depending on the level of expression of different BAP constructs. As seen in the Figure 2, right panel, when scFv-BAP construct was purified using the same protocol, much less protein could be obtained.

Stability and Specificity of BAP

Routinely, BAP-fusion prepared from culture broth or crude periplasmic extract could be used directly for most experiments. This is because the fusion protein is linked to the ompA signal peptide, which can direct the secretion of the protein into periplasm and eventually leaked out of into the culture broth [10]. Even if only a fraction of the fusion protein were present in the culture broth, it is sufficient for most detection assay [15, 24]. The residual activity of alkaline phosphatase in different conditions was measured to determine its stability as shown in Figure 3, upper panel.

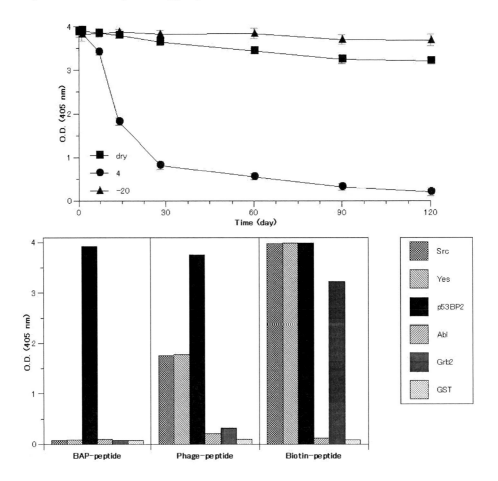

Figure 3. Stability and specificity of bacterial alkaline phosphatase. *Upper panel* illustrates the residual activity of BAP after storage in different conditions at different durations (7, 28, 60, 90, and 120 days). BAP activity was estimated by checking OD_{405} after incubation with pNPP, a chromogenic substrate for BAP. The protein was kept at 4°C in culture supernatant (4), purified and stored at -20°C (-20), or left dry at room temperature on microtiter plate wells (dry). The Intersectin SH3 domain-BAP fusion was used in this study. Different result might be obtained with other constructs. *Bottom panel* illustrates the specificity of the binding of peptide-BAP fusion compared with other formats. Various GST-SH3 fusion proteins (Src, Yes, p53BP2, Abl, and Grb2) were immobilized onto triplicate wells of a microtiter plate and incubated with different formats of p53BP2 SH3 peptide ligand i.e., BAP-fusion, (BAP-peptide), N-terminal fusion to the minor coat protein III of bacteriophage M13 (Phage-peptide), and multivalent biotinylated peptide, pre-complexed with SA-AP (Biotin-peptide). Bound BAP fusion and biotinylated peptides were detected by colorimetric assay with 8 mM pNPP (p-Nitrophenyl Phosphate), whereas bound bacteriophage particles were detected by ELISA [70]. The average OD_{405} values and standard errors are shown. The synthetic peptide ligand of p53BP2 was SGSGSYDASSAPQRPPLPVRKSRPGG, whereas the peptide ligand that fused to BAP and phage coat proteins was WVVDSRPDIPLRRSLP.

We found that less than half of the enzyme activity was left after keeping the enzyme in crude culture broth at 4 °C for 2 weeks, and most of the enzyme activity was completely demised after one month. However, if the enzyme was purified and kept at -20°C, more than 90% of relative activity could be detected after storage for 3 months. Surprisingly, if the enzyme is left dry at room temperature, approximately 80% of the activity was still remained after 3 months. This experimental data was observed when the enzymes were left dry on plastic pins, or on wells of microtiter plate. Thus, bacterial alkaline phosphatase is a relatively stable enzyme, suitable for various assays. It is worthwhile to note that the enzyme is sensitive to buffer containing phosphate [9], thus Tris-buffered saline (TBS) is preferred instead of phosphate-buffered saline (PBS).

Bacterial alkaline phosphatase is a homodimeric enzyme [7, 9, 27]. Its N- terminus protrudes away from the globular body of the protein; therefore, many protein segments or peptides can be fused to the N-terminus of BAP without interfering with the catalytic activity of the enzyme. The homodimeric structure of the bacterial alkaline phosphatase suggested that the peptide or protein that displays as BAP-fusion will interact with its partner in a dimeric form as well. This format allows robust and specific interaction with its partners as demonstrated in Figure 3, low panel. The peptide ligands of p53BP2-SH3 domains (WVVDSRPDIPLRRSLP) [28] were tested for its binding properties with various GST-fusion SH3 domains. The binding specificity was the highest when the peptide ligand was fused to BAP (BAP-peptide) because it didn't cross-react with other SH3 domains. When the same peptide ligand was displayed on the minor coat protein (pIII) of bacteriophage M13 (Phage-peptide), its binding was less specific, as it also cross-reacted with Src and Yes SH3 domain. The least binding specificity was observed when the peptide was chemically synthesized as biotinylated peptide and pre-complexed with streptavidin-conjugated alkaline phosphatase (Biotin-peptide). In this format, the peptide was found to react with most of the SH3 domains tested. Thus, our results suggested that BAP-fusion could be used for specific and robust detection of molecular interactions.

Detection on Microtiter Plate and Estimation of Binding Affinity with BAP Fusions

One of the most common assays to detect molecular interaction is on microtiter plate. The well-known assay on microtiter plate is enzyme-linked immunosorbent assay (ELISA). This technique allows detection of a small amount of a large number of samples at the same time, suitable for high-throughput application. Both qualitative and quantitative analysis can be performed. As demonstrated in Figure 4, application of BAP-fusion for detection of protein-protein interaction on microtiter plate was reported. On the left panel, various peptide ligands of Eps15 homology (EH) domain [29, 30] that have been isolated from phage-displayed peptide library [31] were tested against different GST-fusion proteins as BAP-fusion peptides. Detection of BAP activity could be done by using soluble substrate such as pNPP. Many assays could be done at the same time on microtiter plates. The binding signal is very strong as seen from the OD_{405} values (2.0-4.0), and the back ground was very low. This allows confirmation of the binding of the peptide ligand that has been isolated from phage library.

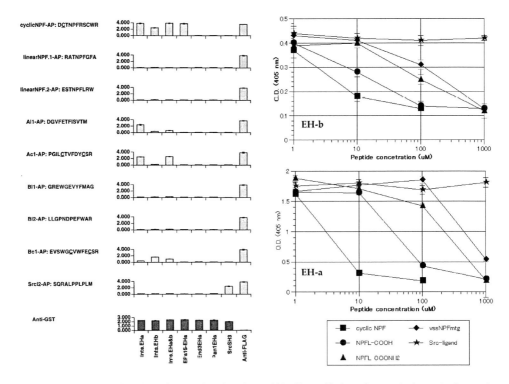

Figure 4. Detection on microtiter plate and estimation of binding affinity of BAP fusions. *Left panel* illustrates the binding characteristic of selected Intersectin EH domain-peptide ligand-BAP fusions. Various GST fusion proteins and the anti-FLAG monoclonal antibody were immobilized onto triplicate wells of a microtiter plat and incubated with different EH domain peptide ligand-BAP fusions. Bound BAP-fusions were detected by colorimetric assay with pNPP. The sequences of the peptides fused to the N-terminal of BAP are shown on the left of each histogram. The target proteins were also detected by ELISA using HRP-conjugated anti-GST antibody and ABTS substrate to demonstrate that equal amounts of proteins were immobilized onto the wells. The average OD_{405} and standard errors are shown. *Right panel* illustrates the estimation of binding affinity by competition of synthetic peptide ligands with the binding of a BAP fusion to the EHa, and EHb domains of Intersectin. Microtiter plate wells, coated with approximately two micrograms of GST fusion protein to EHa (EH-a, bottom) or EHb (EH-b, top), were incubated with a peptide (DCTNPFRSCWR)-BAP fusion in the presence of various concentrations of competitor peptides (μM). After one hour of incubation, the BAP fusion retained in the wells after washing was detected by a colorimetric assay with pNPP. Optical density at 405 nm of the average of triplicate samples and standard deviations are shown. cyclic NPF sequence is DCTNPFRSCWR (with intramolecular disulfide bond), NPFL-COOH and NPFL-COONH2 correspond to the C-terminus (aa 327-335) of RAB/Rip, vssNPFmtg: is the internal NPF motif of RAB/Rip (aa 310-318), and Src-ligand is SGSGILAPPVPPRNTR.

In addition to confirmation of the interaction, BAP-fusion can also be used to estimate the binding affinity as demonstrated in Figure 4, right panel. The EH domain peptide ligand BAP fusion (cyclic-NPF-AP) [31] were added to the wells of microtiter plate that had been coated with GST fusion to two Intersectin EH domains, EH-a and EH-b [31], in the presence of various concentration of different soluble peptides. The binding affinity could be estimated by determination of the concentration of soluble peptide that can inhibit 50% of the binding of BAP fusion, IC_{50}. As shown in our results, the IC_{50} of cyclic NPF to EHa and EHb is approximately, 3 μM and 9 μM, respectively. This analysis also suggested that cyclic NPF can bind with the highest affinity to EH domains, followed by NPFL-COOH motif, whereas the interactions of an internal NPF with EH domain is relatively weak. In addition, it also suggested that the carboxylate group at the C-terminus is involved in the binding, as the IC_{50} of NPFL-COONH2 is higher than that of NPFL-COOH.

Dot Blot/Spot Analysis with BAP-Fusions

BAP fusion proteins can also be used for detecting protein-protein interaction by dot blot analysis. This research demonstrated the application of SH3 domain peptide ligand-BAP fusion [28] and Intersectin EH domain-BAP fusions [31] for the detection of molecular interaction using dot-blot/spot format. GST fusions of various SH3 domains were dot-blotted onto PVDF membrane and incubated with various peptide ligand BAP fusions. The signal can be detected by using precipitated colorimetric or chemiluminescent substrates such as NBT/BCIP or CSPD, respectively.

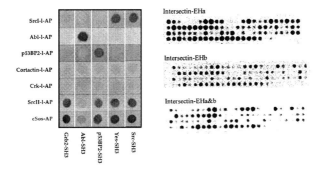

Figure 5. Dot Blot/Spot analysis with BAP fusions. *Left panel* illustrates the binding of different BAP-fusion peptides (peptide ligands of SH3 domains of Src (type I and II), Abl, p53BP2, Cortactin, Crk, and 300 C-terminal amino acids of the mouse Sos protein, cSos) to different GST-SH3 fusion proteins were dot-blotted onto membrane strips. One microgram of each GST-SH3 fusion proteins (Grb2-SH3, Abl-SH3, p53BP2-SH3, Yes-SH3, and Src SH3) was immobilized onto PVDF membrane using dot blot apparatus, cut, and strips were probed with the seven different BAP fusions. The amount of BAP fusion protein bound to the membrane strips was visualized with the chromogenic reagents 5-Bromo-4-Cholor-3-Indolyl Phosphate (BCIP) and Nitro Blue Tetrazolium (NBT). *Right Panel* illustrates the binding of Intersectin EH domain-BAP fusions to synthetic peptide immobilized on membrane. Seven-amino acid-long-peptide were synthesized onto the cellulose membrane by Spot-Synthesis method, and probed with BAP fusions to Intersectin EHa, EHb, or EHaandb. The membrane could be re-used several time, as shown in this figure, the membrane was first probe with EHa, washed, and re-probe with the second and the third BAP constructs. The results showed the binding to ninety six peptides; peptide 1-8 corresponds to alanine scan of TTNPFLL (C-terminus of epsin2/Ibp2, MP90); peptide 9-16 corresponds to alanine scan of NTNPFLL (C-terminus of epsin1, Ibp1); peptide 17-92 corresponds to 19-amino acid replacement of TTNPFLL; peptide 93-94 corresponds to internal NPF motif of epsin2/Ibp2; and peptide 95-96 corresponds to internal NPF motif of epsin1/Ibp1. The bound BAP fusions on membrane were detected with CSPD chemiluminescent substrate and the signal was detected with Fluor-S Multimager.

As shown in Figure 5, left panel, specific interaction of SH3 peptide ligand-BAP with different SH3 domains could be observed by dot-blotted analysis, similar to the result obtained from microtiter plate assay. The type I peptide ligand of Src SH3 domain can interact with both SH3 domains of Src and Yes, because their structures are highly similar [28]. On the contrary, peptide ligand of SH3 domains of Abl or p53BP2 did not cross-react with other SH3 domains. This is because the Abl and p53BP2 are relatively distinct [28]. Src-SH3-peptide ligand type II (SrcII) and cSos were relatively promiscuous in their biding activities, as they were able to cross react with almost every SH3 domains.

In addition to detecting GST fusion protein immobilized on PVDF membrane, BAP fusion could also be efficiently used to detect synthetic peptides that were immobilized on the membrane. As demonstrated in Figure 5, right panel, a large number of different seven-amino acid-long peptides were synthesized onto the cellulose membrane using Spot-Synthesis method [26]. With this method, as much as 425 peptides could be spotted onto one cellulose membrane, allowing rapid and extensive analysis of various molecular interactions between ligand and its target. The membrane is quite durable, as it can be washed with mild detergent (1% TBS + 0.1% Triton X-100) and re-used for a couple of more times. As seen in this report, alanine scan, and peptide mapping for the study of EH domain-ligand interaction could be done using this assay.

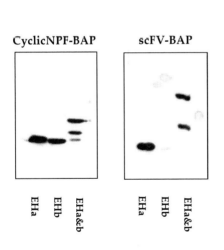
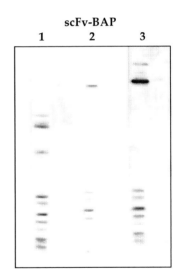

Figure 6. Western blot analysis of BAP fusion. *Left panel* illustrates the detection of GST fusion to different Intersectin EH domains with two BAP fusion probes. GST fusions were resolved by 12% SDS-PAGE and the proteins were blotted onto PVDF membrane before incubating with culture supernatant containing EH-ligand-BAP fusion (CyclicNPF-BAP), or scFv from biopanning of Phage display library with EHa domain (scFV-BAP). The BAP fusion was detected with chemiluminescent substrate (CSPD) and exposed to BioMax Light-1 film (Kodak). The cyclicNPF sequence is DCTNPFSCWR. *Right panel* illustrates western blot analysis of COS-7 cell lysate with BAP fusion to three different scFvs that have been isolated from Phage display library. COS cell lysate was resolved in 12% SDS-PAGE, blotted onto PVDF membrane and probed with 50ng of different scFv-BAP that have been purified by IMAC. The BAP fusion was detected with chemiluminescent substrate (CSPD) and exposed to BioMax Light-1 film (Kodak).

Western Blot Analysis with BAP Fusions

Western blot analysis is commonly used in molecular cell biology research to detect denatured recombinant proteins or proteins from cellular extracts. In this research we demonstrated that BAP-fusion can be used as a convenient one-step detection probe for western blot analysis. EH domain peptide ligand (cyclicNPF-BAP) [31] and single chain variable fragments of monoclonal antibody (scFvs) isolated from phage display antibody library were used to detect GST-EH domain fusion proteins (Figure 6, left panel) or cell lysate (Figure 6, right panel). CyclicNPF-BAP, which is the EH domain peptide ligand [29] could interact with all GST-EH domain fusions; the signal from the binding is robust with no background. When using BAP-fusion to scFv, which has been affinity selected from Phage display library against EHa, only the bands of GST fusion to EHa and EHaandEHb were positive. The result on the right panel illustrated western blot analysis of whole cell lysate using three different scFv-BAPs as probes. Different patterns of protein bands with different scFv-BAPs could be detected. The background remained clear after long exposure (>15min). These results suggested that BAP-fusion is also an efficient reagent for western blot analysis.

Discussion

Numerous examples of application of bacterial alkaline phosphatase (BAP) fusion protein in different aspects of molecular biotechnology research have been reported [2, 3, 13, 14, 16-20, 22, 23, 32-51]. This article demonstrated that BAP-fusion can be used as a versatile molecular probe for detection of molecular interaction in various formats. These include detection on microtiter plate, dot blot, and western blot analysis.

The signal from the detection is strong and the background is very low. The binding specificity is high, and can be used to estimate binding affinity in a competitive inhibition experiment. In addition to the types of assay reported in this study [32, 36, 42, 47, 52-54], there have been previous reports on other formats of detection such as direct cell staining [34], characterization of receptor ligand interaction [13], or protein localization study [2].

The type of protein that is mostly used to fuse with BAP is a variable region of monoclonal antibody; such as scFv or Fab fragments [2, 6, 11, 12, 33, 34, 38-41, 43-46, 48-51, 53-61]. This protein is about 250 amino acid-long. In addition, BAP fusions to peptides [24, 62, 63], and protein domains have also been reported [15, 24]. In this research, peptides ranging from 9 amino acid to 300 amino acid-long (C-terminal of mouse Sos) have been successfully expressed as BAP fusion proteins. Thus, BAP fusion is a flexible system for an expression of a wide variety of proteins.

The *E. coli* alkaline phosphatase is a homodimeric enzyme with its N-terminus protruding out of the globular body of the proteins [5, 64, 65], thus peptide or protein as large as 300 amino acid-long can be fused to the N-terminus of the enzyme without interfering with the enzymatic activity. Moreover, the ten histidine tag and FLAG epitope at the C terminus does not interfere with its catalytic activity, as demonstrated in this report and from previously published work [20], even if the C-terminus of the enzyme doesn't protrude out. These tags are useful for one-step affinity purification, or immuno-detection with anti-FLAG antibody. It has been shown that 6xhistidine tag is sufficient for one-step affinity purification; however we found that in some case, 10xhistidine is more efficient. Even though BAP fusion is a dimmer, it can be used to detect specific molecular interactions and estimate the binding affinity. However, in case it is necessary to determine the binding affinity with monomeric form, such as in BiAcore measurement, a monomeric form of BAP with mutations that inhibit the interaction at the dimeric interface [66] could be used instead.

Bacterial alkaline phosphatase (BAP) that is commonly used came from *E. coli*; however in some case, fusion with human placental alkaline phosphatase has also been reported [2, 23, 45, 62]. BAP is a relatively stable enzyme. As seen in our report that it can be left dry at room temperature without loosing much activity. However, if the enzyme is left in culture broth, it can be kept only for 1 or 2 weeks, this may be the result of proteolysis by some proteases. Thus, fresh sample of culture broth should be prepared every time if it will be used directly for detection.

As seen from this study, the binding signal from an assay on microtiter plate is very strong, OD_{405} values of 2.0-4.0, suggesting that BAP fusion is a very sensitive probe. Recently, there has been a report on the improving *E. coli* alkaline phosphatase efficacy by two additional mutations inside and outside the catalytic pocket (D330N/D153G). This mutant is highly thermostable and the catalytic efficiency (kcat/Km) value was increased by a factor of two, relative to that of the wild type [67]. This mutant should be useful for the assay that requires a very sensitive detection. Further improvement of BAP property such as acid tolerant, solvent, or detergent resistant should also be useful for certain biotechnological applications. Recently, a moderately thermostable alkaline phosphatase from *Geobacillus thermodenitrificans* T2 have been reported [6].

For most binding experiments, BAP fusions in crude culture broth, periplasmic extract, or cell lysate could be used directly as one-step detection probe without further purification. The level of expression and the secretion efficiency depends greatly on the structure of the fused peptide [68]. Thus, it will be beneficial for the researcher to optimize the induction of protein expression in order to obtain the highest amount of the protein. Even if the BAP-fusion is fused to the ompA signal peptide, in most case a large amount of protein was found to accumulate in the cytoplasm. This could be the result of high level of expression of the protein by the *tac* promoter, resulting in the saturation of *E. coli* secretion machinery. Thus, normally, if the BAP fusion will be further purified by IMAC, the whole cell lysate should be used for purification. Purified BAP fusion is useful for long-term storage or certain type of assay. There are a number of BAP-fusion vectors available for generation of BAP fusion constructs [2, 11, 35, 38, 43, 46, 49, 51, 52, 54, 55, 57, 59, 62, 69]. These vectors are different in the multiple cloning site, type of promoter, strains of *E. coli* for expression, and the presence of tag. Recently, a vector for over expression of genes from phage display library using ligation independent cloning (LIC) has been developed [12]. The vector pMY101 that has been used in this study was generated specifically for the expression of BAP fusion of peptide from phage display library with *Xho*I and *Xba*I cloning site, while the vector pMY201 from this study contains larger multiple cloning site and two tags for affinity purification.

Thus it can be concluded that BAP fusion is a versatile molecular probe for detection of various interactions in biotechnological research. It is convenient, and sufficiently sensitive and specific for binding assay in many formats. It is suitable not only for the study of a particular interaction, but highly adaptable for high-throughput analysis.

Acknowledgements

This research was supported by Suranaree University of Technology and National Research Council of Thailand (NRCT). The author would like to thank Miss Podjamas Pansri and Miss Suphap Emrat for technical assistance, and Prof. Dr. Brian K. Kay for invaluable guidance.

References

[1]   Harlow, E. and D. Lane, *Using Antibodies: A Laboratory Manual: Portable Protocol. 1* ed. 1998: Cold Spring Harbor Laboratory Press. 495.

[2]   Brennan, C. and J. Fabes, Alkaline phosphatase fusion proteins as affinity probes for protein localization studies. *Sci. STKE*, 2003. 2003(168): p. PL2.

[3]   Muller, H. and M.J. Soares, Alkaline phosphatase fusion proteins as tags for identifying targets for placental ligands. *Methods Mol. Med.*, 2006. 122: p. 331-40.

[4]   Morton, R.K., The purification of aklaline phosphatases of animal tissues. *Biochem. J.*, 1954. 57(4): p. 595-603.

[5]   Coleman, J.E., Structure and mechanism of alkaline phosphatase. *Annu. Rev. Biophys. Biomol. Struct.,* 1992. 21: p. 441-83.

[6]   Zhang, Y., et al., A Moderately Thermostable Alkaline Phosphatase from Geobacillus thermodenitrificans T2: Cloning, Expression and Biochemical Characterization. *Appl. Biochem. Biotechnol.*, 2008.

[7]   Chang, C.N., W.J. Kuang, and E.Y. Chen, Nucleotide sequence of the alkaline phosphatase gene of Escherichia coli. *Gene*, 1986. 44(1): p. 121-5.

[8]   Le Du, M.H., et al., Artificial evolution of an enzyme active site: structural studies of three highly active mutants of Escherichia coli alkaline phosphatase. *J. Mol. Biol.*, 2002. 316(4): p. 941-53.

[9]   Holtz, K.M., et al., *The mechanism of the alkaline phosphatase reaction: insights from NMR, crystallography and site-specific mutagenesis.*

[10]   Reaction mechanism of alkaline phosphatase based on crystal structures. Two-metal ion catalysis. *FEBS Lett*, 1999. 462(1-2): p. 7-11.

[11]   Yamabhai, M., et al., Secretion of recombinant Bacillus hydrolytic enzymes using Escherichia coli expression systems. *J. Biotechnol.*, 2008. 133(1): p. 50-7.

[12]   Carrier, A., et al., Recombinant antibody-alkaline phosphatase conjugates for diagnosis of human IgGs: application to anti-HBsAg detection. *J. Immunol. Methods,* 1995. 181(2): p. 177-86.

[13]   Han, Z., et al., Accelerated screening of phage-display output with alkaline phosphatase fusions. *Comb. Chem. High Throughput Screen*, 2004. 7(1): p. 55-62.

[14] Kanayama, T., et al., Basis of a high-throughput method for nuclear receptor ligands. *J. Biochem.*, 2003. 133(6): p. 791-7.

[15] Yamabhai, M. and B.K. Kay, Examining the specificity of Src homology 3 domain--ligand interactions with alkaline phosphatase fusion proteins. *Anal. Biochem.*, 1997. 247(1): p. 143-51.

[16] Yamabhai, M., A Convenient Method for the Screening of Compounds that Inhibit Specific Molecular Interactions Using Alkaline Phosphatase Fusion System. *Acta Horticulturae*, 2005. 678: p. 51-57.

[17] Lee, J.C., et al., A mammalian cell-based reverse two-hybrid system for functional analysis of 3C viral protease of human enterovirus 71. *Anal. Biochem.*, 2008. 375(1): p. 115-23.

[18] Lee, J.C., et al., High-throughput cell-based screening for hepatitis C virus NS3/4A protease inhibitors. *Assay Drug Dev. Technol.*, 2005. 3(4): p. 385-92.

[19] Olesen, C.E., et al., Novel methods for chemiluminescent detection of reporter enzymes. *Methods Enzymol*, 2000. 326: p. 175-202.

[20] Buranda, T., et al., Detection of epitope-tagged proteins in flow cytometry: fluorescence resonance energy transfer-based assays on beads with femtomole resolution. *Anal. Biochem.*, 2001. 298(2): p. 151-62.

[21] Hengsakul, M. and A.E. Cass, Alkaline phosphatase-Strep tag fusion protein binding to streptavidin: resonant mirror studies. *J. Mol. Biol.*, 1997. 266(3): p. 621-32.

[22] Mersich, C. and A. Jungbauer, Generic method for quantification of FLAG-tagged fusion proteins by a real time biosensor. *J. Biochem. Biophys Methods*, 2007. 70(4): p. 555-63.

[23] Olin, A.I., et al., The proteoglycans aggrecan and Versican form networks with fibulin-2 through their lectin domain binding. *J. Biol. Chem.*, 2001. 276(2): p. 1253-61.

[24] Stoker, A., Methods for identifying extracellular ligands of RPTPs. *Methods*, 2005. 35(1): p. 80-9.

[25] Yamabhai, M. and B.K. Kay, Mapping protein-protein interactions with alkaline phosphatase fusion proteins. *Methods Enzymol*, 2001. 332: p. 88-102.

[26] Laemmli, U.K., Cleavage of structural proteins during the assembly of the head of bacteriophage T4. *Nature,* 1970. 227(5259): p. 680-5.

[27] Frank, R., The SPOT-synthesis technique. Synthetic peptide arrays on membrane supports--principles and applications. J Immunol Methods, 2002. 267(1): p. 13-26.

[28] Bradshaw, R.A., et al., Amino acid sequence of Escherichia coli alkaline phosphatase. *Proc. Natl. Acad. Sci. USA*, 1981. 78(6): p. 3473-7.

[29] Sparks, A.B., et al., Distinct ligand preferences of Src homology 3 domains from Src, Yes, Abl, Cortactin, p53bp2, PLCgamma, Crk, and Grb2. *Proc. Natl. Acad. Sci. USA*, 1996. 93(4): p. 1540-4.

[30] de Beer, T., et al., Molecular mechanism of NPF recognition by EH domains. *Nat. Struct Biol.,* 2000. 7(11): p. 1018-22.

[31] Santolini, E., et al., The EH network. *Exp. Cell Res.,* 1999. 253(1): p. 186-209.

[32] Yamabhai, M., et al., Intersectin, a novel adaptor protein with two Eps15 homology and five Src homology 3 domains. *J. Biol. Chem.*, 1998. 273(47): p. 31401-7.

[33] Angeles, T.S., et al., Enzyme-linked immunosorbent assay for trkA tyrosine kinase activity. *Anal. Biochem.*, 1996. 236(1): p. 49-55.

[34] Belin, P., et al., Toxicity-based selection of Escherichia coli mutants for functional recombinant protein production: application to an antibody fragment. *Protein Eng. Des. Sel.*, 2004. 17(5): p. 491-500.

[35] Bourin, P., et al., Immunolabeling of CD3-positive lymphocytes with a recombinant single-chain antibody/alkaline phosphatase conjugate. *Biol. Chem.*, 2000. 381(2): p. 173-8.

[36] Cheng, H.J. and J.G. Flanagan, Cloning and characterization of RTK ligands using receptor-alkaline phosphatase fusion proteins. *Methods Mol. Biol.*, 2001. 124: p. 313-34.

[37] Comitti, R., et al., A monoclonal-based, two-site enzyme immunoassay of human insulin. *J. Immunol. Methods*, 1987. 99(1): p. 25-37.

[38] Flanagan, J.G. and H.J. Cheng, Alkaline phosphatase fusion proteins for molecular characterization and cloning of receptors and their ligands. *Methods Enzymol*, 2000. 327: p. 198-210.

[39] Griep, R.A., et al., pSKAP/S: An expression vector for the production of single-chain Fv alkaline phosphatase fusion proteins. *Protein Expr. Purif*, 1999. 16(1): p. 63-9.

[40] Harper, K., et al., A scFv-alkaline phosphatase fusion protein which detects potato leafroll luteovirus in plant extracts by ELISA. *J. Virol. Methods*, 1997. 63(1-2): p. 237-42.

[41] Kobayashi, N., et al., Immunoenzymometric assay for a small molecule,11-deoxycortisol, with attomole-range sensitivity employing an scFv-enzyme fusion protein and anti-idiotype antibodies. *Anal. Chem.*, 2006. 78(7): p. 2244-53.

[42] Kobayashi, N., et al., Generation of a single-chain Fv fragment for the monitoring of deoxycholic acid residues anchored on endogenous proteins. *Steroids*, 2005. 70(4): p. 285-94.

[43] Lobb, R.R., et al., A direct binding assay for the vascular cell adhesion molecule-1 (VCAM1) interaction with alpha 4 integrins. *Cell Adhes Commun.*, 1995. 3(5): p. 385-97.

[44] Mousli, M., M. Goyffon, and P. Billiald, Production and characterization of a bivalent single chain Fv/alkaline phosphatase conjugate specific for the hemocyanin of the scorpion Androctonus australis. *Biochim. Biophys Acta,* 1998. 1425(2): p. 348-60.

[45] Schoonjans, R., et al., A new model for intermediate molecular weight recombinant bispecific and trispecific antibodies by efficient heterodimerization of single chain variable domains through fusion to a Fab-chain. *Biomol. Eng*, 2001. 17(6): p. 193-202.

[46] Sheikholvaezin, A., et al., Construction and purification of a covalently linked divalent tandem single-chain Fv antibody against placental alkaline phosphatase. *Hybridoma* (Larchmt), 2006. 25(5): p. 255-63.

[47] Suzuki, C., et al., Construction, bacterial expression, and characterization of hapten-specific single-chain Fv and alkaline phosphatase fusion protein. *J. Biochem.*, 1997. 122(2): p. 322-9.

[48] Suzuki, C., et al., Open sandwich ELISA with V(H)-/V(L)-alkaline phosphatase fusion proteins. *J. Immunol. Methods,* 1999. 224(1-2): p. 171-84.

[49] Velappan, N., et al., Selection and characterization of scFv antibodies against the Sin Nombre hantavirus nucleocapsid protein. *J. Immunol. Methods*, 2007. 321(1-2): p. 60-9.

[50] Wang, S.H., et al., Construction of single chain variable fragment (ScFv) and BiscFv-alkaline phosphatase fusion protein for detection of Bacillus anthracis. *Anal. Chem.*, 2006. 78(4): p. 997-1004.

[51] Wozniak, G., et al., An ELISA for the detection of TIMP-1 based on recombinant single chain Fv fusion proteins. *Clin. Chim. Acta*, 2003. 335(1-2): p. 49-57.

[52] Yang, D.F., et al., Construction of single chain Fv antibody against transferrin receptor and its protein fusion with alkaline phosphatase. *World J. Gastroenterol.*, 2005. 11(21): p. 3300-3.

[53] Butera, D., et al., Cloning, expression, and characterization of a bi-functional disintegrin/alkaline phosphatase hybrid protein. *Protein Expr. Purif*, 2003. 31(2): p. 286-91.

[54] Mousli, M., et al., Recombinant single-chain Fv antibody fragment-alkaline phosphatase conjugate: a novel in vitro tool to estimate rabies viral glycoprotein antigen in vaccine manufacture. *J. Virol. Methods*, 2007. 146(1-2): p. 246-56.

[55] Muller, B.H., et al., Recombinant single-chain Fv antibody fragment-alkaline phosphatase conjugate for one-step immunodetection in molecular hybridization. *J. Immunol. Methods*, 1999. 227(1-2): p. 177-85.

[56] Boulain, J.C. and F. Ducancel, Expression of recombinant alkaline phosphatase conjugates in Escherichia coli. *Methods Mol. Biol.*, 2004. 267: p. 101-12.

[57] Lindner, P., et al., Specific detection of his-tagged proteins with recombinant anti-His tag scFv-phosphatase or scFv-phage fusions. *Biotechniques*, 1997. 22(1): p. 140-9.

[58] Martin, C.D., et al., A simple vector system to improve performance and utilisation of recombinant antibodies. *BMC Biotechnol.*, 2006. 6: p. 46.

[59] Nakayama, M., D. Neri, and O. Ohara, A new simplified method for preparation of a synthetic phage antibody with practically acceptable detection sensitivity on immunoblots. *Hum. Antibodies*, 2001. 10(2): p. 55-65.

[60] Rau, D., K. Kramer, and B. Hock, Single-chain Fv antibody-alkaline phosphatase fusion proteins produced by one-step cloning as rapid detection tools for ELISA. *J. Immunoassay Immunochem*, 2002. 23(2): p. 129-43.

[61] Tachibana, H., et al., Bacterial expression of a human monoclonal antibody-alkaline phosphatase conjugate specific for Entamoeba histolytica. *Clin. Diagn. Lab. Immunol.*, 2004. 11(1): p. 216-8.

[62] Takazawa, T., et al., Enzymatic labeling of a single chain variable fragment of an antibody with alkaline phosphatase by microbial transglutaminase. *Biotechnol. Bioeng*, 2004. 86(4): p. 399-404.

[63] Voss, S.D., et al., An integrated vector system for cellular studies of phage display-derived peptides. *Anal. Biochem.*, 2002. 308(2): p. 364-72.

[64] Wright, R.M., et al., A high-capacity alkaline phosphatase reporter system for the rapid analysis of specificity and relative affinity of peptides from phage-display libraries. *J. Immunol. Methods*, 2001. 253(1-2): p. 223-32.

[65]   Schlesinger, M.J., The reversible dissociation of the alkaline phosphatase of Escherichia coli. II. Properties of the subunit. *J. Biol. Chem.*, 1965. 240(11): p. 4293-8.

[66]   Stec, B., K.M. Holtz, and E.R. Kantrowitz, A revised mechanism for the alkaline phosphatase reaction involving three metal ions. *J. Mol. Biol.*, 2000. 299(5): p. 1303-11.

[67]   Boulanger, R.R., Jr. and E.R. Kantrowitz, Characterization of a monomeric Escherichia coli alkaline phosphatase formed upon a single amino acid substitution. *J. Biol. Chem.*, 2003. 278(26): p. 23497-501.

[68]   Muller, B.H., et al., Improving Escherichia coli alkaline phosphatase efficacy by additional mutations inside and outside the catalytic pocket. *Chembiochem*, 2001. 2(7-8): p. 517-23.

[69]   Mergulhao, F.J., D.K. Summers, and G.A. Monteiro, Recombinant protein secretion in Escherichia coli. *Biotechnol. Adv.*, 2005. 23(3): p. 177-202. Epub 2005 Jan 8.

[70]   Huang, X., et al., Construction of a high sensitive Escherichia coli alkaline phosphatase reporter system for screening affinity peptides. *J. Biochem. Biophys Methods,* 2007. 70(3): p. 435-9.

[71]   Kay, B.K., J. Kasanov, and M. Yamabhai, Screening phage-displayed combinatorial peptide libraries. *Methods,* 2001. 24(3): p. 240-6.

Index

C

D

H

I

O

P

S

U

V